Molecular Mimicry, Microbes, and Autoimmunity

Molecular Mimicry, Microbes, and Autoimmunity

Edited by

Madeleine W. Cunningham
Department of Microbiology and Immunology
University of Oklahoma Health Sciences Center
Oklahoma City, Oklahoma 73190

Robert S. Fujinami
Department of Neurology
University of Utah School of Medicine
Salt Lake City, Utah 84105

ASM PRESS

Washington, D.C.

ASM Press
American Society for Microbiology
1752 N Street, N.W.
Washington, DC 20036-2084

Library of Congress Cataloging-in-Publication Data

Molecular mimicry, microbes, and autoimmunity / edited by Madeleine Cunningham, Robert S. Fujinami.
p.cm.
Includes index.
ISBN 1-55581-194-9
1. Autoimmune diseases—Etiology. 2. Viral antigens. 3. Commmunicable diseases—Complications. 4. Antigenic determinants. 5. Immunospecificity. I. Cunningham, Madeleine W. II. Fujinami, Robert S.

RC600 .M645 2000
616.97'8—dc21 00-041598

Printed in the United States of America

CONTENTS

CONTRIBUTORS

Heiner Appel • Department of Cancer Immunology and AIDS, Dana-Farber Cancer Institute, Boston, MA 02115

Kurt Bachmaier • Amgen Institute, Ontario Cancer Institute, Department of Medical Biophysics, University of Toronto, 620 University Avenue, Suite 706, Toronto, Ontario M5G 2C1, Canada

David O. Beenhouwer • Departments of Cell Biology and Internal Medicine, Division of Infectious Diseases, Albert Einstein College of Medicine, Bronx, NY 10461

Miri Blank • Research Unit of Autoimmune Diseases, Internal Medicine B, Sheba Medical Center, Tel-Hashomer 52621, Israel

Irun R. Cohen • Department of Immunology, The Weizmann Institute of Science, Rehovot 76100, Israel

Edecio Cunha-Neto • Transplantation Immunology Laboratory, Heart Institute (InCor), University of São Paolo School of Medicine, São Paolo, Brazil

Madeleine W. Cunningham • Department of Microbiology and Immunology, University of Oklahoma Health Sciences Center, Biomedical Research Center Room 219, P.O. Box 26901, Oklahoma City, OK 73190

Gina Cunto-Amesty • Department of Pathology and Laboratory Medicine, University of Pennsylvania, Room 205, John Morgan Building, 36th and Hamilton Walk, Philadelphia, PA 19104-6082

Betty Diamond • Department of Microbiology and Immunology and Department of Medicine, Albert Einstein College of Medicine, 1300 Morris Park Avenue, Bronx, NY 10461

Robert S. Fujinami • Department of Neurology, University of Utah School of Medicine, 30 N 1900 East, Room 3R330, Salt Lake City, UT 84132

Charles J. Gauntt • Department of Microbiology, The University of Texas Health Science Center at San Antonio, 7703 Floyd Curl Drive, San Antonio, TX 78284-7758

Neil S. Greenspan • Institute of Pathology, Biomedical Research Building, Room 927, Case Western Reserve University, 10900 Euclid Avenue, Cleveland, OH 44106-4943

Dawn M. Gross • Department of Pathology, Program in Immunology, Sackler School of Biomedical Sciences, Tufts University School of Medicine, Boston, MA 02111

John B. Harley • Arthritis/Immunology, Oklahoma Medical Research Foundation, 825 N.E. 13th Street, Oklahoma City, OK 73104

Sally A. Huber • Department of Pathology, University of Vermont, 55A South Park Drive, Colchester, VT 05446

Brigitte T. Huber • Department of Pathology, Program in Immunology, Sackler School of Biomedical Sciences, Tufts University School of Medicine, Boston, MA 02111

Judith A. James • Arthritis/Immunology, Oklahoma Medical Research Foundation, 825 N.E. 13th Street, Oklahoma City, OK 73104

Jorge Kalil • Howard Hughes Medical Institute and Department of Medicine and Heart Institute (InCor), University of São Paolo School of Medicine, São Paolo, Brazil

Thomas Kieber-Emmons • Department of Pathology and Laboratory Medicine, University of Pennsylvania, Room 205, John Morgan Building, 36th and Hamilton Walk, Philadelphia, PA 19104-6082

Malak Kotb • Department of Surgery, The University of Tennessee, Memphis, 956 Court Avenue, A202, Memphis, TN 38163

Czeslawa Kowal • Department of Microbiology and Immunology and Department of Medicine, Albert Einstein College of Medicine, 1300 Morris Park Avenue, Bronx, NY 10461

Ilan Krause • Research Unit of Autoimmune Diseases, Internal Medicine B, Sheba Medical Center, Tel-Hashomer 52621, Israel

Rena May • Department of Cell Biology, Albert Einstein College of Medicine, Bronx, NY 10461

Michael B. A. Oldstone • Department of Neuropharmacology, The Scripps Research Institute, 10550 N. Torrey Pines Road, La Jolla, CA 92037

Josef M. Penninger • Amgen Institute, Ontario Cancer Institute, Department of Medical Biophysics, University of Toronto, 620 University Avenue, Suite 706, Toronto, Ontario M5G 2C1, Canada

Clemencia Pinilla • Torrey Pines Institute for Molecular Studies, 3550 General Atomics Court, San Diego, CA 92121

Anthony Quinn • Division of Immune Regulation, La Jolla Institute for Allergy and Immunology, 10355 Science Center Drive, San Diego, CA 92121

Matthew D. Scharff • Department of Cell Biology, Albert Einstein College of Medicine, Bronx, NY 10461

R. Hal Scofield • Arthritis/Immunology, Oklahoma Medical Research Foundation, 825 N.E. 13th Street, Oklahoma City, OK 73104

Eli E. Sercarz • Division of Immune Regulation, La Jolla Institute for Allergy and Immunology, 10355 Science Center Drive, San Diego, CA 92121

Alexander R. Shikhman • Division of Rheumatology, Scripps Clinic-MS 113, 10666 N. Torrey Pines Road, La Jolla, CA 92037

Yehuda Shoenfeld • Research Unit of Autoimmune Diseases, Internal Medicine B, Sheba Medical Center, Tel-Hashomer, 52621 Israel

Philippe Valadon • Department of Cell Biology, Albert Einstein College of Medicine, Bronx, NY 10461

Kumar Visvanathan • 1230 York Avenue, New York, NY 10021

Matthias G. von Herrath • Division of Virology, IMM6, Department of Neuropharmacology, The Scripps Research Institute, 10550 N. Torrey Pines Road, La Jolla, CA 92037

Kai W. Wucherpfennig • Department of Cancer Immunology and AIDS, Dana-Farber Cancer Institute, Boston, MA 02115

John B. Zabriskie • 1230 York Avenue, New York, NY 10021

PREFACE

Over the years, interest in the role of infectious agents in autoimmune diseases has grown until the microbiologists and the immunologists now meet to share new ideas about the mechanisms of immune-mediated diseases. Considerable knowledge about the pathogenesis of bacterial, viral, fungal, and parasitic diseases has led to new thinking about how microbes affect the immune system. Rapid progress in biotechnology has allowed researchers to find answers to questions about the role of molecular mimicry in disease. Explosive growth in the field of microbial immunology, and the realization that pathogens lead to immune responses capable of tissue and organ destruction, have set the framework for the studies described here on molecular mimicry between microbes and host molecules.

The work under the many topics in this book is at the forefront of the field and is described by those who are waging the war against autoimmune disease, a malady which may result in part from a foreign or infectious insult to the immune system. Those who have been tantalized by the intriguing theory that infectious agents break tolerance, or in some fashion trigger the immune system to respond against self, have spent lifetimes researching molecular mimicry and the role of infectious agents in autoimmune diseases. The readers of this book are privileged to be able to read from the best minds in science and medical research.

Introducing molecular mimicry, Michael Oldstone and John Zabriskie write about the past and the future in molecular mimicry and disease. Dr. Oldstone provides a Foreword on the basis of mimicry and its role in autoimmunity and states the challenges for the future. Dr. Zabriskie and Kumar Visvanathan (chapter 1) explain how mimicry in group A streptococcal disease was discovered and investigated as the basis of rheumatic fever. For the theoretical and immunological-minded, Irun Cohen (chapter 2) provides a stimulating treatise on principles of molecular mimicry and autoimmune disease. He brings his new ideas about autoimmune disease together with his knowledge about heat shock proteins, arthritis, and diabetes. These chapters set the stage for succeeding authors to develop the theme of molecular mimicry, microbes, and autoimmunity.

Some of the chapters provide examples of models of molecular mimicry and infectious diseases, including mimicry between viruses and the central nervous system and its potential role in multiple sclerosis, as described by Robert Fujinami (chapter 3). Mimicry between the group A streptococcal M protein and cardiac myosin and the immunopathogenesis of rheumatic fever is reviewed by Madeleine Cunningham (chapter 4), and mimicry between coxsackievirus and heart proteins is discussed by Sally Huber and Charles Gauntt (chapter 5). Some of the latest developments in the role of microbes in atherosclerosis are described in the chapter by Josef Penninger and Kirk Bachmaier (chapter 6) on mimicry between *Chlamydia trachomatis* outer membrane protein and cardiac myosin.

Other chapters explain mimicry from the viewpoint of the immunochemist and immunologist and are outstanding contributions to our understanding of mimicry. Niel Greenspan, Clemencia Pinilla, and Alexander Shikhman (chapter 7) describe striking mimicry between peptides and the group A streptococcal carbohydrate *N*-acetylglucosamine and how it may relate to valvulitis in rheumatic heart disease. The role of superantigens in molecular mimicry and autoimmune diseases is discussed by Malak Kotb (chapter 8). This chapter details how superantigens and molecular mimicry may work together to bring about autoimmunity.

Several chapters approach the issue of mimicry in systemic lupus erythematosus, an important autoimmune disease. Peptide induction of systemic lupus erythematosus is described by John Harley, Hal Scofield, and Judith James (chapter 9). Intriguing is the chapter by Czeslawa Kowal and Betty Diamond (chapter 10) on mimicry between DNA, carbohydrates, and peptides and the implications in systemic lupus erythematosus. Later in the book, a third chapter on mimicry in systemic lupus erythematosus and antiphospholipid syndrome, by Yehuda Shoenfeld and colleagues (chapter 16), describes the role of antiphospholipid antibodies in experimental animal models.

Matthew Scharff and colleagues (chapter 11) describe a beautiful and intriguing study of peptide mimicry of the polysaccharide capsule of *Cryptococcus neoformans*, the fungus which causes cryptococcal meningitis. Autoimmunity in Lyme arthritis in animal models is described by Dawn Gross and Brigitte Huber (chapter 12). They describe mimicry between OspA of *Borrelia burgdorferi* and LFA-1. Another important point of these studies is that peptide mimicry of carbohydrates may lead to strategies for better immunization against carbohydrate epitopes of pathogens. The chapter by Gina Cunto-Amesty and Thomas Kieber-Emmons (chapter 13) looks further at peptide mimicry of carbohydrates.

The importance of T lymphocytes in molecular mimicry and the recognition of self and non-self is recognized by Heiner Appel and Kai Wucherpfennig (chapter 14) in their chapter on the structural basis of cross-reactivity by T cells. The report of the recognition of peptides from infectious microorganisms by human T-cell clones from multiple sclerosis patients was important in delineating the presence of these T cells in human disease. More T-cell immunology and the role of T cells in mimicry and determinant spreading are discussed by Anthony Quinn and Eli Sercarz (chapter 15). Although mimicry might begin the process of breaking tolerance to a self epitope, spreading of immune responses throughout the autoantigen may represent the development of disease.

Matthias von Herrath (chapter 17) writes about diabetes and the role of viruses in development of immune destruction of the islet cells in the pancreas, as well as potential therapeutic strategies and prevention therapies. Finally, molecular mimicry between the Chagas' disease parasite and heart proteins is described elegantly by Edecio Cunha-Neto and Jorge Kalil (chapter 18). These chapters show that there is now much more interest and advanced work in potential mechanisms of molecular mimicry in autoimmune disease.

This book represents years of work in the field of molecular mimicry, autoimmunity, and immunology, but even so, more is yet to be learned about how self-tolerance is broken and how autoimmunity can be prevented or arrested. Novel mechanisms of immune-mediated destruction in animal models are being discovered at a rapid pace in all the autoimmune diseases. Although there are many models of molecular mimicry within this book, the main goal is to understand better the mechanisms which lead to autoimmune disease so that it can be prevented or successfully treated. The potential for induction of tolerance against autoimmune diseases is one important mechanism of prevention, but many other immunotherapies are also under investigation, as different types of immune regulatory mechanisms may be blocked in order to prevent or cure disease.

Although this book contains information about a diverse group of autoimmune diseases, sequelae, and microorganisms, the unifying theme is molecular mimicry. It was impossible to cover all of the microbes implicated in mimicry or all autoimmune diseases; however, we believe that the microbes and diseases chosen represent excellent examples of molecular mimicry in autoimmunity.

Madeleine W. Cunningham
Robert S. Fujinami

Molecular Mimicry, Microbes, and Autoimmunity
Edited by M. W. Cunningham and R. S. Fujinami

Foreword

Molecular Mimicry: an Idea Whose Time Has Come

Michael B. A. Oldstone

Colleagues and I defined molecular mimicry as similar structures shared by molecules from dissimilar genes or by their protein products (5, 10, 12). Either the molecules' linear amino acid sequences or their conformational fit may be shared, even though their origins are as separate as, for example, a virus and a normal host self-determinant. Because guanine-cytosine (GC) sequences and introns designed to be spliced away may provide false hybridization signals and nonsense homologies, respectively, focus on molecular mimicry is necessary at the protein level. Such homologies between proteins have been detected either by use of immunologic reagents, humoral or cellular, that cross-react with two presumably unrelated protein structures or by computer searches to match proteins described in storage banks. Regardless of the methods used for identification, it is now abundantly clear that molecular mimicry between proteins encoded by numerous microbes and host self-proteins occurs and is not uncommon (for a review, see reference 12). Such data are of interest not only in autoimmunity but also as a likely mechanism by which viral proteins are processed inside cells (3).

Examples of molecular mimicry were first described in the early 1980s, when it was found that monoclonal antibodies against viruses reacted with host protein inside the cell (5). Indeed, multiple monoclonal antibodies against a battery of viruses were noted to be cross-reacting with host determinants (for a review, see reference 12). When the frequency of cross-reactivity between viral proteins and host self-antigens was analyzed with more than 800 monoclonal antibodies, it was noted that nearly 5% of the monoclonal antibodies against 15 different viruses, including such commonly found representatives of DNA and RNA viruses as the herpesvirus group, vaccinia virus, myxoviruses, paramyxoviruses, arenaviruses, flaviviruses, alphaviruses, rhabdoviruses, coronaviruses, and human retroviruses, cross-reacted with host-cell determinants expressed on or in uninfected tissues (11, 15). On the basis of needing five to six amino acids to induce a monoclonal antibody response, the probability that 20 amino acids will occur in six identical residues between two proteins is 20^6 or 1 in 128,000,000. Similarly, a variety of T lymphocytes sensitized to cellular proteins (i.e., myelin basic protein, proteolipid protein of myelin, and glutamic decarboxylase [GAD]) were also noted to cross-react (i.e., proliferate, lyse or release, or display cytokines) when they were added to proteins or peptides from selected viruses. The reverse

Michael B. A. Oldstone • Department of Neuropharmacology, The Scripps Research Institute, 10550 N. Torrey Pines Road, La Jolla, CA 92037.

was also noted, in that T lymphocytes sensitized specifically to a virus would cross-react with a host protein or peptide. Collectively, these data indicated that molecular mimicry is not uncommon.

Computer searches revealed interesting sequence homologies that might explain a variety of diseases; for example, the amino acids shared between a number of coagulation proteins and dengue virus or between human immunodeficiency virus and brain proteins could suggest part of the pathogenic mechanism for dengue hemorrhagic shock syndrome and AIDS dementia complex, respectively. Clinical studies have shown a high degree of correlation with the immune response to GAD and other islet antigens in patients who progress to or who have insulin-dependent diabetes. Sequences obtained by computer search revealed identity between a component of GAD amino acids 247 to 279 and other auto-antigens with several viruses (2, 9). Similarly, over the last year evidence has linked chlamydia protein with heart disease and herpes simplex virus with corneal antigens, with the protein and antigen acting as mimics via induction of antibodies and T lymphocytes, respectively; these proteins were implicated as likely causes of the respective diseases (1, 17).

An essential step in validation of the molecular mimicry concept was obtaining biological evidence from animal models that showed that molecular mimicry is more than an epiphenomenon. The first observation used myelin basic protein and allergic encephalomyelitis (4). The myelin basic protein sequences that cause allergic encephalitis are known, and the encephalitogenic site of 8 to 10 amino acids had been mapped for several different animal species. With the use of computer-assisted analysis, several viral proteins that showed significant homology with the encephalitogenic site of myelin basic protein were uncovered, including homologous fits between the myelin basic protein and nucleoprotein of the hemagglutinin of influenza virus, the core protein of adenovirus, the EC-LF2 protein of Epstein-Barr virus, and the hepatitis B virus polymerase (HBVP), as well as others. However, the best fit occurred between the myelin basic protein encephalitogenic site in the rabbit and HBVP. It was then shown that inoculation of the HBVP peptide peripherally into rabbits caused perivascular infiltration localized to the central nervous system, reminiscent of disease induced by inoculation of whole myelin basic protein or the peptide component of the encephalitogenic site of myelin basic protein (4). Furthermore, a specific immune response, both cellular and humoral, to myelin basic protein occurred (4). Thereafter, several other animal models that showed that cross-reactivity between a microbe and self-antigen leads to disease were developed (for reviews, see references 11 and 12). Interesting examples are presented in publications from the laboratories of Harvey Cantor and Priscilla Schaffer (17) for herpes simplex virus and corneal antigens and from the laboratory of Larry Steinman (13) for viral peptides and experimental autoimmune encephalomyelitis (EAE). In the study of Steinman et al. (13), it was noted that a major epitope of myelin basic protein at amino acids 87 to 99 (VHFFKNIVTPRTP) induces EAE. VHFFK was mapped as containing the major residues necessary for the binding of this self-myelin molecule to the T-cell receptor and to the major histocompatibility complex (MHC). Peptides from papillomavirus strains that contain the motif VHFFK induce EAE. In contrast, peptides from papillomavirus type 40 that contain VHFFR and a peptide from papillomavirus type 32 that contains VHFFH prevented EAE. While $CD4^+$ T-cell lines that produce experimental allergic encephalomyelitis following adoptive transfer produce gamma interferon and tumor necrosis factor alpha, those that prevent or suppress disease produce interleukin 4. Thus, in this study, microbial peptides that differ from the core motif of the myelin basic protein self-antigen at amino acids 87 to 99 by a single residue either could cause disease or, in contrast, could function as altered peptide ligands and could behave as T-cell receptor antagonists and modulate autoimmune disease.

The most difficult step is to definitively prove the relevance of molecular mimicry to naturally occurring human disease. Various correlations ranging from those that are reasonably convincing to those that are less so have been published (for reviews, see references 9, 10, and 12). Two examples are selected for brief mention here. The majority of patients with the autoimmune disease myasthenia gravis have antibodies against the acetylcholine receptor (AChR). Purification of antibodies from patients with myasthenia gravis with the human AChR α-subunit from amino acids 157 to 170 provided immunoglobulin G antibodies that bound to native AChR and that inhibited the binding of α-bungarotoxin to the receptor. The human AChR α-subunit from amino acids 160 to 167 showed specific immunological cross-reactivity with a shared homologous domain on herpes simplex virus glycoprotein D, residues 286 to 293, by both specific binding and inhibition assays. Antibodies to the human AChR α-subunit bound to herpes simplex virus-infected cells. The data on the immunological cross-reactivity of the AChR "self-epitope" with herpes simplex virus and the presence of cross-reactive antibodies in the sera of patients with myasthenia gravis suggest that this virus may be associated with the initiation of some cases of myasthenia gravis (14).

In another study, Wucherpfennig and Strominger (16) imposed a structural requirement for molecular mimicry searches as did Hammer et al. (7). Wucherpfennig and Strominger (16) used the known structures for MHC class II disease-associated molecules with peptide binding and the T-cell receptor for a known immunodominant myelin basic protein peptide. By a database search, a panel of 129 peptides from microbes that matched the molecular mimicry motif were obtained, and these were tested with several T-cell clones obtained from the cerebrospinal fluid of multiple sclerosis patients (16). Eight peptides (seven of viral origin and one of bacterial origin) were found to efficiently activate three of these clones, whereas only one of the eight peptides would have been identified as an appropriate molecular mimic by sequence alignment. These observations indicated that a single T-cell receptor could recognize several distinct but structurally related peptides from multiple pathogens, suggesting more permissivity for the T-cell receptor than has been previously appreciated, an observation supported by many recent reports (6, 8). Hammer et al. (7) analyzed the possible peptide motifs that bound to the HLA-DR β chain associated with autoimmune disease. The influence of single key residues was tested by using site-directed mutations. A selection of peptides that bound to rheumatoid arthritis-linked DR allotypes was shown to be critical for amino acid position 71 of DR β and amino acid position 57 of DQ β. Extension of these findings to insulin-dependent diabetes mellitus suggested that the insulin-dependent diabetes mellitus "autoimmune" peptide would bear a negative charge at P9 in order to bind preferentially to the diabetes-associated DQ allele. In toto, these findings should assist in identification of potential peptides (self?, viral?, microbe?) associated with the pathogenesis of autoimmune disease. One variation on this theme that helps to explain the linkage of MHC with autoimmune disease and infectious agents is that MHC-derived self-peptides prominent in selecting the T-cell repertoire may be mutated and/or may share immunological cross-reactive epitopes with microbial antigens. These findings, in conjunction with the analysis in the study by Ruiz et al. (13) described above, raise the possibility that products from viral or microbial agents may be used to block autoimmune diseases.

In summary, molecular mimicry is but one mechanism by which autoimmune diseases can occur in association with infectious agents. The concept of molecular mimicry remains a viable hypothesis for framing questions and approaches to uncovering the initiating infectious agent as well as recognizing the "self"-determinant, understanding the pathogenic mechanism(s) involved, and designing strategies for the treatment and prevention of autoimmune disorders. The Oxford Dictionary defines hypothesis as "A supposition or conjecture put forward to account for certain facts and used as a basis for further investigation by which it may be proved or disproved." In many instances hard data dictate molecular mimicry as a mechanism for disease causation. For others, additional information is required before molecular mimicry can be accepted or rejected. The availability of computer data banks, structural information on specific MHC alleles, and MHC maps for particular autoimmune diseases and the ability to identify the anchoring and flanking sequences of a peptide that binds to that MHC allele or to the T-cell receptor provide the opportunity to critically evaluate and identify the microbial causes of autoimmune diseases. The application and use of transgenic models designed to evaluate molecular mimicry offer the opportunity to understand the sequence of events that lead to the pathology as well as to design specific and unique therapies in order to reverse or prevent the autoimmune process and disease.

REFERENCES

1. **Bachmaier, K., N. Neu, L. M. de la Maza, S. Pal, A. Hessel, and J. M. Penninger.** 1999. Chlamydia infections and heart disease linked through antigenic mimicry. *Science* **283:**1335–1339.
2. **Brusic, V., G. Rudy, and L. Harrison.** 1997. Molecular mimicry—from hypothesis towards evidence. *Immunol. Today* **18:**95–96.
3. **Dales, S., R. S. Fujinami, and M. B. A. Oldstone.** 1983. Infection with vaccinia favors the selection of hybridomas synthesizing auto-antibodies against intermediate filaments, among them one cross-reacting with the virus hemagglutinin. *J. Immunol.* **131:**1546–1553.
4. **Fujinami, R. S., and M. B. A. Oldstone.** 1985. Amino acid homology between the encephalitogenic site of myelin basic protein and virus: mechanism for autoimmunity. *Science* **230:**1043–1045.
5. **Fujinami, R. S., M. B. A. Oldstone, Z. Wroblewska, M. E. Frankel, and H. Koprowski.** 1983. Molecular mimicry in virus infection: crossreaction of measles virus phosphoprotein or of herpes simplex virus protein with human intermediate filaments. *Proc. Natl. Acad. Sci. USA* **80:**2346–2350.
6. **Garcia, K. C., M. Degano, R. L. Stanfield, A. Brunmark, M. R. Jackson, P. A. Peterson, L. Teyton, and I. A. Wilson.** 1996. The αβ T cell receptor structure at 2.5 Å and its orientation in the TCR-MHC complex. *Science* **274:**209–219.
7. **Hammer, J., F. Gallazzi, E. Bono, R. W. Karr, J. Guenot, P. Valsasnini, Z. A. Nagy, and F. Sinigaglia.** 1995. Peptide binding to HLA-DR4 molecules: correlation with rheumatoid arthritis association. *J. Exp. Med.* **181:**1847–1855.
8. **Kersh, G. J., and P. M. Allen.** 1996. Essential flexibility in the T-cell recognition of antigen. *Nature* **380:**495–498.
9. **Maclaren, N., and J. Atkinson.** 1999. Insulin dependent diabetes mellitus: hypothesis of molecular mimicry between islet cell antigens and microorganisms. *Mol. Med. Today* **Feb:**76–83.
10. **Oldstone, M. B. A.** 1987. Molecular mimicry and autoimmune disease. *Cell* **50:**819–820.
11. **Oldstone, M. B. A.** 1989. Molecular mimicry as a mechanism for the cause and as a probe uncovering etiologic agent(s) of autoimmune disease. *Curr. Top. Microbiol. Immunol.* **145:**127–135.
12. **Oldstone, M. B. A.** 1998. Molecular mimicry and immune mediated diseases. *FASEB J.* **12:**1255–1265.

13. **Ruiz, P. J., H. Garren, D. L. Hirschberg, A. M. Langer-Gould, M. Levite, M. V. Karpuj, S. Southwood, A. Sette, P. Conlon, and L. Steinman.** 1999. Microbial epitopes act as altered peptide ligands to prevent experimental autoimmune encephalomyelitis. *J. Exp. Med.* **189:**1275–1283.
14. **Schwimmbeck, P. L., T. Dyrberg, D. Drachman, and M. B. A. Oldstone.** 1989. Molecular mimicry and myasthenia gravis: an autoantigenic site of the acetylcholine receptor α-subunit that has biologic activity and reacts immunochemically with herpes simplex virus. *J. Clin. Investig.* **84:**1174–1180.
15. **Srinivasappa, J., J. Saegusa, B. S. Prabhakar, M. K. Gentry, M. J. Buchmeier, T. J. Wiktor, H. Koprowski, M. B. A. Oldstone, and A. L. Notkins.** 1986. Molecular mimicry: frequency of reactivity of monoclonal antiviral antibodies with normal tissues. *J. Virol.* **57:**397–401.
16. **Wucherpfennig, K. W., and J. L. Strominger.** 1995. Molecular mimicry in T-cell mediated autoimmunity: viral peptides activate human T-cell clones specific for myelin basic protein. *Cell* **80:**695–705.
17. **Zhao, Z.-S., F. Granucci, L. Yeh, P. Schaffer, and H. Cantor.** 1997. Molecular mimicry by herpes simplex virus-type 1: autoimmune disease after viral infection. *Science* **279:**1344–1347.

Molecular Mimicry, Microbes, and Autoimmunity
Edited by M. W. Cunningham and R. S. Fujinami

Chapter 1

An Overview: Molecular Mimicry and Disease

Kumar Visvanathan and John B. Zabriskie

Some diseases are initiated by quite normal immune responses to foreign antigens, such as microbes, but the antibodies or T cells that are stimulated happen to recognize a similar (cross-reactive) self-protein. Molecular sequencing techniques have revealed numerous short stretches of homology between various microbial antigens and self-antigens. This homology is called "molecular mimicry" and is postulated to be one reason why immune responses against foreign antigens can lead to reactivity against self. It is possible that this is the initiating mechanism in a large number of autoimmune diseases.

The term *molecular mimicry* was originally coined by Damian (13) in his studies on parasitic organisms, but it is now clear that the sharing of antigenic determinants between host and microbe is a quite common event and probably occurs far more frequently than the evidence has so far indicated. It is our belief that these cross-reactions are occurring constantly, and, as we shall see later, it is perhaps only in the genetically programmed individual that immunologically relevant pathological damage occurs.

In order for these potentially harmful events to occur, at least three conditions must be in place:

1. significant antigenic mimicry between microbe and host,
2. an abnormal cellular and humoral response on the part of the host to the microbial antigens cross-reactive with tissue antigens, and
3. genetic factors that favor an abnormal host response to the cross-reactive antigens.

Table 1 is a partial listing of the known cross-reactions between microbe and host, and as we continue to know more about the genomes of bacteria and the human genome, this list will certainly expand. It is the purpose of this chapter to discuss these microbe-host interactions in more detail by using several different microbe-host disease states to illustrate the importance and nature of these cross-reactions in the genetically programmed host.

GROUP A STREPTOCOCCI AND DISEASE STATES

Figure 1 is a schematic drawing of the various cellular structures of the organism, and one can see that there are numerous cross-reactions between this organism and host tis-

Kumar Visvanathan and John B. Zabriskie • 1230 York Avenue, New York, NY 10021.

Table 1. A selection of cross-reactions between microbes and mammalian tissues

Organism	Tissue	Possible disease association
Streptococcus pyogenes	Heart, brain, joints	Rheumatic fever
Typanosoma cruzi (Cruzin)	High-density lipoprotein	Chagas' disease
Coxsackie B virus	Cardiac myosin	Cardiomyopathy
Shigella	HLA-B27 lymphocytes	Ankylosing spondylitis
Escherichia coli	Colon tissue	Ulcerative colitis
T. bacillus	Cartilage proteoglycans	Rheumatoid arthritis

sues. The molecular details of these reactions will be discussed in greater depth in other chapters. We will concentrate on those reactions as they relate to disease states.

The concept of autoimmunity playing a role in this disease was introduced more than 50 years ago by a number of investigators when antibodies to the heart were noted in the sera of patients with acute rheumatic fever (ARF) and/or rheumatic heart disease. The origin of these antibodies was better defined when Kaplan and Frengley (30) noted that immunization of rabbits with group A streptococci induced antibodies that bound to the human heart in a manner strikingly similar to that observed with patient sera. Further experiments revealed that the cross-reacting antigen being identified was similar to (but not identical to) M protein and was also present in a limited number of group A streptococcal strains. More recently, Fischetti (22) identified the complete chemical structure of the M protein and demonstrated a close homology of the M-protein moiety with cardiac

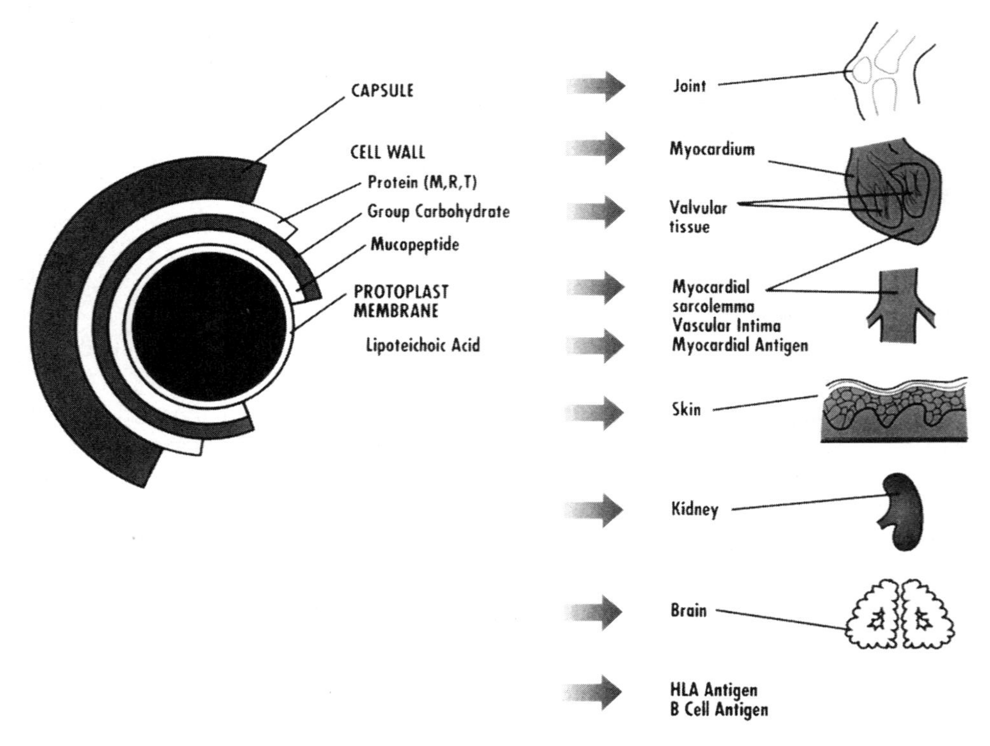

Figure 1. Schematic representation of the various structures of the group A streptococcus. Note the wide variety of cross-reactions between its antigens and mammalian tissues.

cytoskeletal proteins. Further studies by Sargent and coworkers (50) and by Cunningham et al. (12) clearly demonstrated that the M-protein moiety contains numerous epitopes cross-reactive with certain cardiac proteins (e.g., tropomyosin and myosin) including those unrelated to the cytoskeletal proteins.

Similar studies in our own laboratory demonstrated the presence of a second cross-reactive antigen that resides in the streptococcal membrane and that is apparently unrelated to M protein. This antigen was present in all streptococcal throat strains tested and has been purified to a series of four peptides that range in molecular mass from 23 to 22 kDa (60). An important feature of this protein was the demonstration by van de Rijn et al. (60) that the heart-reactive staining pattern of sera from patients with ARF was completely abolished by preabsorption of the sera with these closely spaced peptides. Further studies by Read et al. (47) with lymphocytes from ARF patients revealed enhanced cellular reactivity to this membrane antigen but not the cell wall antigen. This reactivity was confined to membrane preparations from streptococcal strains isolated from the throat but did not occur with membrane preparations from streptococcal strains isolated from patients with skin-associated infections.

Autoimmune mechanisms are also thought to play a role in another major clinical manifestation of ARF, namely, rheumatic chorea. Through the work of Husby et al. (27) it was clear that the sera of patients with active chorea contained an antibody that bound to the cytoplasm of the cells of the caudate nuclei and that was specific for these cells. The presence of the antibody correlated with the clinical activity of the disease and could be absorbed by streptococcal antigens. Furthermore, the antibody was detected in the cerebrospinal fluid of five patients with active chorea (unpublished data). The importance of this antibody in the actual disease state is evidenced by the fact that serial plasmapheresis and removal of the antibody have resulted in marked clinical improvement in a small number of patients (S. Swedo, A. Aron, and J. B. Zabriskie, unpublished data).

Turning to the cellular reactivity in patients with rheumatic fever (RF), lymphocytes from the RF valvular lesions have been an area of active investigation in our laboratory. For example, using a panel of monoclonal antibodies specific for various cell surface markers, Kemeny et al. (32) demonstrated that the valvular lesions contained equal numbers of macrophages and T cells. Helper T cells predominated in the more active cardiac lesions; the ratio of T helper cells to T suppressor cells in patients with chronic carditis closely resembled that seen in normal peripheral blood. Recently, Yoshinaga et al. (66) extended these experiments one step further and were able to isolate and clone T-cell lines from these valvular specimens. When these T-cell clones were stimulated with various streptococcal antigens, only streptococcal membrane antigens isolated from RF-associated strains elicited a reaction in proliferation assays. Antigens isolated from nephritogenic strains or antigens from other streptococcal groups were nonreactive. One unanticipated finding was that none of these T-cell clones reacted with cardiac antigens or a variety of purified M-protein antigens. In contrast to these findings, Guilherme and colleagues (25) have isolated individual T-cell clones from valvular specimens from patients with RF, and the clones reacted with both streptococcal antigens and mammalian cytoskeletal proteins. The exact nature of these reactive antigens is under investigation.

The question of whether or not these heart-reactive antibodies or activated T cells isolated from patients are actually cytotoxic for the relevant tissue antigens has been difficult to prove. The experiments of Yang and colleagues (65) clearly demonstrated that guinea pigs sensitized to streptococcal membranes produced activated T cells specifically cytotoxic for guinea pig neonatal myofibers but not for other tissues, whereas immunization with cell walls did not. However, the addition of heart-reactive antibodies obtained from the sera of these animals did not enhance the cytotoxicity, indicating that the observed cytotoxicity was antibody independent. The few experiments performed to date with cells obtained from patients with inactive RF also show increased cytotoxicity for human atrial myofibers. Once again, the addition of heart-reactive antibody from the sera of these patients did not enhance the cytotoxicity. These results would suggest that cellular autoreactivity by activated T cells (21) may play a more important role in the rheumatic cardiac lesion than was previously thought.

In this context, recent observations by Cunningham and coworkers (12) are worth mentioning and may correlate with observations made by Rose and colleagues (48) in an experimental model of cardiomyopathy (see below). Using streptococcus-induced monoclonal antibodies that cross-react with M protein and human cardiac myosin, they could neutralize coxsackievirus groups B3 and B4. The virus-neutralizing antibodies were also cytotoxic for heart and fibroblast cell lines and reacted with viral capsid proteins on Western blots.

Given the evidence of an abnormal host immune response to streptococcal antigens cross-reactive with relevant target antigens, the tantalizing and as yet unresolved question is whether there is a genetic basis for this response in patients with RF. Beginning with Cheadle's work (9) 100 years ago, numerous investigators have presented evidence for and against this concept. Recent articles have suggested that susceptibility to ARF following streptococcal infection might be related to alleles of the human major histocompatibility complex (MHC), but the association with a given phenotype has been either controversial or inconclusive.

To date the strongest association of a cell marker with RF has been the identification of a B-cell marker apparently unrelated to MHC by Patarroyo et al. in Colombia and our group in New York City (43). The relative risk of contracting the disease if one bore the marker was 12.9. Because of the limited supply of the original allo-antiserum used to identify the marker, we embarked on a series of studies in which we attempted to reproduce this marker using hybridoma technology and immunization of mice with B cells isolated from patients with RF. Of a number of monoclonal antibodies, we have recently isolated one called D8/17 which identifies all patients with RF tested in a number of widely different geographic areas of the world (33). The antibody is also not related to MHC, and studies with families indicate that this marker is inherited in an autosomal recessive fashion. The exact nature of the antigen is as yet unknown, but our experiments suggest that it is not present on other lymphoid cells. The antigen is, however, also expressed on other tissues, most notably, the heart and smooth muscle cells. What role this antigen plays in the disease process is unknown at present, but its expression on B cells from all patients with RF strongly suggests its direct involvement in the etiopathogenesis of RF.

CHAGAS' DISEASE

Although rarely seen in the more temperate climates, Chagas' disease is endemic in many parts of the world, most notably in South America, where an estimated 11 million or more people are infected with *Trypanosoma cruzi.* The acute disease is the result of multiplication of the organism in the reticuloendothelial system followed by direct invasion of muscle tissue, the liver, and occasionally the meninges. In contrast, in chronic Chagas' cardiomyopathy, the organism is absent but there is a progressive inflammatory cardiac myofibrillar degeneration with necrosis. An important clue to the etiology of the chronic form of the disease was provided by Cossio and coworkers (11), who noted that the sera from a high percentage of these patients contained antibodies reactive with components of cardiac sarcolemma as well as with the valvular endothelial cells. A causal role for this antibody in disease pathogenesis was suggested when cardiac-specimen biopsy from patients with Chagas' disease demonstrated immunoglobulin bound to cardiac myofibers. Antibody eluted from this tissue bound to myofibers of unaffected heart tissue in a pattern similar to that seen in the patient.

In addition to the presence of immunoglobulin in these biopsy specimens, numerous investigators have also commented on the presence of chronic lymphocytic infiltrates in these tissues, suggesting that abnormalities in cellular reactivity might also be involved in the initiation and/or perpetuation of target organ damage. von Kreuter and Santos-Buch (63) examined cellular immunity in a mouse model of Chagas' disease and noted that sensitized cells from these animals exhibited cytotoxic activity for mouse heart cells in culture. As with RF (see above), the addition of heart-reactive antibody from these animals did not enhance the cytotoxicity, again suggesting that cellular autoimmunity to heart antigens might play the primary role in this disease.

As with the chorea of RF, a new chapter is being written on the relationship of the *T. cruzi* antigens to structures of the parasympathetic nervous system that cause the neurological manifestations of the disease. Several investigators have now reported antigenic cross-reactions between *T. cruzi* and neuronal and glial antigens. Using cloning techniques and recombinant technology, Van Voorhis and Eigen (62) have shown that a 160-kDa *T. cruzi* flagellar protein cross-reacts with a 48-kDa human neuronal antigen. Immunoperoxidase studies of brain sections revealed that at least part of the reaction is related to a neuronal protein of the mysenteric plexus (also found in the neurons in the ileum), perhaps explaining the observed megacolon and megaesophagus so often seen in patients with Chagas' disease. Interestingly, this cross-reaction was also localized to the Purkinje's fibers of the brain.

In summary, autoimmune phenomena have been reported and implicated in the etiology of progressive cardiac damage in both experimentally infected animals and humans with chronic Chagas' disease (11, 24). It has been suggested that both changes in the T-cell repertoire (38) and molecular mimicry (62) might be implicated in the pathogenesis of the autoimmune phenomena seen following *T. cruzi* infection. Yet another intriguing possibility is a chronic immune response to local parasites that may persist in cardiac tissue during the chronic stage of the disease (29). d'Imperio Lima et al. (14) have described persistent B-cell activation and immunoglobulin production in the murine model of *T. cruzi* infection, even when injected parasites are no longer detected in the blood. Finnegan et al. (21) have reported that autoreactive $CD4^+$ T-cell clones are capable of polyclonal activation of immunoglobulin production by B cells in both MHC-restricted and noncognate fashions.

Recently, Freire-de-Lima and colleagues (23) have reported that splenic $CD4^+$ T cells from chronically infected mice have markedly enhanced autoreactivity toward normal syngeneic but not allogenic resting B cells. This suggests that enhanced $CD4^+$ T-cell autoreactivity can be a potential cause of persistent B-cell activation and autoantibody production in patients with chronic Chagas' disease.

CARDIOMYOPATHIES

It has long been known that numerous microbes, parasites, and viruses can cause acute myocarditis. In these clinical syndromes the organism can be readily demonstrated in the tissues of the individual. In contrast, there is an ill-defined group of cardiomyopathies in which the offending organism that caused the acute inflammation is no longer present, yet the cardiomyopathy continues to progress. These cases are distinguished by the presence of lymphocytic infiltrations and, in a majority of cases, increased titers of heart-reactive antibodies. Among the agents considered to be important causal agents in this group of cardiomyopathies have been the group B coxsackieviruses. These agents frequently cause acute cardiac inflammation in humans, and 10 to 12% of patients go on to develop a more chronic phase of the disease. Because chronic and progressive myocarditis is restricted to a small number of patients, it has also been suggested that there might be a genetic basis for the disease.

This concept has been greatly strengthened by the elegant work of Rose and colleagues (48) with an animal model of coxsackievirus group B3-induced myocarditis. Using genetically different strains of mice, they could induce a chronic myocarditis primarily in *H-2* congenic strains bearing an A background (i.e., ACA and ASW mice), whereas C57BL/10 *H-2* mice were resistant. A key finding in those studies was that only those mice with heart-reactive antibodies in their sera went on to develop chronic cardiomyopathy; these antibodies were primarily directed against the cardiac isoform of myosin. A causal role for myosin in the etiology of this cardiomyopathy was shown when injection of myosin alone was able to induce severe immunologically mediated myocarditis in certain genetic strains of mice.

Hoping to demonstrate that myosin and the coxsackie B virus shared common antigenic determinants, Rose and colleagues (48) were disappointed in their inability to show that heart-reactive antibodies in their coxsackievirus group B3-induced cardiomyopathy cross-reacted with viral antigens. In contrast, Saegusa et al. (49a), in preparing a series of different monoclonal antibodies against coxsackie B4 virus, identified one which both neutralized the coxsackie B4 virus and bound to rabbit and mouse heart tissues. An important observation was that this antibody did not bind to human heart tissue, suggesting that both the strain of virus and the myosin of a given species may be crucial for the induction of cardiac disease. Whether or not viruses other then coxsackie B virus are involved in chronic human cardiomyopathies and whether this progression is on a genetic basis, although unknown at present, are particularly fertile fields for future study.

KLEBSIELLA AND ANKYLOSING SPONDYLITIS

One of the most intriguing "experiments of nature" was recorded by Noer (41), who investigated an outbreak of *Shigella* dysentery involving 607 men among 1,276 sailors aboard an aircraft carrier. Long-term follow-up of these men revealed that 10 men developed Reiter's syndrome. Four of five (80%) of these individuals were HLA-B27 positive.

This extraordinary story ushered in the concept of microbe-host mimicry involving *Klebsiella* and *Shigella* antigens and specific antigens of MHC. The first observations were noted by Ebringer and Ebringer and colleagues (15–18), who reported that antisera prepared against HLA-B27 lymphocytes reacted with *Klebsiella pneumoniae* antigens and was more common in the feces of ankylosing spondylitis (AS) and acute anterior uveitis patients and in those individuals with active disease than in the feces of patients with inactive disease. On a more molecular level, Ogawasara and colleagues (42) used an HLA-B27 monoclonal antibody and showed that it was reactive against the 60- and 80-kDa antigens of *K. pneumoniae* which were not seen in other gram-negative bacteria. Subsequent studies showed that six amino acids (OTDRED) were shared between HLA-B27.1 and the nitrogenase reductase enzyme of *K. pneumoniae* (53). Most recently, Ewing et al. (20) synthesized 74 overlapping peptides from the *K. pneumoniae* nitrogenase reductase (residues 181 to 199) and from the HLA-B27.1 molecule (residues 65 to 85), which were then tested against sera from patients with AS and compared to sera from B27-positive and B27-negative close relatives of the patients with AS. The strongest reactivity was seen to the *Klebsiella* peptide NSRQTDR, while the same sera reacted to the HLA-B27 peptide KAKAQTDR. This reactivity was not seen in the individuals who did not have AS. In summary, it appears that sera from AS patients contain antibodies that react to both *K. pneumoniae* nitrogenase peptides and HLA-B27.1 peptides. Furthermore, it appears that there are at least two epitopes on HLA-B27.1 in the α-1 domain in the MHC genome region that are autoantigenic in AS patients.

MYCOBACTERIUM TUBERCULOSIS ANTIGENS

Since the pathogenic mechanisms underlying rheumatoid arthritis remain unclear, many investigators have turned to experimental models of arthritis in an effort to further understand the disease in humans. Among these models, the adjuvant arthritis (AA) model with heat-killed *Mycobacterium tuberculosis* has been studied extensively (8, 44). The antigenic epitope recognized by arthritogenic T-cell clone A2B was identified as residues 180 to 188 of the 65-kDa mycobacterial heat shock protein (hsp) (61). Attempts to use the 65-kDa protein to induce AA failed, but immunization with the protein induced a state of resistance to AA induction by the large hsp of *M. tuberculosis* (7). An interesting and potentially important therapeutic observation was that tolerance induced by hsp65 conferred protection against the other types of rat experimental arthritis such as that induced by streptococcal cell walls (59) and collagen type II (58).

Mycobacterial hsp65 belongs to the hsp60 family of hsps. It is highly conserved and shares 48% amino acid identity with the mammalian homologue PI or hsp60 (28). It also cross-reacts with antigens present on synovial fluid membranes obtained from juvenile chronic arthritis patients (5). The level of expression of this protein has also been shown to be elevated in the inflamed synovia and subcutaneous nodules of rheumatoid arthritis patients (31). More recently, Anderton and colleagues (2) prepared a series of peptides that spanned the hsp65 protein and found that nine hsp65 epitopes had MHC class II-restricted T-cell responses. These observations were followed by experiments in which they tested each hsp65 epitope for its ability to offer protective immunity against experimental arthritis. Only one peptide (amino acids 256 to 270) was able to confer protection against AA. Of interest, this same peptide also protected against arthritis induced by cp 20961, a synthetic aminolipid (3). In contrast to the belief that cross-reactive antigens to self might induce more active disease, these studies indicate that cross-reactivity between bacterial hsp65 and self hsp60 might maintain a protective self-reactive T-cell population.

INTESTINAL ANTIGENS

There has been a long and somewhat controversial history of cross-reactions between *Escherichia coli* antigens and antigens of the mammalian intestinal tract. Perlmann et al. (45) showed years ago that the sera of patients with ulcerative colitis contain antibodies that bind to both intestinal and *E. coli* antigens. An important observation was that the cross-reactions were primarily seen with fetal intestinal antigens (37). The fact that the observed cross-reactions have been seen in other inflammatory conditions does not necessarily detract from these observations. As pointed out above, many cross-reactions between microbes and hosts exist. It is only when these cross-reactions occur in the genetically susceptible host that tissue damage to the target organ may result.

From these somewhat simplified beginnings, evidence from more recent investigations has led to the concept of a much more complicated interaction between host and intestinal flora in which numerous nonspecific toxins, cell wall peptidoglycans, and lipopolysaccharides derived from a variety of organisms can initiate and trigger an inflammatory response in the bowel.

It is also clear that lymphokines and cytokines play an important role in this inflammatory response. For example, deletion of interleukin 10 (IL-10) or transforming growth factor β genes in mice results in chronic intestinal inflammation (35, 36). Genetic deletion of IL-2 or T-cell receptors also leads to spontaneous colitis (39, 40). These experiments imply that disturbing T-lymphocyte regulatory circuits can have detrimental effects on the bowel. An important point to remember is that the presence of the normal intestinal flora is also important in the regulation of the disease. Thus, Lewis rats with normal pathogen-free intestinal flora show chronic midbowel, small ulcers after indomethacin treatment, while littermates raised under germfree conditions do not (52). IL-2 "knockout" mice (49) and HLA-B27 transgenic rats (56) that develop active colitis under normal intestinal flora conditions have no evidence of intestinal inflammation when raised in a germfree environment.

RHEUMATOID ARTHRITIS

The experimental models of such diseases as streptococcal cell wall and AA strongly suggest that cross-reactive hsps of microbial and/or mammalian origin might be involved in the disease in humans. Alternatively, the peptidoglycan fractions of various organisms might directly activate cytokine cascades such as tumor necrosis factor alpha (TNF-α) and IL-1. The latter concept is particularly attractive in that the peptidoglycan moiety is very similar in all gram-positive organisms (including mycobacteria), with modifications mainly in the side chains of the molecules. The remarkable conformational similarity between lipopolysaccharide and the peptidoglycan structures suggests that both gram-positive and gram-negative structures (Fig. 2) are capable of binding to a single specific 70-kDa site on macrophages, thereby releasing TNF-α and other cytokines (46). As noted above, all these molecules are capable of initiating an inflammatory cascade.

PEPTIDOGLYCAN

HO CH2 HO CH2 O O O NH CO CH3 H O H— C O CH4 NH CO CH3 CO

L- Ala
D-Glu-NH2
L-Lys — L- Ala-L-Ala — D-Ala
D — Ala L-Lys
(D— Ala) D-Glu
L- Ala
(MurNAc-GlcNAc)

LIPID A

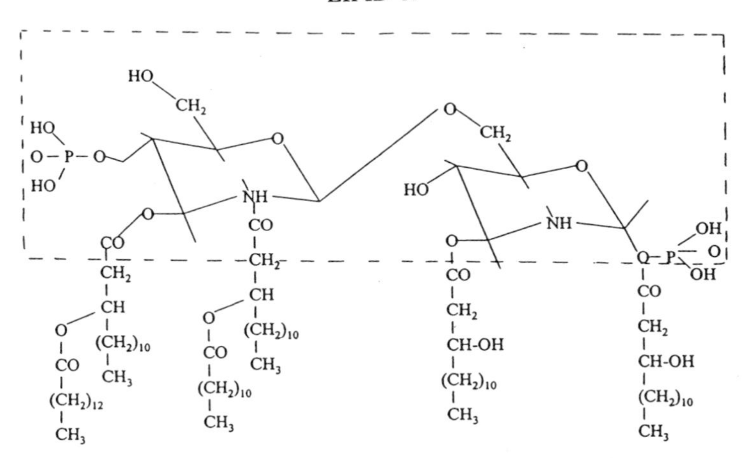

Figure 2. Schematic drawing of the two-dimensional structure of lipid A of gram-negative organisms and the peptidoglycan structure of gram-positive organisms. While there are obvious differences in their structures, the stenciled areas of each structure are quite similar.

Since microbes can enter the body in a variety of ways, including the intestine, and given the fact that the peptidoglycan moieties are remarkably similar in many organisms, these molecules could provide the intermittent stimulus for the waxing and waning symptoms seen in patients with rheumatoid arthritis. This could occur either directly on target cells (macrophages) or as a result of being carried via macrophages to the synovial tissue. These molecules could also cause the reactive arthritis seen in both patients with Crohn's disease and patients with ulcerative colitis. In this context, it is important to note that in experimental arthritis, animals raised in a germfree environment do not develop arthritis (56).

Numerous investigators have reported on the presence of autoantibodies in rheumatoid arthritis patients, and these antibodies have reacted with various molecules including histones, keratin, collagen, cardiolipins, gliadin of the intestinal mucosa, and self-immunoglobulin G (self-IgG) (1). Exactly what role these autoantibodies play in the disease process remains unclear, and as for all autoantibodies, there is always a question of whether they are primary or secondary to tissue damage. In terms of pathogenesis, perhaps the most interesting of these autoantibodies is the rheumatoid factor, which is a complex of immunoglobulin M (IgM) antibodies directed to the Fc portion of IgG. Work by Bokisch et al. (4) a number of years ago suggested that the binding of IgG to microbial antigens induced a conformational change in the IgG molecule, thus allowing the IgM molecules to bind to the altered form of the IgG molecule. These complexes of antibodies (perhaps containing antigen) could migrate to the synovium and thus induce inflammation in the joint either as a complex or via direct stimulation of synovial cells by the antigen(s) (10). The levels of these high-affinity B-cell rheumatoid factors do correlate with disease severity, and the production of cytokines and complement deposition also show a high degree of correlation. Clinically, perhaps the cytokine most released in the joint is TNF-α (7), and the recent clinical success in using the soluble TNF-α receptors in patients with rheumatoid arthritis suggests that this may be the most inflammatory molecule in the disease. Notably, blockage of this cytokine results in disease remission both in humans (40) and in the streptococcal cell wall arthritis model (S. Kerwar, unpublished data).

While not completely relevant to this chapter, a recent publication by Bradley and coworkers (6) points out the significance of the MHC in an animal model of collagen antibodies. On the basis of the knowledge that certain HLA-DR4 subtypes predispose an animal to rheumatoid arthritis while other DR4 haplotypes do not, they generated mice transgenic for human HLA-DQ6, an allele associated with the nonsusceptible haplotype. These mice were resistant to collagen antibodies, while mice that expressed both DQ6 and DQ8 developed moderate levels of collagen antibodies. In contrast, DQ8 mice developed severe arthritis. These studies point out the importance of the genetic background in this model and also emphasize that HLA-DQ polymorphism may also turn out to play an important role in rheumatoid arthritis.

SPONDYLOARTHROPATHIES

An association between an environmental agent and certain components of the MHC now appears to be clear for the spondyloarthropathies, most notably, AS and Reiter's syndrome. For example, when one examines certain North American Indian tribes such as the Pima, Chippewa, and Bella Coola, the prevalence of AS ranges between 27 and 93/1,000 population, whereas its prevalence is 0.5/1,000 population in England. If one also looks at HLA-B27 by racial or ethnic group, the frequency of HLA-B27 in Pima or Chippewa Indians ranges between 17 and 23%, whereas it is 8% among Caucasians in England and the United States (26).

As noted in the section on cross-reactivity, there appears to be a clear molecular association between structures of the HLA-B27 allele and antigens of *Klebsiella*. This is a clear case of molecular mimicry between host and microbes, and the sera of AS patients contain antibodies that are cross-reactive with both molecules (16). Experimental data underline the importance of these observations. Taurog and Hammer (54) developed rats transgenic for HLA-B27 and human β_2-microglobulin. Of five lines produced, two (lines 21-44 and 33-3) developed spontaneous inflammatory disease closely resembling the B27-associated disease in humans. Furthermore, the disease-prone lines had higher levels of expression of B27 thymic mRNA and splenic cell surface protein by the time of disease onset. Thus, disease susceptibility appeared to correlate with both the gene copy number and the quantity of B27 in lymphoid cells (55).

The crucial and as yet unresolved question is: What are the underlying mechanisms involved in this relationship that actually initiate the disease process? Do antibodies to the MHC domains suppress the normal immune response to a microbial stimulus, thereby increasing the autoreactive response to the target tissues? Does the MHC binding in some manner initiate an inflammatory cascade in the joints with direct stimulation by the microbial antigens? All of these questions need to be explored further. On the experimental side, the observations from studies with transgenic mice raise interesting questions regarding the microbe-host relationship. If these animals are raised in a germfree environment, do they get the disease? Recent studies suggest that the answer is that they do not (56). At what point in the evolution of disease in those animals do we see antibodies to microbial antigens that cross-react with HLA-B27? Obviously, much additional work needs to be done. Yet, the observation that a human HLA-B27 gene produces an inflammatory disease in transgenic mice opens up a large new field of investigation, especially with respect to the arthritogenic properties of the antigens involved.

ULCERATIVE COLITIS AND CROHN'S DISEASE

Once again, there would appear to be in the host a genetic component which creates an abnormal immune or inflammatory response to certain microbes. In this case, it is the intestinal antigens which provide the inflammatory stimulus. Clinical evidence for a genetic factor may be found in racial differences of increased susceptibility in Caucasians compared to the susceptibilities of African-Americans and Asians, the increased susceptibilities of Jewish individuals compared to those of non-Jewish individuals, and the increased incidence of Crohn's disease compared to that of ulcerative colitis observed in monozygotic twins (51). Genetic regulation of immunoresponsiveness has also been involved, especially with regard to HLA class II genes. Statistically significant associations have been established for HLA-DR2 in ulcerative colitis and HLA-DR1 DQW5 in Crohn's disease (57, 64).

While it is almost impossible to identify a given intestinal microbe as the main cause of these diseases, a common denominator in these diseases would appear to be the peptidoglycan-polysaccharide complexes (PPCs) common to many intestinal organisms. Recently, by using a monoclonal antibody directed against this molecule, PPCs were detected intracellularly in the mucosa and submucosa of Crohn's disease patients, whereas they were not detected in controls. At a cellular level, the lymphocytes of patients with active Crohn's disease had elevated responses to the PPCs compared to the responses of lymphocytes of controls (34). This fact, coupled with the observation that the lipopolysaccharide structures of gram-negative organisms and peptidoglycans are quite similar and the fact that each can stimulate TNF-α (and probably other cytokine) production, could account for the necessary inflammatory stimulus.

Thus, these molecules may either directly or when they are attached to macrophages provide the arthritogenic stimulus so often seen in these two diseases. Examination of the synovial fluids for local production of cytokines as well as immunological evidence of intracellular antigens in macrophages would be an important step in understanding this concept.

SIGNIFICANCE

It is now clear that many microbes share antigenic determinants not only with other microbes but also with a variety of human tissues. Cross-reactions between streptococci and heart tissue, *E. coli* antigens and colonic antigens, HLA-B27 and *Shigella* antigens, and gram-negative bacteria and blood groups are but a few examples of these cross-reactions seen in nature. The majority of these cross-reactions are generally harmless and may serve the purpose of protecting humans from a wide variety of microbes which might be pathogenic. However, the combination of the right microbe infecting a genetically susceptible host may result in serious autoimmune consequences.

As indicated above, perhaps the two best examples of microbially induced autoimmune cardiomyopathies are RF and Chagas' disease. The cross-reactions between host and relevant target organs have been well-defined for those diseases; there is an abnormal host cellular and humoral response to tissue and microbial antigens. Furthermore, in patients with RF there appears to be a genetic predisposition to the disease. Whether this genetic predisposition allows breakage of tolerance on the part of the host, an abnormal response to the specific cross-reactive determinant, or production of anti-idiotypic antibodies that bear the cross-reactive epitopes is unknown and will require further investigation.

In terms of the immune response of the host, it is becoming increasingly apparent that the cellular response to cross-reactive antigens may play a more significant role in the observed pathological damage then was previously thought. In both RF and Chagas' disease, T cells are specifically cytotoxic for the target organ, and the addition of antibody does not enhance the cytotoxicity, suggesting a more crucial role for cellular immunity in the disease process. Similar studies with experimental models of autoimmune heart disease point to the same conclusion. To date, these studies have not been done for either the PPS syndrome or the idiopathic cardiomyopathies.

In more general terms, what is the significance of heart-reactive antibodies in these disease states? It is well known that the levels of organ-specific antibodies increase with age, appear in a variety of other disease states (e.g., lupus and rheumatoid arthritis), and in many instances do not appear to cause damage. Even for RF and Chagas' disease, is it a question of "what came first: the chicken or the egg?"

In other words, are these cross-reactive antibodies directly involved in the disease process or are they merely a reflection of prior heart damage and a fortuitous cross-reaction with a given microbe. The evidence in RF suggests that the latter concept is not true since these heart-reactive antibodies are absorbed by both streptococcal and cardiac antigens, while heart-reactive antibodies in PPS are absorbed only by cardiac tissue. Thus, in RF the antiheart antibody appears to relate to the infection, whereas in postpericardiotomy syndrome the antibody appears to relate to tissue damage (19).

Whether or not these heart-reactive antibodies (from whatever origin) play a direct role in the disease process is difficult to ascertain with certainty. They do correlate with the disease state and disappear during convalescence; they are seen in the pathological tissue specimens, but they do not apparently kill target organ cells or enhance specific cellular cytotoxicity in vitro. Perhaps complement activation and/or other factors are needed in situ to produce the observed immunoglobulin staining seen in the lesions of several cardiomyopathies. To our knowledge, these experiments have not been attempted in vitro.

What is becoming increasingly clear not only in cardiac autoimmunity but also in other autoimmune diseases such as rheumatic arthritis is that cellular activation specific for a given target tissue appears to be important in the pathogenesis of disease. Cytotoxic T cells specific for cardiac myofibers are seen in both RF and Chagas' disease, and it is our impression that they may be important in the pathogenesis of other cardiomyopathies as well. Obviously, it will be important to define the exact nature of the cross-reactive epitopes recognized by these active cells and their role in causing cytotoxicity to human tissue.

Finally, the specific mechanisms operating in the interaction between host genetic factors and the environment that result in the induction of autoimmune disease of the heart are still not clear. In RF there appears to be a marker called D8/17 which is inherited in an autosomal fashion and which is preferentially increased in the B cells of patients with RF. However, it is well known that not all strains of group A streptococci cause RF, even in the genetically susceptible individual. The most plausible explanation for this discrepancy is that only certain strains exhibit the epitopes that activate the immune system to produce tissue cross-reactive antibodies and activated T cells. Then, only in the context of the immune response (a heightened response, break in tolerance, etc.) will the disease occur. A similar mechanism can be postulated for the other cardiomyopathies, as it appears that only select individuals will develop progressive autoimmune disease following active infection. Future studies of each of the cardiopathies mentioned above not only should concentrate on the immune response of the host to a given cross-reactive antigen (either in the microbe or in the host), but should examine more carefully whether or not a particular microbe exhibits a specific epitope.

Obviously, the field is ripe for application of the modern tools of molecular biology to the study of these intriguing host-microbe relationships in cardiac disease.

REFERENCES

1. **Albani, S., and D. A. Carson.** 1997. Etiology and pathogenesis of rheumatoid arthritis, p. 985. *In* W. J. Koopman (ed.), *Arthritis and Allied Conditions,* vol. 1 The Williams & Wilkins Co., Baltimore, Md.
2. **Anderton, S. M., R. van der Zee, A. Noordzij, and W. van Eden.** 1994. Differential mycobacterial 65-kDa heat shock protein T cell epitope recognition after adjuvant arthritis-inducing or protective immunization protocols. *J. Immunol.* **152:**3656–3664.
3. **Anderton, S. M., R. van der Zee, B. Prakken, A. Noordzij, and W. van Eden.** 1995. Activation of T cells recognizing self 60-kD heat shock protein can protect against experimental arthritis. *J. Exp. Med.* **181:**943–952.
4. **Bokisch, V. A., D. Bernstein, and R. M. Krause.** 1972. Occurrence of 19S and 7S anti-IgGs during hyperimmunization of rabbits with streptococci. *J. Exp. Med.* **136:**799–815.

5. **Boog, C. J., E. R. de Graeff-Meeder, M. A. Lucassen, R. van der Zee, M. M. Voorhorst-Ogink, P. J. van Kooten, H. J. Geuze, and W. van Eden.** 1992. Two monoclonal antibodies generated against human hsp60 show reactivity with synovial membranes of patients with juvenile chronic arthritis. *J. Exp. Med.* **175:**1805–1810.
6. **Bradley, D. S., G. H. Nabozny, S. Cheng, P. Zhou, M. M. Griffiths, H. S. Luthra, and C. S. David.** 1997. HLA-DQB1 polymorphism determines incidence, onset, and severity of collagen-induced arthritis in transgenic mice. Implications in human rheumatoid arthritis. *J. Clin. Investig.* **100:**2227–2234.
7. **Brennan, F. M., R. N. Maini, and M. Feldmann.** 1992. TNF alpha—a pivotal role in rheumatoid arthritis? *Br. J. Rheumatol.* **31:**293–298.
8. **Chang, Y. H., C. M. Pearson, and C. Abe.** 1980. Adjuvant polyarthritis IV induction by a synthetic adjuvant: immunological, histopathologic, and other studies. *Arthritis Rheum.* **23:**735–741.
9. **Cheadle, W. B.** 1889. Harvean lectures on the various manifestations of the rheumatic state as exemplified in childhood and early life. *Lancet* **i:**821–827.
10. **Chien, Y. H., and M. M. Davis.** 1993. How alpha beta T-cell receptors "see" peptide/MHC complexes. *Immunol. Today* **14:**597–602.
11. **Cossio, P. M., R. P. Laguens, C. Diez, A. Szarfman, A. Segal, and R. M. Arana.** 1974. Chagasic cardiopathy. Antibodies reacting with plasma membrane of striated muscle and endothelial cells. *Circulation* **50:**1252–1259.
12. **Cunningham, M. W., S. M. Antone, J. M. Gulizia, B. M. McManus, V. A. Fischetti, and C. J. Gauntt.** 1992. Cytotoxic and viral neutralizing antibodies crossreact with streptococcal M protein, enteroviruses, and human cardiac myosin. *Proc. Natl. Acad. Sci. USA* **89:**1320–1324.
13. **Damian, R. T.** 1964. Molecular mimicry: antigen sharing by parasite and host and its consequences. *Am. Nat.* **98:**129–149.
14. **d'Imperio Lima, M. R., H. Eisen, P. Minoprio, M. Joskowicz, and A. Coutinho.** 1986. Persistence of polyclonal B cell activation with undetectable parasitemia in late stages of experimental Chagas' disease. *J. Immunol.* **137:**353–356.
15. **Ebringer, A.** 1983. The cross-tolerance hypothesis, HLA-B27 and ankylosing spondylitis. *Br. J. Rheumatol.* **22**(4 Suppl. 2)**:**53–66.
16. **Ebringer, A., P. Cowling, N. Ngwa-Suh, D. C. O. James, and R. Ebringer.** 1976. Cross-reactivity between Klebsiella aerogenes species and B27 lymphocyte antigens as an aetiological factor in ankylosing spondylitis, p. 27. *In* J. Dausset and A. Svejgaard (ed.), *HLA and Disease,* vol. 8. Institut National de la Santé et da la Recherch Médicale, Paris, France.
17. **Ebringer, R., D. Cawdell, and A. Ebringer.** 1979. Klebsiella pneumoniae and acute anterior uveitis in ankylosing spondylitis. *Br. Med. J.* **1**(6160)**:**383.
18. **Ebringer, R. W., D. R. Cawdell, P. Cowling, and A. Ebringer.** 1978. Sequential studies in ankylosing spondylitis. Association of Klebsiella pneumoniae with active disease. *Ann. Rheum. Dis.* **37:**146–151.
19. **Engle, M. A., J. C. McCabe, P. A. Ebert, and J. Zabriskie.** 1974. The postpericardiotomy syndrome and antiheart antibodies. *Circulation* **49:**401–406.
20. **Ewing, C., R. Ebringer, G. Tribbick, and H. M. Geysen.** 1990. Antibody activity in ankylosing spondylitis sera to two sites on HLA B27.1 at the MHC groove region (within sequence 65–85), and to a Klebsiella pneumoniae nitrogenase reductase peptide (within sequence 181–199). *J. Exp. Med.* **171:**1635–1647.
21. **Finnegan, A., B. W. Needleman, and R. J. Hodes.** 1990. Function of autoreactive T cells in immune responses. *Immunol. Rev.* **116:**15–31.
22. **Fischetti, V. A.** 1989. Streptococcal M protein: molecular design and biological behavior. *Clin. Microbiol. Rev.* **2:**285–314.
23. **Freire-de-Lima, C., L. M. Pecanha, and G. A. Dos Reis.** 1996. Chronic experimental Chagas' disease: functional syngeneic T-B-cell cooperation in vitro in the absence of an exogenous stimulus. *Infect. Immun.* **64:**2861–2866.
24. **Gazzinelli, R. T., M. J. Morato, R. M. Nunes, J. R. Cancado, Z. Brener, and G. Gazzinelli.** 1988. Idiotype stimulation of T lymphocytes from Trypanosoma cruzi-infected patients. *J. Immunol.* **140:**3167–3172.
25. **Guilherme, L., E. Cunha-Neto, V. Coelho, R. Snitcowsky, P. M. Pomerantzeff, R. V. Assis, F. Pedra, J. Neumann, A. Goldberg, M. E. Patarroyo, et al.** 1995. Human heart-infiltrating T-cell clones from rheumatic heart disease patients recognize both streptococcal and cardiac proteins. *Circulation* **92:**415–420.

26. **Hochberg, M. C.** 1984. Epidemiology, p. 1–42. *In* A. Calvin (ed.), *Spondyloarthropathies*. Grune and Stratton, New York, N.Y.
27. **Husby, G., I. van de Rijn, J. B. Zabriskie, Z. H. Abdin, and R. C. Williams, Jr.** 1976. Antibodies reacting with cytoplasm of subthalamic and caudate nuclei neurons in chorea and acute rheumatic fever. *J. Exp. Med.* **144:**1094–1110.
28. **Jindal, S., A. K. Dudani, B. Singh, C. B. Harley, and R. S. Gupta.** 1989. Primary structure of a human mitochondrial protein homologous to the bacterial and plant chaperonins and to the 65-kilodalton mycobacterial antigen. *Mol. Cell. Biol.* **9:**2279–2283.
29. **Jones, E. M., D. G. Colley, S. Tostes, E. R. Lopes, C. L. Vnencak-Jones, and T. L. McCurley.** 1992. A Trypanosoma cruzi DNA sequence amplified from inflammatory lesions in human chagasic cardiomyopathy. *Trans. Assoc. Am. Physicians* **105:**182–189.
30. **Kaplan, M. H., and J. D. Frengley.** 1969. Autoimmunity to the heart in cardiac disease. Current concepts of the relation of autoimmunity to rheumatic fever, postcardiotomy and postinfarction syndromes and cardiomyopathies. *Am. J. Cardiol.* **24:**459–473.
31. **Karlsson-Parra, A., K. Soderstrom, M. Ferm, J. Ivanyi, R. Kiessling, and L. Klareskog.** 1990. Presence of human 65 kD heat shock protein (hsp) in inflamed joints and subcutaneous nodules of RA patients. *Scand. J. Immunol.* **31:**283–288.
32. **Kemeny, E., G. Husby, R. C. Williams, Jr., and J. B. Zabriskie.** 1994. Tissue distribution of antigen(s) defined by monoclonal antibody D8/17 reacting with B lymphocytes of patients with rheumatic heart disease. *Clin. Immunol. Immunopathol.* **72:**35–43.
33. **Khanna, A. K., D. R. Buskirk, R. C. Williams, Jr., A. Gibofsky, M. K. Crow, A. Menon, M. Fotino, H. M. Reid, T. Poon-King, P. Rubinstein, et al.** 1989. Presence of a non-HLA B cell antigen in rheumatic fever patients and their families as defined by a monoclonal antibody. *J. Clin. Investig.* **83:**1710–1716.
34. **Klasen, I. S., M. J. Melief, A. G. van Halteren, W. R. Schouten, M. van Blankenstein, G. Hoke, H. de Visser, H. Hooijkaas, and M. P. Hazenberg.** 1994. The presence of peptidoglycan-polysaccharide complexes in the bowel wall and the cellular responses to these complexes in Crohn's disease. *Clin. Immunol. Immunopathol.* **71:**303–308.
35. **Kuhn, R., J. Lohler, D. Rennick, K. Rajewsky, and W. Muller.** 1993. Interleukin-10-deficient mice develop chronic enterocolitis. *Cell* **75:**263–274.
36. **Kulkarni, A. B., C. G. Huh, D. Becker, A. Geiser, M. Lyght, K. C. Flanders, A. B. Roberts, M. B. Sporn, J. M. Ward, and S. Karlsson.** 1993. Transforming growth factor beta 1 null mutation in mice causes excessive inflammatory response and early death. *Proc. Natl. Acad. Sci. USA* **90:**770–774.
37. **Lagercrantz, R., S. Hammarstrom, P. Perlmann, and B. E. Gustafsson.** 1968. Immunological studies in ulcerative colitis. IV. Origin of autoantibodies. *J. Exp. Med.* **128:**1339–1352.
38. **Minoprio, P., S. Itohara, C. Heusser, S. Tonegawa, and A. Coutinho.** 1989. Immunobiology of murine T. cruzi infection: the predominance of parasite-nonspecific responses and the activation of TCRI T cells. *Immunol. Rev.* **112:**183–207.
39. **Mombaerts, P., E. Mizoguchi, M. J. Grusby, L. H. Glimcher, A. K. Bhan, and S. Tonegawa.** 1993. Spontaneous development of inflammatory bowel disease in T cell receptor mutant mice. *Cell* **75:**274–282.
40. **Moreland, L. W., S. W. Baumgartner, M. H. Schiff, E. A. Tindall, R. M. Fleischmann, A. L. Weaver, R. E. Ettlinger, S. Cohen, W. J. Koopman, K. Mohler, M. B. Widmer, and C. M. Blosch.** 1997. Treatment of rheumatoid arthritis with a recombinant human tumor necrosis factor receptor (p75)-Fc fusion protein. *N. Engl. J. Med.* **337:**141–147.
41. **Noer, H. R.** 1966. An "experimental" epidemic of Reiter's syndrome. *JAMA* **198:**693–698.
42. **Ogawasara, M., D. H. Kono, and D. T. Y. Yu.** 1986. Mimicry of human histocompatibility HLA-B27 antigens by *Klebsiella pneumoniae. Infect. Immun.* **51:**901–908.
43. **Patarroyo, M. E., R. J. Winchester, A. Vejerano, A. Gibofsky, F. Chalem, J. B. Zabriskie, and H. G. Kunkel.** 1979. Association of a B-cell alloantigen with susceptibility to rheumatic fever. *Nature* **278:**173–174.
44. **Pearson, C. M.** 1956. Development of arthritis, periarthritis and periostitis in rats given adjuvant. *Proc. Soc. Exp. Biol. Med.* **91:**95–101.
45. **Perlmann, P., S. Hammarstrom, R. Lagercrantz, and B. E. Gustafsson.** 1965. Antigen from colon of germfree rats and antibodies in human ulcerative colitis. *Ann. N. Y. Acad. Sci.* **124:**377–394.

46. **Rabin, R. L., M. M. Bieber, and N. N. Teng.** 1993. Lipopolysaccharide and peptidoglycan share binding sites on human peripheral blood monocytes. *J. Infect. Dis.* **168:**135–142.
47. **Read, S. E., V. A. Fischetti, V. Utermohlen, R. E. Falk, and J. B. Zabriskie.** 1974. Cellular reactivity studies to streptococcal antigens. Migration inhibition studies in patients with streptococcal infections and rheumatic fever. *J. Clin. Investig.* **54:**439–450.
48. **Rose, N. R., K. W. Beisel, A. Herskowitz, M. Neu, L. J. Wolfgram, F. Alvarez, M. D. Traystman, and S. W. Craig.** 1987. Cardiac myosin and autoimmune myocarditis. *CIBA Found. Symp.* **29:**3–24.
49. **Sadlack, B., H. Merz, H. Schorle, A. Schimpl, A. C. Feller, and I. Horak.** 1993. Ulcerative colitis-like disease in mice with a disrupted interleukin-2 gene. *Cell* **75:**253–261.
49a. **Saegusa, J., B. S. Prabhakar, K. Essani, P. R. McClintock, Y. Fukuda, V. J. Ferrans, and A. L. Notkins.** 1986. Monoclonal antibody to coxsackievirus B4 reacts with myocardium. *J. Infect. Dis.* **153**: 372-373. (Letter.)
50. **Sargent, S. J., E. H. Beachey, C. E. Corbett, and J. B. Dale.** 1987. Sequence of protective epitopes of streptococcal M proteins shared with cardiac sarcolemmal membranes. *J. Immunol.* **139:**1285–1290.
51. **Sartor, R. B.** 1995. Current concepts of the etiology and pathogenesis of ulcerative colitis and Crohn's disease. *Gastroenterol. Clin. N. Am.* **24**(3)**:**475-507.
52. **Sartor, R. B.** 1994. Cytokines in intestinal inflammation: pathophysiological and clinical considerations. *Gastroenterology* **106:**533–539.
53. **Schwimmbeck, P. L., D. T. Yu, and M. B. Oldstone.** 1987. Autoantibodies to HLA B27 in the sera of HLA B27 patients with ankylosing spondylitis and Reiter's syndrome. Molecular mimicry with Klebsiella pneumoniae as potential mechanism of autoimmune disease. *J. Exp. Med.* **166:**173–181.
54. **Taurog, J. D., and R. E. Hammer.** 1996. Experimental spondyloarthropathy in HLA-B27 transgenic rats. *Clin. Rheumatol.* **15**(Suppl. 1)**:**22–27.
55. **Taurog, J. D., S. D. Maika, W. A. Simmons, M. Breban, and R. E. Hammer.** 1993. Susceptibility to inflammatory disease in HLA-B27 transgenic rat lines correlates with the level of B27 expression. *J. Immunol.* **150:**4168–4178.
56. **Taurog, J. D., J. A. Richardson, J. T. Croft, W. A. Simmons, M. Zhou, J. L. Fernandez-Sueiro, E. Balish, and R. E. Hammer.** 1994. The germfree state prevents development of gut and joint inflammatory disease in HLA-B27 transgenic rats. *J. Exp. Med.* **180:**2359–2364.
57. **Toyoda, H., S. J. Wang, H. Y. Yang, A. Redford, D. Magalong, D. Tyan, C. K. McElree, S. R. Pressman, F. Shanahan, S. R. Targan, et al.** 1993. Distinct associations of HLA class II genes with inflammatory bowel disease. *Gastroenterology* **104:**741–748.
58. **Trentham, D. E., A. S. Townes, and A. H. Kang.** 1977. Autoimmunity to type II collagen an experimental model of arthritis. *J. Exp. Med.* **146:**857–868.
59. **van den Broek, M. F., E. J. Hogervorst, M. C. Van Bruggen, W. Van Eden, R. van der Zee, and W. B. van den Berg.** 1989. Protection against streptococcal cell wall-induced arthritis by pretreatment with the 65-kD mycobacterial heat shock protein. *J. Exp. Med.* **170:**449–466.
60. **van de Rijn, I., J. B. Zabriskie, and M. McCarty.** 1977. Group A streptococcal antigens cross-reactive with myocardium. Purification of heart-reactive antibody and isolation and characterization of the streptococcal antigen. *J. Exp. Med.* **146:**579–599.
61. **van Eden, W., J. E. Thole, R. van der Zee, A. Noordzij, J. D. van Embden, E. J. Hensen, and I. R. Cohen.** 1988. Cloning of the mycobacterial epitope recognized by T lymphocytes in adjuvant arthritis. *Nature* **331:**171–173.
62. **Van Voorhis, W. C., and H. Eisen.** 1989. F1-160. A surface antigen of Trypanosoma cruzi that mimics mammalian nervous tissue. *J. Exp. Med.* **169:**641–652.
63. **von Kreuter, B. F., and C. A. Santos-Buch.** 1986. Pathoimmune polymyositis induced in C3H/HeJ mice by Trypanosoma cruzi infection. *Clin. Exp. Rheumatol.* **4:**83–89.
64. **Yang, H., J. I. Rotter, H. Toyoda, C. Landers, D. Tyran, C. K. McElree, and S. R. Targan.** 1993. Ulcerative colitis: a genetically heterogeneous disorder defined by genetic (HLA class II) and subclinical (antineutrophil cytoplasmic antibodies) markers. *J. Clin. Investig.* **92:**1080–1084.
65. **Yang, L. C., P. R. Soprey, M. K. Wittner, and E. N. Fox.** 1977. Streptococcal-induced cell-mediated-immune destruction of cardiac myofibers in vitro. *J. Exp. Med.* **146:**344–360.
66. **Yoshinaga, M., F. Figueroa, M. R. Wahid, R. H. Marcus, E. Suh, and J. B. Zabriskie.** 1995. Antigenic specificity of lymphocytes isolated from valvular specimens of rheumatic fever patients. *J. Autoimmun.* **8:**601–613.

Molecular Mimicry, Microbes, and Autoimmunity
Edited by M. W. Cunningham and R. S. Fujinami

Chapter 2

Principles of Molecular Mimicry and Autoimmune Disease

Irun R. Cohen

The role of molecular mimicry in the host-parasite relationship is complex, and different investigators may have different ideas in mind when they consider the subject. On the one hand, parasitologists interested in the host-parasite relationship have called attention to molecular mimicry as a strategy deployed by some parasites to evade rejection by the host's immune system (16, 17); by expressing host-like antigens, a parasite might masquerade, at least in part, as the host self and so enjoy the protection from immune attack afforded by "self-tolerance." On the other hand, immunologists have become interested in molecular mimicry as a possible cause of autoimmune disease (32, 36); an immune response to a parasite that expresses a host-like antigen could break tolerance to the mimicked self-antigen. My aim here is to examine the immunological principles behind the phenomenon of host-parasite mimicry and autoimmune disease. Which intrinsic features of the immune system make molecular mimicry possible? How can molecular mimicry trigger autoimmune disease? How does the immune system protect itself against the dangers of molecular mimicry? I shall organize the discussion around several topics: autoimmune disease, physiological autoimmunity, the immunological homunculus, receptor degeneracy, creation of specificity, activation of autoimmune disease, and the immune dialogue.

AUTOIMMUNE DISEASE

An autoimmune disease is caused by the malfunction of otherwise healthy components of the body caused by the individual's own immune system. Four conditions must be fulfilled for autoimmunity to be expressed as a disease:

1. The individual's repertoire of T cells and/or B cells must include the precursors of autoimmune clones, cells that bear antigen receptors capable of recognizing self-molecules.
2. The autoimmune clones must be triggered into activation.
3. Damaging effector mechanisms must be generated by the ensuing immune response.
4. The noxious effector response must persist or must at least recur at a frequency and magnitude sufficient to hurt the patient.

Irun R. Cohen • Department of Immunology, The Weizmann Institute of Science, Rehovot, Israel 76100.

In short, an autoimmune disease is the result of the unregulated activation of existing autoimmune lymphocytes that leads to harmful inflammation. Why, however, should autoimmune lymphocytes exist at all?

Antigen recognition, including self-recognition, requires preexisting T cells and B cells that bear antigen receptors complementary to the antigen, be it a foreign antigen or a self-antigen; this is the basic tenet of the clonal selection theory of acquired immunity (9). So, an autoimmune disease would not be possible unless autoimmune lymphocytes preceded the outbreak of the disease. In other words, molecular mimicry of the host by the parasite, to endanger the host, presupposes the host's potential for autoimmunity. A parasite that deploys a host-mimicking antigen cannot be blamed for creating host self-recognition; the parasite is to be blamed only for unleashing the disease. To understand the parasite's role in pathogenesis, it does not suffice merely to know how a parasite molecule looks like a host molecule; the host is supplied at the outset with plenty of genuine self-molecules and with the receptors needed to recognize at least some of them (12). Investigators need to know what it is that the parasite supplies, in addition to a mimicry antigen, that activates the harmful attack. Before the activation issue is dealt with, however, consider the natural self-recognition that must precede activation.

PHYSIOLOGICAL AUTOIMMUNITY ONE

It is now generally accepted that all the functional T cells patrolling one's body have undergone positive selection for their ability to bind to antigen epitopes present in the thymus; such epitopes normally can arise only from self-antigens (45). It goes like this. Newborn T cells die in the thymus in two ways and survive in one way. They die, for lack of stimulation, if they fail to find to any self-antigen epitope, or they die of apoptosis if they are too strongly stimulated by a self-antigen to which they do bind. The newborn thymus T cell survives to leave the thymus if it happens to meet a self-antigen epitope that stimulates it "just right." In molecular terms, "just right" is interpreted to mean self-recognition with a binding affinity (or better, avidity) that is neither too low nor too high: too low means death by understimulation, and too high means death by overstimulation (34, 42). Thus, thymic T-cell education refers to the selection for life of T cells that bear anti-self-receptors of affinity (avidity) that is low, but not too low. B cells and the T cells maturing outside of the thymus may undergo other types of developmental selections.

How, upon leaving the thymus, however, do the self-selected T cells manage to recognize the foreign antigens, which, after all, is their specialty? The key point is that the thymus positively selects newborn T cells that bind to self-epitopes with a relatively low affinity; once a T cell has left the thymus, binding of such low affinity is thought to be insufficient to trigger an effector response. It is assumed that the mature T cell will no longer respond to the self-epitope that saved its life when it was a newborn in the thymus. In other words, most self-epitopes outside of the thymus are seen with such a low affinity that, metaphorically speaking, they are overlooked. Only foreign antigens that happen to bind autoimmune receptors with a relatively high affinity will activate the receptor-bearing clones. The foreign antigens that look like self, but with sufficiently altered affinity, serve as ligands for effective recognition. Thus, any foreign antigen that elicits an immune response must do so by mimicking a self-epitope. A foreign epitope is an altered self-ligand. From this point of view, all immunity is based on mimicry.

Note that this concept of mimicry, like much that seems at the moment to be logical in biology, is tentative; the immune system is so complex that the obvious and the unimaginable often change places. Mimicry, like all things biological, is a matter of degree; affinity is relative, by definition, and so the borders between self and not-self are intrinsically fuzzy. This is a key idea, and I will enlarge on it when I discuss degeneracy below.

In summary, one can conclude that low-affinity autoimmunity is physiological and serves as the formative principle of the T-cell repertoire directed in practice to foreign antigens. This is *physiological autoimmunity one.*

PHYSIOLOGICAL AUTOIMMUNITY TWO: THE IMMUNOLOGICAL HOMUNCULUS

In addition to the lymphocytes that bind to self-epitopes with a low affinity, there are sets of T cells and B cells, at a high frequency, that seem to bind to particular self-antigens with a high affinity. It appears that healthy people and healthy rodents, too, express autoantibodies to a small but relatively fixed set of self-antigens (26). Viewed in their totality, these autoantibodies, both immunoglobulin M (IgM) and IgG, form distinct and stable patterns of autoreactivity. Particular autoantibodies and their patterns may change during immunization and in autoimmune disease (18), but the autoantibody patterns do not arise from experience with foreign antigens; they appear to have been selected by the individual's self-antigens (21).

In addition to autoantibodies, humans are also outfitted with T cells reactive with a limited set of dominant self-antigens (12, 15). It is quite feasible to raise from the peripheral blood of healthy persons lines of T cells that react strongly to certain self-antigens, which include myelin basic protein (MBP) and other antigens targeted in autoimmune diseases (29). This natural repertoire of T-cell self-reactivity develops in the thymus (45) and can be detected outside the thymus from birth. The cord blood of newborn infants, for example, contains a high frequency of T cells reactive with the 60-kDa heat shock protein (hsp60) (19). The combined set of autoimmune T cells and B cells creates, as it were, the immune system's internal picture of certain self-molecules. This special set of self-reactivity has been termed the "immunological homunculus," the immune system's image of the individual (11, 12, 15).

If rodents and humans feature immunological homunculi, then the positive selection of strong autoimmunity to at least some self-antigens has been conserved during evolution and so is likely to be advantageous. It is now known that cytokines and other immune molecules have important roles in wound healing, angiogenesis, the repair of bones and connective tissues, and other processes needed to maintain the body (14). Recent experiments indicate that autoimmunity to MBP can help maintain the central nervous system in the face of traumatic injury (30). Thus, some of the agents of autoimmunity expressed in the immunological homunculus might function in the ongoing maintenance of the self (37). Other autoimmune reactivities might assist immunity to foreign antigens. For example, autoimmunity to self-hsp60 can help fight infectious agents (25) and regulate allograft rejection (7). Thus, autoimmunity can be of benefit, at least under certain conditions. Be that as it may, for molecular mimicry to trigger an autoimmune disease, the target self-antigen would have to be recognizable in the periphery by mature T cells; this suggests that homuncular self-antigens ought to be prominent in clinically significant mimicry. Could the limited number of self-antigens in the homuncular set explain the immunological regularity of the major autoimmune diseases (12, 15)? A few diseases, such as rheumatoid arthritis, multiple sclerosis, type I diabetes, and lupus, account for much of the clinical spectrum; perhaps preexisting homuncular autoimmunity molds the stereotypic patterns of clinical disease. In any case, physiological autoimmunity is a foundation for pathological mimicry.

MOLECULAR MIMICRY AND RECEPTOR DEGENERACY

Two molecules can be said to mimic one another as antigens if both molecules, irrespective of their chemical similarity or dissimilarity, can be bound by the same antigen receptor. Mimicry between foreign and self, as discussed above, is a fundamental principle in the T-cell repertoire; a T cell first undergoes positive selection for a self-antigen in the thymus and is again selected by a foreign antigen in the periphery. Antigen mimicry is common in the periphery, too. A single T-cell clone can be demonstrated to recognize many different peptide epitopes (2, 3, 6, 31).

The very existence of antigen mimicry should jolt the thinking about the adaptive immune system. The adaptive immune system specializes in creating specificity. It is the only biological system known to manufacture its own receptors, millions and millions of them, somatically in the lifetime of the individual. It also uses these receptors in the process of making exceedingly fine distinctions between antigens. The jolt comes when it is discovered that this fine specificity is not inherent in the antigen receptors; the antigen receptors, like all other biological receptors, are fundamentally degenerate. Antigens are prone to mimic one another.

The experts may argue about just how degenerate a T cell or antibody may be (28). However, whether one receptor can bind to only a handful or to a thousand different ligands does not negate the conclusion that the fine specificity of the immune response must be created outside of the primary interaction between the receptor and its ligand. In a word, the specificity of an immune response cannot be *reduced* to the specificity of one antigen receptor.

A lack of ligand fidelity is intrinsic to receptors, not only to antigen receptors but to biological receptors generally (24, 27). Any receptor can bind to more than one specific ligand, an attribute called "degeneracy." Receptors are degenerate by nature because their binding sites interact with their ligands by means of multiple noncovalent contacts. The strength by which a ligand is bound, its affinity, depends greatly on the collection of contact points with which the ligand interacts with the receptor. One can readily imagine how different ligands might use different points of contact and be bound with the same affinity to the same receptor. Obviously, receptors can bind to different ligands with different affinities. Indeed, any ligand that can fit a receptor binding site with a sufficient strength of interaction will bind to and could be recognized by the receptor. Thus, a potential for functional mimicry is built into biological recognition at the molecular level. There is no way to get around it.

In fact, one does not need to lament receptor degeneracy; receptor degeneracy is functional. Pharmacological intervention, for example, is based on the interactions of ligands, called drugs, with receptors that evolved to interact with other, more physiological ligands. Viruses and bacteria use host receptors for their maintenance in the host niche; parasites even mimic host cytokines (20). Antigens are only one example of a universal mimicry. The immune system, however, is clearly able to discriminate among antigens. How can the immune system act specifically when its antigen receptors are degenerate?

THE CREATION OF SPECIFICITY

The degeneracy question is not only hard to answer; the degeneracy question has been hard to ask. Investigators and scientists have been taught by the clonal selection theory of acquired immunity to equate the specificity of the immune response with the specificity of the antigen receptors (9, 10). The classical formulation of the theory taught that immune specificity was achieved automatically. Specificity was postulated to be inherent in the lymphocyte's antigen receptor and had to be realized only by the binding of the receptor to the antigen. Receptor specificity was the starting point of clonal selection; specificity was thought to be a given. On the contrary, antigenic mimicry and receptor degeneracy indicate that the world of the immune system is not as simple as that originally postulated by the founding fathers (11). Specificity is not given; it must be taken (14). But how?

It is clear that the specificity of an immune response cannot be reduced to the specificity of a single receptor. Immune specificity emerges from the collective of interactions that form the response. Consider that evolution does not care so much about recognition as it does about response. Recall that the immune system, like all biological systems, has evolved to adapt, to produce meaningful responses. Survival depends on what you do with what you see. Seeing alone (i.e., binding) has little meaning (13, 14). The immune system is adaptive when it succeeds in producing responses appropriate to the situation (11, 38). Degenerate receptors do not matter, as long as the response can discriminate between right and wrong. Immune specificity is functionally related to the actual response of the system; immune specificity emerges from a chain of events involving many different cells and molecules. The nature of the effector response is rarely governed by the tyranny of a single degenerate receptor. A polyclonal serum, for example, is usually much more specific than any one of the component clones of antibodies. (Experimentalists often complain that their pet monoclonal antibodies bind to all kinds of junk antigens.)

Macrophages, T cells, and B cells each sense a unique aspect of the immune encounter, and, at the same time, these cells mutually exchange information (by way of cytokines, chemokines, antibodies, and other cell interaction molecules) that is obtained from their different views. In other words, each cell type is influenced by the views of the others. Macrophages sense and report on contexts of damage and inflammation; T cells see epitopes processed from the innards of antigens; B cells can see molecular conformation. The cytokines and other interaction molecules exchanged between these various agents help the cells modify their responses; macrophages, T cells, and B cells help (or suppress) one another in a network of mutual interaction. In short, the immune system responds not only to antigens but also to its own response.

The specificity of the immune response emerges from such collective interactions. I have termed this collective interaction "co-respondence" (14). The many mechanisms responsible for co-respondence and the emergence of specificity are issues beyond the scope of this chapter, and interested persons can read more elsewhere (14). The relevant points for the present discussion are that antigenic mimicry can induce an autoimmune disease only if the mimicry leads to a damaging response, that receptor-ligand binding by itself is not a disease, that molecular mimicry of self-antigens is only a part of the story, and that specificity is generated as the immune response unfolds.

ACTIVATION OF AUTOIMMUNE DISEASE

What is required of an infectious agent immunologically to transform physiological autoimmunity into an autoimmune disease? The infectious agent supplies a collective of signals for co-respondence. Antigen mimicry is not enough to cause damage. To induce an autoimmune disease, an infectious agent must activate physiological autoimmunity unnaturally. Three experimental examples will help define the principles of activation.

One requirement for activation is that the infectious agent supply adjuvant signals (22). An experimental example will be one of contrived mimicry. A transgenic mouse constructed so that a viral antigen is expressed by its pancreatic beta cells tends to relate to the viral antigen immunologically as a self-antigen. Although the viral antigen is expressed in the cells that secrete insulin, the mouse will not develop autoimmune diabetes automatically, even if the mouse has many T cells that bear receptors for the viral antigen. The mouse, however, can be induced to attack its beta cells by infecting the mouse with the particular virus (43, 44). The viral self-antigen is there in the beta cells, and the T cells that can see it are in the mouse, but that does not suffice; the development of autoimmune diabetes needs the infection. Why? Because the infection supplies adjuvant signals. An antigen without adjuvant signals is not a sufficient stimulus for an effector response. Accessory ligands produced by antigen-presenting cells and tissue cells are required to tell the T cells and B cells how to respond. Co-respondence is essential. One may conclude that the immune system has evolved to respond not to antigens but to antigens in context (1, 11, 13). Infectious agents and the inflammation that they produce generate the co-response network that leads to an effector response. An autoimmune disease may emerge when a relevant self-molecule is targeted in the context of the infection.

It is beyond the scope of this chapter to delve into a detailed description of all the accessory signals that may be generated by an infectious agent. Suffice it to say that the innate arm of the immune response, carried in the germ line of the species, can recognize many of the invariant molecules carried by microbes and the tissue signals generated by infected or damaged body cells (22). The first principle of autoimmune mimicry, therefore, is that the infectious agent supplies sufficient adjuvant signals; mimicking molecules are not enough. Indeed, even a genuine self-molecule will not induce an experimental autoimmune disease in a mouse unless immunization is done with an adjuvant.

The second requirement for pathogenic mimicry relates to the fine structure of the mimic itself. An example is experimental autoimmune encephalomyelitis (EAE). EAE (5) can be induced in many species of animals, including humans, by immunization to epitopes of the antigen MBP. The induction of the disease, of course, requires complete Freund's adjuvant (CFA; dead mycobacteria in oil) or some other array of accessory signals. (MBP without CFA will not induce EAE; in fact, the administration of MBP alone will likely induce resistance to EAE [4]). Given the CFA, however, the MBP itself also matters. Immunization with rat MBP can induce EAE in the susceptible Lewis rat strain, but guinea pig MBP is much more encephalitogenic; guinea pig MBP will cause more severe EAE than that caused by 10-fold more rat MBP (4). Felix Mor, Michael Kantorowitz, and I recently investigated the encephalitogenicity of the dominant MBP peptide sequence (positions 71 to 87) in the Lewis rat and compared the rat and guinea pig peptides, which differ by one conservative amino acid substitution at position 78: the guinea pig peptide features serine (S) in place of the threonine (T) in the rat peptide. The alteration is biologically significant; the S-containing peptide, compared to the T-containing peptide, induced more severe EAE and influenced the Vβ usage and the CDR3 sequence of the T-cell receptors of the encephalitogenic T cells (30a). According to a model of the binding motif of the rat IA molecule (33), we concluded that the position of the S-T substitution interacted with a peptide anchor position P6 in the cleft of the rat IA molecule. The S alteration could be seen to have a better fit in the P6 pocket.

Thus, the sequence alteration of a mimicking antigen epitope can be critical. The autoimmune disease EAE, the example here, arose with greater vigor when the mimicry between the foreign (guinea pig) and the self (rat) was less than perfect. Here, the altered peptide ligand gained affinity for the major histocompatibility complex (MHC). Thus, even an increase in affinity for an MHC molecule and not only a change in affinity for the antigen receptor can lead to a change in the effector (cytokine) phenotype and, thus, in the pathogenicity of the response. Of course, antigen alterations similarly could affect, for better or worse, other processes involved in immune activation including antigen uptake and processing. The chemical nature of the mimicry can have wide-ranging effects.

The third principal element involved in autoimmune activation by mimicry is the immune history of the subject; co-respondence has a memory (14). A mimicking self-antigen associated with an effective adjuvant can elicit different responses in genetically identical individuals. Each immune system is uniquely fashioned by its history of past immunological experience. An example is adjuvant arthritis (AA). AA can be induced in rats by immunization with CFA enriched for its content of mycobacteria (23). In the case of AA, the mycobacteria provide more than adjuvant signals alone; along with adjuvant signals, an antigen of mycobacteria, apparently related to hsp65 (41), appears to mimic an epitope in cartilage proteoglycan (40). Nathan Karin and I investigated the differences in AA induced in purebred Lewis rats raised under germfree conditions compared to that induced in genetically identical Lewis rats raised in contaminated quarters (the contamination was not defined). We found that mycobacterial immunization produced AA of markedly different characteristics in the two colonies (unpublished data). The contaminated rats developed acute AA that spontaneously remitted after 2 weeks, and these rats acquired long-term resistance to attempts to reinduce AA. The clean rats, in contrast, developed chronic AA that lasted at least 9 months. Moreover, the clean rats repeatedly manifested exacerbations of AA each time that they were challenged with mycobacteria. In short, the previous exposure of the rats to infectious agents had a significant effect on the response to pathogenic mycobacterial mimicry.

AA shows that contact with infectious agents can not only trigger autoimmunity but can also protect against autoimmunity. Resistance to AA can be acquired by a previous bout of AA, provided the rats have had an enriched immunological experience generally. The ability of infection to protect against autoimmunity has been documented in the case of the autoimmune diabetes developing in NOD mice (8). Although the diabetes is spontaneous (it has no known trigger), the disease can be prevented by experience of the NOD mice with a variety of infectious agents. These experimental examples of resistance to autoimmunity induced by infection support the idea that the current increases in the incidences of some autoimmune diseases and of allergies in developed countries may be due to too much hygiene, to a lack of experience with infections (35). Mimicry apparently can lead to regulation of autoimmune processes as well as to dysregulation. Nevertheless, repeated infections with group A streptococcal strains can induce repeated bouts of acute rheumatic fever and cumulative heart damage (39). The role of immune experience in autoimmune activation is complex indeed.

IMMUNE DIALOGUE

Mimicry is a complex subject because the immune system is a complex system, and complex systems just do not have simple solutions. The world is filled with mimicry; just submit your favorite peptide sequence into a protein sequence bank and see how many homologies turn up. The potential for mimicry inherent in the evolution and coevolution of hosts and parasites is compounded by the intrinsic degeneracy of biological receptors. To protect itself against receptor degeneracy and universal mimicry, the immune system generates functional specificity by using adjuvant signals, collective interactions of co-responding cells, and the immune history of the individual. Immune specificity can be likened to a cellular dialogue (1). Specificity is thus an emergent property of a series of interactions (14). This conclusion is not of much comfort to those who like to have specificity for free, but comfort is never for free.

SUMMARY: ANTIGENIC MIMICRY AND AUTOIMMUNE DISEASE

Let me summarize the roles of infectious agents and antigenic mimicry in the induction of autoimmune disease. The four conditions outlined at the outset will be the framework:

1. Physiological autoimmunity, including the immunological homunculus, lays the foundation for parasite-induced autoimmune disease. Here, the host, not the parasite, is responsible.
2. The host's autoimmune clones are activated through the co-respondence of a combination of mimicking antigen epitopes (with suitable alterations) together with adjuvant signals. The infectious agent is responsible for the co-respondence combination. Antigen mimicry alone will not suffice.
3. The adjuvant signals generated by the parasite not only activate the system, they instruct the immune system to produce a damaging type of response.
4. The roles of mimicry and infection in recurrence and chronicity are not well understood. Immunomodulatory cytokines produced by the parasite may play a role, along with signals generated by host tissue damage. Antigen mimicry alone is not responsible.

REFERENCES

1. **Atlan, H., and I. R. Cohen.** 1998. Immune information, self-organization and meaning. *Int. Immunol.* **10:**711–717.

2. **Ausubel, L. J., C. K. Kwan., A. Sette, V. Kuchroo, and D. A. Hafler.** 1996. Complementary mutations in an antigenic peptide allow for crossreactivity of autoreactive T-cell clones. *Proc. Natl. Acad. Sci. USA* **93:**15317–15322.
3. **Baum, H., H. Davis, and M. Peakman.** 1996. Molecular mimicry in the MHC: hidden clues to autoimmunity? *Immunol. Today* **17:**64–70.
4. **Ben-Nun, A., and I. R. Cohen.** 1982. Spontaneous remission and acquired resistance to autoimmune encephalomyelitis (EAE) are associated with suppression of T cell reactivity: suppressed EAE effector T cells recovered as T cell lines. *J. Immunol.* **128:**1450–1457.
5. **Ben-Nun, A., H. Wekerle, and I. R. Cohen.** 1981. The rapid isolation of clonable antigen-specific T lymphocyte lines capable of mediating autoimmune encephalomyelitis. *Eur. J. Immunol.* **11:**195–199.
6. **Bhardwaj, V., V. Kumar, H. M. Geysen, and E. E. Sercarz.** 1993. Degenerate recognition of a dissimilar antigenic peptide by myelin basic protein-reactive T cells. Implications for thymic education and autoimmunity. *J. Immunol.* **151:**5000–5010.
7. **Birk, O. S., S. L. Gur, D. Elias, R. Margalit, F. Mor, P. Carmi, J. Bockova, D. M. Altmann, and I. R. Cohen.** 1999. The 60-kDa heat shock protein modulates allograft rejection. *Proc. Natl. Acad. Sci. USA* **96:**5159–5163.
8. **Bowman, M. A., E. H. Leiter., and M. A. Atkinson.** 1994. Prevention of diabetes in the NOD mouse: implications for therapeutic intervention in human disease. *Immunol. Today* **15:**115–120.
9. **Burnet, F. M.** 1959. *The Clonal Selection Theory of Acquired Immunity.* Cambridge University Press, Cambridge, United Kingdom.
10. **Burnet, F. M.** 1969. *Self and Not-Self.* Cambridge University Press, Cambridge, United Kingdom.
11. **Cohen, I. R.** 1992. The cognitive principle challenges clonal selection. *Immunol. Today* **13:**441–444.
12. **Cohen, I. R.** 1992. The cognitive paradigm and the immunological homunculus. *Immunol. Today* **13:**490–494.
13. **Cohen, I. R.** 1995. Language, meaning and the immune system. *Isr. J. Med. Sci.* **31:**36–37.
14. **Cohen, I. R.** 2000. *Tending Adam's Garden: Evolving the Cognitive Immune Self.* Academic Press, London, United Kingdom.
15. **Cohen, I. R., and D. B. Young.** 1991. Autoimmunity, microbial immunity and the immunological homunculus. *Immunol. Today* **12:**105–110.
16. **Damian, R. T.** 1964. Molecular mimicry: antigen sharing by parasite and host and its consequences. *Am. Nat.* **98:**129–149.
17. **Damian, R. T.** 1997. Parasite immune evasion and exploitation: reflections and projections. *Parasitology* **115:**S169–S175.
18. **Fesel, C., and A. Coutinho.** 1998. Dynamics of serum IgM autoreactive repertoires following immunization: strain specificity, inheritance and association with autoimmune disease susceptibility. *Eur. J. Immunol.* **28:**3616–3629.
19. **Fischer, H. P., C. E. Sharrock, and G. S. Panayi.** 1992. High frequency of cord blood lymphocytes against mycobacterial 65-kDa heat-shock protein. *Eur. J. Immunol.* **22:**1667–1669.
20. **Grencis, R. K., and G. M. Entwistle.** 1997. Production of an interferon-gamma homologue by an intestinal nematode: functionally significant or interesting artifact? *Parasitology* **115:**S101–S106.
21. **Haury, M., A. Sundblad, A. Grandien, C. Barreau, A. Coutinho, and A. Nobrega.** 1997. The repertoire of serum IgM in normal mice is largely independent of external antigenic contact. *Eur. J. Immunol.* **27:**1557–1563.
22. **Hoffmann, J. A., F. C. Kafatos, C. A. Janeway, and R. A. Ezekowitz.** 1999. Phylogenetic perspectives in innate immunity. *Science* **284:**1313–1318.
23. **Holoshitz, J., Y. Naparstek, A. Ben-Nun, and I. R. Cohen.** 1983. Lines of T lymphocytes induce or vaccinate against autoimmune arthritis. *Science* **219:**56–58.
24. **Klotz, I. M., and D. L. Hunston.** 1975. Protein interactions with small molecules. Relationships between stoichiometric binding constants, site binding constants, and empirical binding parameters. *J. Biol. Chem.* **250:**3001–3009.
25. **Konen-Waizmann, S., A. Cohen, M. Fridkin, and I. R. Cohen.** 1999. Self heat-shock protein (hsp60) peptide serves in a conjugate vaccine against a lethal pneumococcal infection. *J. Infect. Dis.* **179:**403–413.
26. **Lacroix-Desmazes, S., S. V. Kaveri, L. Mouthon, A. Ayouba, E. Malanchere, A. Coutinho, and M. D. Kazatchkine.** 1998. Self-reactive antibodies (natural autoantibodies) in healthy individuals. *J. Immunol. Methods* **216:**117–137.

27. **Lancet, D., E. Sadovsky, and E. Seidemann.** 1993. Probability model for molecular recognition in biological receptor repertoires: significance to the olfactory system. *Proc. Natl. Acad. Sci. USA* **90:**3715–3719.
28. **Mason, D.** 1998. A very high level of crossreactivity is an essential feature of the T-cell receptor. *Immunol. Today* **19:**395–404.
29. **McLaurin, J. A., D. Hafler, and J. P. Antel.** 1995. Reactivity of normal T-cell lines to MBP isolated from normal and multiple sclerosis white matter. *J. Neurol. Sci.* **128:**205–211.
30. **Moalem, G., R. Leibowitz-Amit, E. Yoels, F. Mor, I. R. Cohen, and M. Schwartz.** 1999. Autoimmune T cells protect neurons from secondary degeneration after central nervous system axotomy. *Nat. Med.* **5:**49–55.
30a**Mor, F., M. Kantorowitz, and I. R. Cohen.** Selection of anti-MBP T-cell lines in the Lewis rat: V-beta 8.2 dominance and CDR-3 motifs are dependent on serine at position 78 of MBP. *J. Neuroimmunol.*, in press.
31. **Nanda, N. K., K. K. Arzoo, H. M. Geysen, and E. E. Sercarz.** 1995. Recognition of multiple peptide cores by a single T cell receptor. *J. Exp. Med.* **182:**531–539.
32. **Oldstone, M. B. A.** 1998. Molecular mimicry and immune-mediated diseases. *FASEB J.* **12:**1255–1265.
33. **Reizis, B., F. Mor, M. Eisenstein, H. Schild, S. Stefanoviic, H. G. Rammensee, and I. R. Cohen.** 1996. The peptide binding specificity of the MHC class II I-A molecule of the Lewis rat, RT1.BI. *Int. Immunol.* **8:**1825–1832.
34. **Robey, E., and B. J. Fowlkes.** 1994. Selective events in T cell development. *Annu. Rev. Immunol.* **12:**675–705.
35. **Rook, G. A., and J. L. Stanford.** 1998. Give us this day our daily germs. *Immunol. Today* **19:**113–116.
36. **Rowley, D., and C. R. Jenkin.** 1962. Antigenic cross-reaction between host and parasite as a possible cause of pathogenicity. *Nature* **193:**151–154.
37. **Segel, L. A., R. L. Bar-Or.** 1999. On the role of feedback in promoting conflicting goals of the adaptive immune system. *J. Immunol.* **163:**1342–1349.
38. **Schwartz, M., G. Moalem, R. Leibowitz-Amit, and I. R. Cohen.** 1999. Innate and adaptive responses can be beneficial for CNS repair. *Trends Neurosci.* **22:**295–299.
39. **Stollerman, G. H.** 1998. The changing face of rheumatic fever in the 20th century. *J. Med. Microbiol.* **47:**655–657.
40. **van Eden, W., J. Holoshitz, Z. Nevo, A. Frenkel, A. Klajman, I. R. Cohen.** 1985. Arthritis induced by a T-lymphocyte clone that responds to Mycobacterium tuberculosis and to cartilage proteoglycans. *Proc. Natl. Acad. Sci. USA* **82:**5117–5120.
41. **van Eden, W., J. E. Thole, R. van der Zee, A. Noordzij, J. D. van Embden, E. J. Hensen, and I. R. Cohen.** 1988. Cloning of the mycobacterial epitope recognized by T lymphocytes in adjuvant arthritis. *Nature* **331:**171–173.
42. **Von Boehmer, H.** 1990. Developmental biology of T cells in T cell-receptor transgenic mice. *Annu. Rev. Immunol.* **8:**531–556.
43. **von Herrath, M. G., S. Guerder, H. Lewicki, R. A. Flavell, and M. B. Oldstone.** 1995. Coexpression of B7-1 and viral ("self") transgenes in pancreatic beta cells can break peripheral ignorance and lead to spontaneous autoimmune diabetes. *Immunity* **3:**727–738.
44. **von Herrath, M. G., C. F. Evans, M. S. Horwitz, and M. B. Oldstone.** 1996. Using transgenic mouse models to dissect the pathogenesis of virus-induced autoimmune disorders of the islets of Langerhans and the central nervous system. *Immunol. Rev.* **152:**111–143.
45. **Wekerle, H., M. Bradl, C. Linington, G. Kaab, and K. Kojima.** 1996. The shaping of the brain-specific T lymphocyte repertoire in the thymus. *Immunol. Rev.* **149:**231–243.

Molecular Mimicry, Microbes, and Autoimmunity
Edited by M. W. Cunningham and R. S. Fujinami

Chapter 3

Molecular Mimicry and Central Nervous System Autoimmune Disease

Robert S. Fujinami

The initial events that lead to the loss of tolerance to self-antigens that results in autoimmune disease are not known. Infections have been associated with the initiation and/or exacerbation of multiple sclerosis (MS), an autoimmune disease of the central nervous system (CNS). It is suspected that tolerance to self-CNS antigens is broken by several linked events. This results in the initiation or induction of anti-CNS immune responses, which lead to inflammation and demyelination.

An example of a virus that induces a CNS inflammatory demyelinating disease is measles virus (18). In most cases of measles, the virus causes an acute respiratory infection in which the measles virus is able to disseminate to the lymphatic system. Viral replication in lymphoid tissues allows large amounts of virus to enter the blood, which leads to the subsequent seeding of other organs. As an antiviral immune response is generated, virus titers in tissues such as skin, liver, and lung drop dramatically and measles virus is eventually cleared (8,17,43). The appearance of the characteristic rash occurs approximately 2 weeks after the initial infection and is due to the clearance of virus-infected cells in the skin by anti-measles virus T cells. Postinfectious encephalomyelitis is a complication of measles and is generally observed 4 to 5 days following the rash. It occurs in approximately 1/1,000 individuals with measles. It has been reported that myelin basic protein (MBP)-specific T-cell proliferative responses are increased in patients with measles virus-induced postinfectious encephalomyelitis (19). In addition, in the CNS there are perivascular mononuclear cell lesions similar to those observed in an experimental CNS autoimmune model known as experimental allergic encephalomyelitis (EAE). MBP is a major component of CNS myelin, and when it is emulsified in adjuvant it can be used to induce EAE when it is injected into animals (for a review, see reference 20). Johnson et al. (19) found that 8 of 17 patients (47%) had lymphocytes that proliferated in response to MBP, whereas 6 of 40 (15%) individuals with measles without encephalomyelitis had such lymphocytes. None of four lymphocyte cultures from healthy children proliferated in response to MBP. No intrathecal anti-measles virus antibodies were detected, suggesting that measles virus-induced postinfectious encephalomyelitis is not dependent on virus replication within the CNS. Thus, acute viral infections can induce autoimmune diseases of the CNS. However,

Robert S. Fujinami • Department of Neurology, University of Utah School of Medicine, 30 N 1900 East, Room 3R330, Salt Lake City, UT 84132.

the actual mechanism(s) is not known, but molecular mimicry provides an interesting hypothesis.

DEVELOPMENT OF THE CONCEPT OF MOLECULAR MIMICRY

Damian (5) first used the term "molecular mimicry" in the context of "eclipsed" antigen, in which "an antigenic determinant of parasite origin resembles an antigenic determinant of its host." This was a means for parasites to "resemble" antigenic determinants of their hosts such that they would not elicit immune responses. He suggested that the concept that shared antigens initiated autoimmunity could not be supported on evolutionary or immunological grounds. Damian did predict that host polymorphisms could have evolved in response to the eclipsed antigens contained in microbes. Interestingly, this could explain one of the driving forces of polymorphisms in the major histocompatibility complex (MHC) and is still a topic of discussion. Excellent reviews on this topic are available elsewhere (6, 7) and will not be further discussed here.

In 1983, in studying the specificities of monoclonal antibodies to measles virus and herpesvirus that were generated, colleagues and I reported that some of these virus-specific antibodies also reacted with self-proteins. At that time we suggested that this was a viable mechanism for the generation of cross-reacting immune responses by viruses, which could lead to autoimmune disease (13). It was later noted that cross-reacting monoclonal antibodies were relatively frequent (approximate frequency, 3%) and were not restricted to a particular type of virus. Both DNA and RNA viruses were represented (38). Some of the monoclonal antibodies to measles virus that were generated were also found to react with human T cells (1). Such antibodies have the potential to bind to the costimulatory molecule on the surfaces of self-reactive T cells that provide sufficient signals for activation. This observation could help explain the spontaneous proliferation of peripheral mononuclear cells seen during natural measles virus infection (44). Thus, cross-reacting antibodies between viruses and self-antigens are relatively frequent and may be responsible for the occurrence of "natural" antibodies found in healthy individuals (26, 32, 37) or the maintenance of some autoreactive T cells.

As mentioned above, EAE can be induced in a variety of species with various myelin proteins or epitopes derived from these myelin proteins. These include MBP, myelin proteolipid protein (PLP), myelin oligodendrocyte protein (MOG), myelin-associated glycoprotein, S100 protein, and myelin oligodendrocyte basic protein (2, 25). In 1985, in extending the molecular mimicry studies, colleagues and I scanned protein databases in an attempt to identify common amino acid sequences with the known encephalitogenic peptides of MBP. We found that the encephalitogenic epitope for the rabbit (amino acids 66 to 25) shared sequence similarity with a portion of the hepatitis B virus polymerase (amino acids 589 to 598) (12). The viral peptide was biochemically synthesized and was injected into rabbits with complete Freud's adjuvant (CFA). What was observed was that most of these animals made antibodies to the viral peptide that cross-reacted with MBP. Interestingly, half of the rabbits sensitized with the viral peptide had peripheral blood mononuclear cells that could recognize MBP, as measured by in vitro proliferation assays. What was surprising was that some of the rabbits had lesions in the CNS consistent with those seen in EAE. This was the first example that a viral peptide that cross-reacted with a self-CNS protein could induce a cross-reacting immune response that led to autoimmune disease. This observation had profound effects on the way in which investigators viewed the concept of molecular mimicry. It is important to say that the hepatitis B virus polymerase has a shared epitope with a rabbit encephalitogenic site but not with any potential human encephalitic sites. Hepatitis B virus does not induce EAE or any EAE-like diseases in humans. In addition, the current recombinant hepatits B virus vaccine does not contain the viral polymerase protein.

Recently, Wucherpfennig and colleagues extended the concept of molecular mimicry and autoimmune CNS disease. They have elegantly shown that the cross-reacting epitope between virus microbe and self-CNS protein does not need to have identical amino acids in order for T-cell recognition to occur. These studies are further described in chapter 14. More recently, Gran et al. (16) described the use of combinatorial libraries to study mimicry between the MBP peptide from positions 87 to 99 and other microbial peptides. They found that some of the peptides that were able to stimulate MBP-specific T-cell clones did not share any common amino acids. In addition, some of the peptides were able to stimulate the T cells more effectively than the MBP peptide could, suggesting that these non-self-peptides were even more potent agonists than MBP. Thus, due to the degeneracy of the T-cell receptor, the term "molecular mimicry" has been broadened to include nonhomologous epitopes.

MODELS OF MIMICRY FOR CNS AUTOIMMUNE DISEASE

In autoimmune disease and particularly in autoimmune diseases of the CNS with the exception of the postinfectious encephalomyelopathies, no virus has been identified as the etiological agent. In autoimmune diseases in general, the autoantigens are not known, and if molecular mimicry is the mechanism for the breaking of tolerance, the cross-reacting epitope may not be obvious, as discussed above. Therefore, colleagues and I chose a different tact to explore this concept. We reasoned that during virus infection, viral proteins similar to self-proteins are targeted for degradation through the TAP (transporter associated with antigen processing) pathway into the proteasomes. Here, proteins are degraded into smaller peptides, which associate with MHC class I molecules for presentation on the surfaces of infected cells (for a review, see reference 30). Recombinant vaccinia viruses that encoded self-CNS antigens were constructed. The first recombinant virus contained the entire coding region for myelin PLP and was designated VVplp. Mice were infected with VVplp and were monitored for the development of CNS autoimmune disease. None of the mice infected with VVplp or a control recombinant vaccinia virus that encoded β-galactosidase (VVsc11) developed any clinical signs or pathological changes in the CNS. However, when mice were infected with VVplp and then sensitized with $PLP_{139-151}$, they were found to have enhanced EAE with an earlier onset of symptoms compared to that for mice infected with control VVsc11. The clinical disease and the inflammatory disease in the CNS were seen earlier in VVplp-infected mice than in control virus-infected mice (3). We also observed enhanced disease when mice were first infected with VVplp and challenged with $PLP_{104-117}$. $PLP_{104-117}$ represents a cryptic epitope, and $PLP_{104-117}$-infected mice generally develop EAE at later times with little or no acute disease. Disease caused by another encephalitogenic PLP epitope, $PLP_{178-191}$, which induces acute disease with a relapsing-remitting disease course, was enhanced the least. Thus, there appears to be a hierarchy in the enhancement that is observed, and this hierarchy is epitope specific. These data suggest that virus infections with common determinants can prime individuals for autoimmune disease.

EAE induced by $PLP_{139-151}$ in SJL/J mice as stated above has an acute phase that is followed by a relapsing-remitting clinical disease. Infection of mice with VVplp prior to challenge with $PLP_{139-151}$ in CFA results in enhanced acute disease, but these mice do not go on to develop the relapsing-remitting disease. This is in contrast to the situation for mice infected with VVsc11, in which the mice undergo an acute phase but later develop a relapsing-remitting disease. At later times, when the CNS of the mice that were infected with control virus (VVsc11) or no virus and that were undergoing relapses is examined, extensive inflammatory demyelinating lesions are evident. However, in mice infected with VVplp and challenged with $PLP_{139-151}$, the CNS parenchyma is clear of lesions, but meningitis is prominent, suggesting that the inflammatory cells are unable to enter the parenchyma. Delayed-type hypersensitivity (DTH) and proliferative responses to PLP are also reduced in VVplp-infected and -challenged mice compared with the control virus-infected mice.

Colleagues and I further investigated the effects of infection on a different autoantigen, MBP. PL/J mice are $H\text{-}2^u$, and the first 11 amino acids of MBP represent a major encephalitogenic peptide for this strain of mouse. Using a different recombinant vaccinia virus that encodes amino acids 1 to 23 of MBP and that contains the encephalitogenic antigen for the PL/J mouse (VV_{M1-23}), we found a much different scenario (4). Mice infected with VV_{M1-23} did not develop any clinical or pathological evidence of EAE. Interestingly, when these mice were challenged with ac-MBP1-23 or whole MBP, mice were protected from developing EAE. VV_{M1-23}-infected mice had reduced ear-swelling (DTH) responses to MBP, and proliferative responses were decreased as well. SJL/J mice ($H\text{-}2^s$), which have a different MHC background and which therefore respond to different encephalitogenic peptides, were not protected, nor were PL/J mice infected with VV_{M1-23} and challenged with a whole spinal cord homogenate that contained multiple CNS autoantigens. This protection did not appear to be mediated by $CD8^+$ T cells since mice depleted of $CD8^+$ cells were still protected. Lymph node cells from protected (VV_{M1-23}-infected) mice could not transfer EAE to naive mice, whereas mice infected with VVsc11 and challenged with MBP could transfer disease. These data indicate that anergy to MBP is favored by infection. Thus, prior exposure can skew the immune repertoire toward unresponsiveness rather than activation. This may depend on the self-antigen or mimic that is presented by the microbe. These data favor a different interpretation of epidemiological and migration studies in MS. The observation is that northern Europeans who spend their first 15 years of life in northern Europe and then migrate to areas of low MS risk (closer to the equator) keep their high-risk phenotype (21, 24). Similarly, the reverse is true, in which individuals who, after age 15, move to northern Europe from an area with a low MS risk keep their low-risk phenotype. This has been attributed to the types and kinds of infections that a genetically susceptible individual was exposed to early in life (23). These infections could prime an individual for MS (discussed further later in this chapter). Another interpretation is that infections in regions where the prevalence of MS is low or where it is less prevalent may actually protect individuals from disease.

In the studies described above, a recombinant virus was used to mirror what may be occurring with viruses that encode cross-reacting determinants. Another way to mimic an infection with a virus that encodes self-proteins is with antigens encoded by plasmid DNA.

These cDNAs are then injected into the host and are used as effective vaccines (45). Colleagues and I speculated that this technique could be another way to deliver self-antigens that mimic host CNS proteins in a manner similar to that in which an infection would. Therefore, to complement the studies with VVplp described above, colleagues and I constructed a cDNA sequence that encoded whole PLP driven by a cytomegalovirus (CMV) promoter or that encoded encephalitogenic regions of PLP, $PLP_{139-151}$, and $PLP_{178-191}$. As a control for these experiments we used a pCMV-β sequence that encodes β-galactosidase. Mice were injected in the gastrocnemius muscles with the cDNAs. Two weeks following the last injection the mice were injected with $PLP_{139-151}$ or $PLP_{178-191}$ in CFA or with phosphate-buffered saline in CFA. The mice were then monitored for the development of EAE. No clinical signs were observed in mice not challenged with encephalitogenic peptides derived from PLP. However, antibody to PLP could be detected. Injection of mice with any of the cDNA constructs that encoded PLP itself or PLP epitopes enhanced the clinical symptoms of EAE over those observed in control plasmid-injected mice. Both the acute and chronic relapsing-remitting phases of the disease were augmented. Lymphoproliferative responses were increased for $PLP_{139-151}$-immunized mice compared with those found for control pCMV-immunized mice. In mice injected with $PLP_{139-151}$ plasmid DNA, proliferative responses were evident in cells obtained from draining lymph nodes even without $PLP_{139-151}$ challenge (41). Thus, injection of mice with cDNAs that encode PLP or encephalitogenic regions can prime individuals for autoimmune CNS disease, but by themselves they do not elicit disease. Recently, Ruiz et al. (35), using similar PLP cDNA constructs, found that mice injected with the constructs instead of being primed for autoimmune disease had suppressed responses to $PLP_{139-151}$. The data of my colleagues and I (41) and those of Ruiz et al. (35) still need to be reconciled.

Interestingly, it has been found that plasmid DNA that does not encode self-antigens could exacerbate EAE. This is most likely due to the CpG motifs contained in bacterially derived DNA. This DNA was shown to enhance gamma interferon and interleukin 6 production. In addition, it was found that natural killer (NK) cells were activated in treated mice. Clinical signs of relapsing-remitting EAE as well as pathological score (demyelination and inflammation) were enhanced (42).

More recently, Whitton and colleagues (33, 34) have demonstrated that ubiquitination of peptides and proteins can enhance their ability to generate $CD8^+$ cytotoxic T lymphocytes (CTLs) with little or no antibody production. Using this scheme, colleagues and I constructed cDNAs that encode a ubiquitinated form of PLP (pCMVUPLP). We speculated that this should drive PLP more efficiently into the proteasomes and allow better processing. From epidemiological studies, as stated before, the first 15 years of residence is important for imprinting a resistance or susceptibility phenotype in individuals genetically susceptible to MS (22). This has been attributed to the types or kinds of infections acquired during this period of time. In addition, there is a genetic component to MS. For example, HLA DRB1*1501 is associated with northern Europeans and is a prevalent marker for MS in this population (9). Besides infections that contribute to MS (10, 27, 31, 36), there is an age susceptibility (23) that could also contribute to disease. Therefore, the following experiment was conducted. Young (age 3 to 4 weeks, just prior to puberty) SJL/J mice (northern European equivalent) were injected three times with pCMVUPLP, which mimicked early infections that may be occurring early in life in a genetically susceptible population. Mice were then challenged with just adjuvant (CFA). The rationale for this is that while many microbes have been reported to be the "MS agent," no one organism has yet been determined to be the cause. However, infections have been reported to initiate disease or are associated with exacerbations (10, 27, 31, 36). Therefore, a nonspecific immunological stimulus was given to the mice. It was observed that the mice developed mild clinical signs of EAE, and infiltrates were seen in the CNSs of the animals. Proliferative responses to $PLP_{139-151}$ were observed. Thus, infections (in this instance, cDNA encoding PLP) early in life may prime an individual for autoimmune disease and a nonspecific event may trigger the attack.

The outset of this discussion presented data that indicated that the mimics were presented to the animal in a manner analogous to what may be occurring during a viral infection. Therefore, there appears to be a dilemma. Viral proteins are favored to be presented via MHC class I, although as virus-infected cells die and lyse, viral proteins can be presented by antigen-presenting cells such as macrophages that have ingested the debris. The models of autoimmune CNS disease described above are mediated by $CD4^+$ MHC class II-restricted T cells, and most autoimmune diseases are associated with certain HLA class II-restricting elements. All epitopes reported in these systems are presented by MHC class II molecules. Furthermore, investigation is required to determine the interplay between infections and disease induction and the role of $CD8^+$ versus $CD4^+$ T cells.

Evans et al. (11) have established an interesting model for virus-induced CNS autoimmune disease. They established transgenic mice that express lymphocytic choriomeningitis virus (LCMV) nucleoprotein or glycoprotein in oligodendrocytes (11). The viral nucleoprotein or glycoprotein would be treated in these mice as a self-CNS protein. The transgenic mice were then infected with LCMV. Infection was by the intraperitoneal route. This route does not lead to CNS LCMV infection. They observed that transgenic and nontransgenic mice made perfectly good immune responses to LCMV. This included CTL and humoral anti-LCMV immune responses. The CNSs of both transgenic and nontransgenic mice had increased levels of $CD8^+$ and $CD4^+$ T cells after infection. However, in the CNSs of transgenic mice, lymphocytes were retained and were present predominately in the white matter. $CD8^+$ T cells were evident as long as a year postinfection. Functionally, lymphocytes obtained from the CNS were cytotoxic for LCMV-infected or recombinant vaccinia virus-encoding nucleoprotein-infected target cells. As a consequence of infection with lymphocytes that infiltrate the CNS, there was a marked up-regulation of MHC class I and II molecules. In addition, focal areas of demyelination were present within inflammatory lesions. Those investigators went on to doubly infect these transgenic mice with LCMV or other related and nonrelated viruses. When transgenic mice were doubly infected with LCMV, they observed an enhancement of inflammatory cells in the CNS. Interestingly, transgenic mice previously infected with LCMV and then infected with vaccinia virus (nonrelated) or Pichinde virus (a distantly related arenavirus) also had increased CNS pathology after infection. These data indicate that the potentiation of autoimmune disease initially induced by molecular mimicry need not be with the same or original virus.

Talbot, Antel, and colleagues (40) have been putting together an interesting story concerning molecular mimicry, human coronavirus, and MS. They generated MBP- and human coronavirus-specific T-cell lines from patients with MS and healthy donors. Most of these T-cell lines were $CD4^+$ T cells. In analyzing the lines they found that almost 30% of the T-cell lines from MS patients showed reactivity to both MBP and coronavirus antigen. Less than 2% of the T cells from control healthy donors showed dual reactivity between virus and MBP. Reciprocal reactivities were seen in some T-cell lines from MS patients but not those from healthy donors. The shared epitopes have yet to be identified. Those investigators support the hypothesis that the cross-reacting immune responses may be involved in MS and other CNS inflammatory demyelinating diseases.

CONTRIBUTION OF ANTIBODY TO DISEASE

Antibodies can have profound effects that relate to autoimmune disease (29). During persistent viral infections antibody-antigen immune complex deposition can lead to glomerulonephritis (29). Colleagues and I proposed that since, during infections, cross-reacting antibodies are very prevalent, some of these immune complexes could comprise not only antibodies and viral proteins but also host proteins. Along these lines we have demonstrated that cross-reacting antibodies can have effects on the pattern of CNS disease that is observed.

Theiler's murine encephalomyelitis virus (TMEV) infection of mice leads to a chronic demyelinating disease that has similarities to the human demyelinating disease MS. In investigating the immune response to TMEV infection, we found that mice produced cross-reacting antibodies that reacted with TMEV protein 1 (VP1) and galactocerebroside (GC), a component of myelin. A monoclonal antibody against VP1 that also reacted with GC was obtained (14). This monoclonal antibody could inhibit the binding of antibodies from TMEV-infected mice to GC and TMEV, indicating that antibodies with similar specificities were present in sera from infected mice. The monoclonal antibody reacted with oligodendrocytes in vitro in newborn mouse brain cell cultures. By immunohistochemistry this monoclonal antibody bound to myelinated regions in the CNS. In addition, the monoclonal antibody was directed against a neutralizing epitope. Interestingly, passive transfer of this monoclonal antibody to naive uninfected mice did not result in demyelinating disease, but when this monoclonal antibody was transferred into mice that had been sensitized for EAE, the extent of the inflammatory demyelinating lesions was enhanced 10-fold over that in EAE-sensitized mice given no antibody (46). Thus, in this instance cross-reacting antibody could participate in lesion development but could not initiate development of the lesion itself.

In other studies colleagues and I explored another cross-reacting monoclonal antibody that defined a common epitope between astrocytes and the gp41 protein of human immunodeficiency virus (HIV) (46). The monoclonal antibody was produced to the HIV gp41 amino acid sequence Leu-Gly-Ile-Trp-Gly-Cys-Ser-Gly-Lys-Leu-Ile-Cys. This antibody not only reacted with HIV gp41 but also reacted with or activated astrocytes. Binding to reactive astrocytes could be inhibited with the HIV peptide. The protein in reactive astrocytes to which the monoclonal antibody bound had an apparent molecular mass of 43 kDa. It was found that the CSF from some patients with AIDS dementia contained antibodies that reacted with activated astrocytes. These types of antibodies found in HIV-infected patients could contribute to some of the CNS complications found in AIDS, such as dementia.

The interest in antibodies to CNS antigens in MS has waxed and waned, similar to the relapsing-remitting course that MS can take. Recently, Bronstein and colleagues (39) have identified a myelin-specific protein called oligodendrocyte-specific protein (OSP). This protein represents 7% of the proteins represented in myelin. They found that SJL/J mice sensitized with various peptides from OSP developed EAE. Mononuclear cell infiltrates with focal demyelination were evident in the CNSs of sensitized mice. In addition, that group (39) found that antibodies to OSP were present in the CSF of patients with relapsing-remitting MS, whereas they were not found in control subjects with other neurological diseases. A region from OSP, positions 114 to 120, was found to be the antibody-reactive site. This site cross-reacted with 27 different pathogens. It was found that all relapsing-remitting MS patients had antibodies that reacted to many of the cross-reactive (microbial) peptides. This suggested that the humoral response to $OSP_{114–120}$ was quite specific for relapsing-remitting MS.

In other studies, Genain et al. (15) found that myelin-specific antibodies were bound to areas of myelin vesiculation in early MS lesions. These MOG-specific antibodies were seen in areas of immunoglobulin deposition. Similar antibodies were seen in nonhuman primates sensitized with MOG or myelin in adjuvant. They presented a scenario in which binding of myelin antibodies or other soluble mediators could initiate the early vacuolation. These changes or myelin laminar separations would lead to additional myelin antigens becoming accessible for binding of additional antibody with complement deposition and macrophage accumulation. Macrophages would aid in myelin disruption through Fc-mediated uptake of antibody-antigen immune complexes. While they stated that the origins of the different antibodies in MS tissue remain unclear, one could speculate that a potential source of the initial antibodies could be through the generation of cross-reacting antibodies by molecular mimicry.

In more recent studies, Mokhtarian et al. (28) made comparisons between Semliki Forest virus (SFV) sequences and known CNS myelin proteins. SFV infection of C57BL6/J mice leads to encephalomyelitis and inflammatory demyelinating disease. They found that lymphocytes from mice infected with SFV proliferated in response to SFV E2 peptides and to a peptide derived from MOG (amino acids 18 to 32 cross-react with E2 amino acids 115 to 129). When mice were immunized with $MOG_{18–32}$ or $E2_{115–129}$, mice developed a late-onset chronic form of EAE. The pathological disease was similar to that (demyelination) found during SFV infection. Antibodies to several myelin proteins in addition to MOG were observed. They speculated that cross-reactive T cells as well as antibodies to MOG contribute to the demyelination observed during SFV infection.

SUMMARY AND CONCLUSION

Virus infections generally occur first in the periphery (Fig. 1) and then, if the CNS is involved, later spread by the hematogenous route or by axonal transport. In the case of virus-induced autoimmunity, viruses most likely infect cells in the periphery, leading to infection of antigen-presenting cells as well as other cells. Infection of nonimmune cells can lead to cell death by several pathways. Early after infection NK cells can kill infected cells as part of the innate immune response. Viruses can cause direct lysis of infected cells through either apoptotic or necrotic pathways. Once an antiviral immune response develops, antiviral antibodies can bind to the surfaces of infected cells, leading to the activation of the complement cascade that eventually kills the infected cell. In addition, as the cellular immune response is mounted $CD8^+$ T cells and, in some instances, $CD4^+$ T cells can lyse virus-infected cells. These three pathways lead to destruction of virus-infected cells. Cell debris can then be ingested by macrophages or other professional antigen-presenting

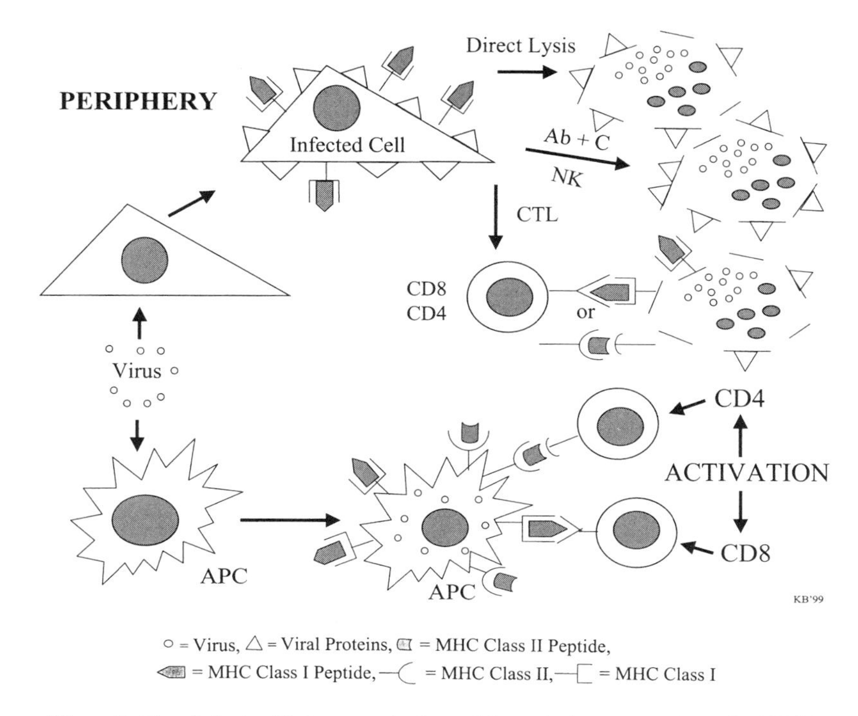

Figure 1. Virus infection likely occurs in the periphery. Antigen-presenting cells (APCs) are infected and can present antigen in the MHC class I to $CD8^+$ T cells. In addition, antigen-presenting cells can ingest and process infected cell debris, leading to presentation of epitopes via class II and activation of $CD4^+$ T cells. Infected cells can be killed by several mechanisms. Virus can directly kill cells, leading to release of additional progeny virus. Activated NK cells can kill infected cells early after infection, leading to a reduction of virus progeny. As an immune response is mounted antibody and complement can kill infected cells. In many instances antibodies need to recognize the viral glycoproteins expressed on the surfaces of the infected cell. $CD8^+$ and, in some cases, $CD4^+$ T cells can lyse virus-infected cells through either perforin or Fas-mediated pathways. Ab, antibody; C, complement.

cells and viral peptides presented via MHC class II. Infection of the antigen-presenting coils leads to the generation of activated $CD4^+$ T cells that potentiate B-cell differentiation and $CD8^+$ T-cell activation.

Generation of cross-reacting T cells and/or antibodies between viral epitopes and CNS determinants can lead to CNS pathology. Autoreactive cells generated in the periphery can then migrate through capillaries into the CNS parenchyma (Fig. 2). There plasma cells can produce antimyelin antibodies that aid in the breakdown of myelin. Macrophages can bind to the Fc portion of the immunoglobulin molecule bound to myelin and initiate myelin degradation. In addition, with complement, antibody can initiate oligodendrocyte lysis and death. Since these cells can myelinate 50 to 200 sheaths, loss of an oligodendrocyte can result in considerable myelin destruction. Also, once they are in the CNS, T cells can release proinflammatory cytokines such as tumor necrosis factor, which has myelinotoxic activity, resulting in further demyelination.

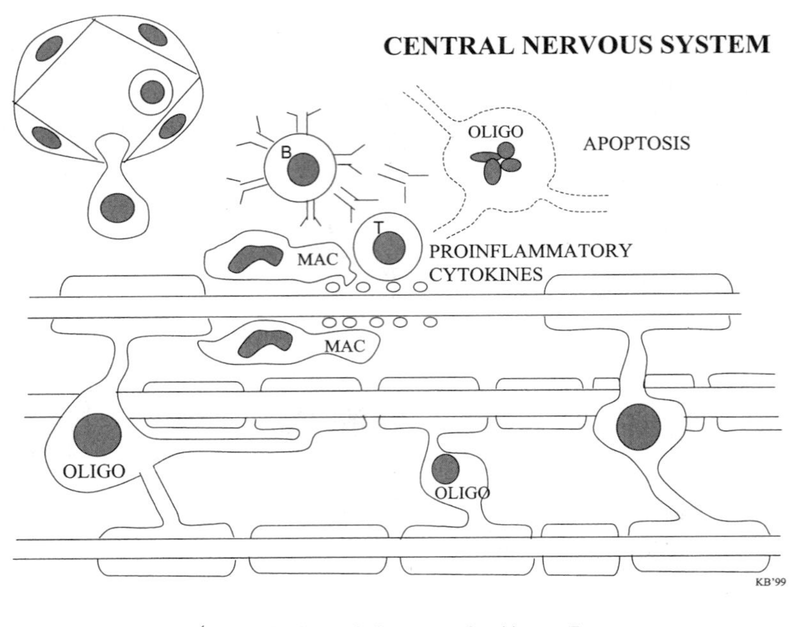

= Antibody, Proinflammatory Cytokines = Tumor Necrosis Factor, Lymphotoxin.

Figure 2. Once $CD4^+$ or $CD8^+$ T cells are activated they can readily cross the blood-brain barrier. There these cells recognize common epitopes between virus and self-CNS antigens and release proinflammatory cytokines. Many of these are myelinotoxic and can kill the myelin-producing cell, the oligodendrocyte (OLIGO). In addition, macrophages (MAC) and plasma cells are drawn into the areas of inflammation. Plasma cells locally produce cross-reacting antibodies that bind to the myelin and/or myelin-supporting cells. Macrophages can recognize antibody produced by the plasma cells bound to myelin through Fc receptors. As myelin is eroded nerve conduction diminishes, leading to clinical signs of disease.

With molecular mimicry, there is an abundance of evidence that cross-reactions can occur at many levels. Many viruses have been implicated in autoimmune diseases of the human CNS, but as yet no cause-and-effect association has been identified. Therefore, molecular mimicry remains a potential mechanism.

Acknowledgments. I thank Kathleen Borick for preparation of the manuscript. This work was supported by grant AI42525 from the National Institutes of Health.

REFERENCES

1. **Bahmanyar, S., J. Srinivasappa, P. Casali, R. S. Fujinami, M. B. A. Oldstone, and A. L. Notkins.** 1987. Antigenic mimicry between measles virus and human T lymphocytes. *J. Infect. Dis.* **156:**526–527.
2. **Barnett, L. A., and R. S. Fujinami.** 1997. Autoantigens of the nervous system, p. 333–363. *In* R. W. Keane and W. F. Hickey (ed.), *Immunology of the Nervous System.* Oxford University Press, New York, N.Y.
3. **Barnett, L. A., J. L. Whitton, Y. Wada, and R. S. Fujinami.** 1993. Enhancement of autoimmune disease using recombinant vaccinia virus encoding myelin proteolipid protein. *J. Neuroimmunol.* **44:**15–25.

4. **Barnett, L. A., J. L. Whitton, L. Y. Wang, and R. S. Fujinami.** 1996. Virus encoding an encephalitogenic peptide protects mice from experimental allergic encephalomyelitis. *J. Neuroimmunol.* **64:**163–173.
5. **Damian, R. T.** 1964. Molecular mimicry: antigen sharing by parasite and host and its consequences. *Am. Nat.* **XCVIII:**129–149.
6. **Damian, R. T.** 1988. Parasites and molecular mimicry, p. 211. *In* Å. Lernmark, T. Dyrberg, L. Terenius, and B. Hökfelt (ed.), *Molecular Mimicry in Health and Disease.* Elsevier Science Publishers B.V. (Biomedical Division), Amsterdam, The Netherlands.
7. **Damian, R. T.** 1989. Molecular mimicry: parasite evasion and host defense, *Curr. Top. Microbiol. Immunol.* **145:**101–115.
8. **Dhib-Jalbut, S., and S. Jacobson.** 1994. Cytotoxic T cells in paramyxovirus infection of humans. *Curr. Top. Microbiol. Immunol.* **189:**109–121.
9. **Ebers, G. C., and D. A. Dyment.** 1998. Genetics of multiple sclerosis. *Semin. Neurol.* **18:**295–299.
10. **Edwards, S., M. Zvartau, H. Clarke, W. Irving, and L. D. Blumhardt.** 1998. Clinical relapses and disease activity on magnetic resonance imaging associated with viral upper respiratory tract infections in multiple sclerosis. *J. Neurol. Neurosurg. Psychiatry* **64:**736–741.
11. **Evans, C. F., M. S. Horwitz, M. V. Hobbs, and M. B. A. Oldstone.** 1996. Viral infection of transgenic mice expressing a viral protein in oligodendrocytes leads to chronic central nervous system autoimmune disease. *J. Exp. Med.* **184:**2371–2384.
12. **Fujinami, R. S., and M. B. A. Oldstone.** 1985. Amino acid homology between the encephalitogenic site of myelin basic protein and virus: mechanism for autoimmunity. *Science* **230:**1043–1045.
13. **Fujinami, R. S., M. B. A. Oldstone, Z. Wroblewska, M. E. Frankel, and H. Koprowski.** 1983. Molecular mimicry in virus infection: crossreaction of measles virus phosphoprotein or of herpes simplex virus protein with human intermediate filaments. *Proc. Natl. Acad. Sci. USA* **80:**2345–2350.
14. **Fujinami, R. S., A. Zurbriggen, and H. C. Powell.** 1988. Monoclonal antibody defines determinant between Theiler's virus and lipid-like structures. *J. Neuroimmunol.* **20:**25–32.
15. **Genain, C. P., B. Cannella, S. L. Hauser, and C. S. Raine.** 1999. Identification of autoantibodies associated with myelin damage in multiple sclerosis. *Nat. Med.* **5:**170–175.
16. **Gran, B., B. Hemmer, and R. Martin.** 1999. Molecular mimicry and multiple sclerosis—a possible role for degenerate T cell recognition in the induction of autoimmune responses. *J. Neural Transm. Suppl.* **55:**19–31.
17. **Jacobson, S., J. R. Richert, W. E. Biddison, A. Satinsky, R. J. Hartzman, and H. F. McFarland.** 1994. Measles virus-specific T4+ human cytotoxic T cell clones are restricted by class II HLA antigens. *J. Immunol.* **133:**754–757.
18. **Johnson, R. T.** 1994. The virology of demyelinating diseases. *Ann. Neurol.* **36**(Suppl.)**:**S54–S60.
19. **Johnson, R. T., D. E. Griffin, R. L. Hirsch, J. S. Wolinsky, S. Roedenbeck, I. Lindo-de-Soriano, and A. Vaisberg.** 1984. Measles encephalomyelitis—clinical and immunologic studies. *N. Engl. J. Med.* **310:**137–141.
20. **Kies, M. W., R. E. Martenson, and G. E. Deibler.** 1972. Myelin basic proteins. *Adv. Exp. Med. Biol.* **32:**201–214.
21. **Kurtzke, J. F.** 1976. Multiple sclerosis among immigrants. *Br. Med. J.* **1:**1527–1528.
22. **Kurtzke, J. F.** 1977. Geography in multiple sclerosis. *J. Neurol.* **215:**1–26.
23. **Kurtzke, J. F.** 1993. Epidemiologic evidence for multiple sclerosis as an infection. *Clin. Microbiol. Rev.* **6:**382–427.
24. **Kurtzke, J. F., L. T. Kurland, and I. D. Goldberg.** 1971. Mortality and migration in multiple sclerosis. *Neurology* **21:**1186–1197.
25. **Maatta, J. A., M. S. Kaldman, S. Sakoda, A. A. Salmi, and A. E. Hinkkanen.** 1999. Encephalitogenicity of myelin-associated oligodendrocytic basic protein and 2′,3′-cyclic nucleotide 3′-phosphodiesterase for BALB/c and SJL mice. *Immunology* **95:**383–388.
26. **Marchalonis, J. J., S. F. Schluter, L. Wilson, D. E. Yocum, J. T. Boyer, and M. M. B. Kay.** 1993. Natural human antibodies to synthetic peptide autoantigens: correlations with age and autoimmune disease. *Gerontology* **39:**65–79.
27. **Metz, L. M., S. D. McGuinness, and C. Harris.** 1998. Urinary tract infections may trigger relapse in multiple sclerosis. *Axon* **19:**67–70.
28. **Mokhtarian, F., Z. Zhang, Y. Shi, E. Gonzales, and R. A. Sobel.** 1999. Molecular mimicry between a viral peptide and a myelin oligodendrocyte glycoprotein peptide induces autoimmune demyelinating disease in mice. *J. Neuroimmunol.* **95:**43-54.

29. **Oldstone, M. B. A., and F. J. Dixon.** 1971. Immune complex disease in chronic viral infections. *J. Exp. Med.* **134:**32s–40s.
30. **Pamer, E., and P. Cresswell.** 1998. Mechanisms of MHC class I–restricted antigen processing. *Annu. Rev. Immunol.* **16:**323–358.
31. **Rapp, N. S., J. Gilroy, and A. M. Lerner.** 1995. Role of bacterial infection in exacerbation of multiple sclerosis. *Am. J. Phys. Med. Rehabil.* **74:**415–418.
32. **Rodman, T. C., S. E. To, J. J. Sullivan, and R. Winston.** 1997. Innate natural antibodies. Primary roles indicated by specific epitopes. *Hum. Immunol.* **55:**87–95.
33. **Rodriguez, F., L. L. An, S. Harkins, J. Zhang, M. Yokoyama, G. Widera, J. T. Fuller, C. Kincaid, I. L. Campbell, and J. L. Whitton.** 1998. DNA immunization with minigenes: low frequency of memory cytotoxic T lymphocytes and inefficient antiviral protection are rectified by ubiquitination. *J. Virol.* **72:**5174–5181.
34. **Rodriguez, F., J. Zhang, and J. L. Whitton.** 1997. DNA immunization: ubiquitination of a viral protein enhances cytotoxic T-lymphocyte induction and antiviral protection but abrogates antibody induction. *J. Virol.* **71:**8497–8503.
35. **Ruiz, P. J., H. Garren, I. U. Ruiz, D. L. Hirschberg, L.-V. T. Nguyen, M. V. Karpuj, M. T. Cooper, D. J. Mitchell, C. G. Fathman, and L. Steinman.** 1999. Suppressive immunization with DNA encoding a self-peptide prevents autoimmune disease: modulation of T cell costimulation. *J. Immunol.* **162:**3336–3341.
36. **Sibley, W. A., C. R. Bamford, and K. Clark.** 1985. Clinical viral infections and multiple sclerosis. *Lancet* **i:**1313–1315.
37. **Spalter, S. H., S. V. Kaveri, E. Bonnin, J. C. Mani, J. P. Cartron, and M. D. Kazatchkine.** 1999. Normal human serum contains natural antibodies reactive with autologous ABO blood group antigens. *Blood* **93:**4418–4424.
38. **Srinivasappa, J., J. Saegusa, B. S. Prabhakar, M. K. Gentry, M. Buchmeier, T. J. Wiktor, H. Koprowski, M. B. A. Oldstone, and A. L. Notkins.** 1986. Molecular mimicry: frequency of reactivity of monoclonal antiviral antibodies with normal tissues. *J. Virol.* **57:**397–401.
39. **Stevens, D. B., K. Chen, R. S. Seitz, E. E. Sercarz, and J. M. Bronstein.** 1999. Oligodendrocyte-specific protein peptides induce experimental autoimmune encephalomyelitis in SJL/J mice. *J. Immunol.* **162:**7501–7509.
40. **Talbot, P. J., J. S. Paquette, C. Ciurli, J. P. Antel, and F. Ouellet.** 1996. Myelin basic protein and human coronavirus 229E cross-reactive T cells in multiple sclerosis. *Ann. Neurol.* **39:**233–240.
41. **Tsunoda, I., L.-Q. Kuang, N. D. Tolley, J. L. Whitton, and R. S. Fujinami.** 1998. Enhancement of experimental allergic encephalomyelitis (EAE) by DNA immunization with myelin proteolipid protein (PLP) plasmid DNA. *J. Neuropathol. Exp. Neurol.* **57:**758–767.
42. **Tsunoda, I., N. D. Tolley, D. J. Theil, J. L. Whitton, H. Kobayashi, and R. S. Fujinami.** 1999. Exacerbation of viral and autoimmune animal models for multiple sclerosis by bacterial DNA. *Brain Pathol.* **9:**481–493.
43. **van Binnendijk, R. S., M. C. Poelen, K. C. Kuijpers, A. D. Osterhaus, and F. G. Uytdehaag.** 1990. The predominance of $CD8^+$ T cells after infection with measles virus suggests a role for $CD8^+$ class I MHC-restricted cytotoxic T lymphocytes (CTL) in recovery from measles. Clonal analyses of human $CD8^+$ class I MHC-restricted CTL. *J. Immunol.* **144:**2394–2399.
44. **Ward, B. J., R. T. Johnson, A. Vaisberg, E. Jauregui, and D. E. Griffin.** 1990. Spontaneous proliferation of peripheral mononuclear cells in natural measles virus infection: identification of dividing cells and correlation with mitogen responsiveness. *Clin. Immunol. Immunopathol.* **55:**315–326.
45. **Whitton, J. L., F. Rodriguez, J. Zhang, and D. E. Hassett.** 1999. DNA immunization: mechanistic studies. *Vaccine* **17:**1612–1619.
46. **Yamada, M., A. Zurbriggen, M. B. A. Oldstone, and R. S. Fujinami.** 1991. Common immunologic determinant between human immunodeficiency virus type I gp41 and astrocytes. *J. Virol.* **65:**1370–1376.

Molecular Mimicry, Microbes, and Autoimmunity
Edited by M. W. Cunningham and R. S. Fujinami

Chapter 4

Molecular Mimicry between Streptococcal M Protein and Cardiac Myosin and the Immunopathogenesis of Rheumatic Fever

Madeleine W. Cunningham

Rheumatic fever is one of the most serious diseases caused by group A streptococci. *Streptococcus pyogenes* is the etiological agent of many suppurative diseases such as pharyngitis, erysipelas, cellulitis, and impetigo, all of which are treatable with penicillin. Rheumatic fever occurs as an autoimmune sequela following streptococcal pharyngitis (8, 9, 67). Treatment of pharyngitis with penicillin immediately eradicates the bacterium, lowers or prevents the immune response against the streptococcus, and lowers the incidence of rheumatic fever. After the discovery of penicillin (32), the rate of rheumatic fever in nations with access to antibiotic therapy dropped steadily over the years until a low incidence of rheumatic fever was seen in the United States and Europe (65). However, recent outbreaks of rheumatic fever have been reported in the United States, with a primary focus of disease in the Salt Lake City, Utah, area since the early 1980s (71, 72). While the discovery of penicillin changed the face of rheumatic fever in the United States and Europe, rheumatic carditis has remained the major cause of acquired heart disease in children worldwide (2, 5, 39, 66–68).

Historically, rheumatic fever is a disease that has been reported in the medical literature since very early times (11, 55), and autoantibodies against the heart were also recognized in the disease (10, 40). Although rheumatic fever was associated with group A streptococcal infection, immunologic cross-reactivity between group A streptococci and heart tissues was not identified until the 1960s (41, 73–75). The identification of cross-reactive sera led investigators to the hypothesis of molecular mimicry between the group A streptococcus and human tissues, including the heart (73–75), joints (6), brain (38), and skin (48). The manifestations of acute rheumatic fever include carditis, the most serious manifestation; arthritis, the most common manifestation; erythema marginatum, a circinate skin rash; subcutaneous nodules; and Sydenham's chorea, a nervous-movement disorder.

Madeleine W. Cunningham • Department of Microbiology and Immunology, University of Oklahoma Health Sciences Center, P.O. Box 26901, Oklahoma City, OK 73190.

Autoantibodies against tissues present in patients with acute rheumatic fever were found to be removed from the sera after absorption with group A streptococci or its membranes or cell walls (42, 69, 74). The specific antigens of the streptococcus and host tissues were not identified in these studies with human and animal sera. It was also unclear if there were tissue-cross-reactive antigens in the wall or membrane of the streptococci, and the results of some of the studies appeared to be in conflict. Due to the polyclonal nature of sera, even the idea that antibodies could be cross-reactive was challenged. In the early 1980s, my laboratory produced human and mouse cross-reactive monoclonal antibodies (MAbs) which have identified the cross-reactive antigens on the streptococcus and in human tissues. In addition, the studies established that cross-reactivity was a feature of some antistreptococcal antibodies which were autoreactive (19, 24, 44).

From the perspective of the microbiologist, the cross-reactive MAbs identified the streptococcal cross-reactive antigens as M protein, the antiphagocytic virulence factor of group A streptococci, and *N*-acetylglucosamine, the immunodominant epitope of the group A carbohydrate. In the eyes of the immunologist, the antibodies were polyreactive and recognized a group of related host proteins including cardiac and skeletal myosin, keratin, vimentin, and tropomyosin (16, 20, 24, 29). The antistreptococcal and antimyosin MAbs were some of the first polyreactive antibodies reported in the literature (24). Although the variable region genes of all of the mouse and human cross-reactive MAbs have been sequenced, there is no immediately identifiable characteristic or sequence that renders them polyreactive (1; N. M. J. Mertens, J., J. E. Galvin, and M. W. Cunningham, submitted for publication). However, one could speculate that certain combinations of V-gene sequences may be important for the conformation necessary for the polyreactivity. There are several recent comprehensive reviews on the studies of the heart- and tissue-cross-reactive MAbs that have been studied over the past 15 years (13–15; M. W. Cunningham, submitted for publication).

To summarize, the MAbs definitively recognized myosin (19) and streptococcal M protein (24, 29) as well as other alpha-helical coiled-coil molecules (29), and in the mouse, a subset recognized DNA (24) and another antibody subset recognized *N*-acetyl-β-D-glucosamine (62). In humans with rheumatic carditis, the antimyosin and antistreptococcal MAbs recognized *N*-acetylglucosamine (1, 60, 61). The role of the cross-reactive antibodies in disease is under investigation. Only a few of the mouse and human cross-reactive MAbs have been found to be cytotoxic and potentially pathogenic (1, 3, 4, 17). Cytotoxicity was associated with the expression of laminin on the cell lines tested. Laminin may be an important link between the myocardium and the heart valve. The myocardium contains laminin around each of the myocyte membranes, while in the valve laminin is present in the basement membrane adjacent to the surface endothelium. Antimyosin MAbs have been shown to deposit in heart tissues in DBA/2 mice and to produce myocarditic lesions (47). Antibody deposition may be due to the immunoglobulin G (IgG) subclass of MAbs and to the expression of myosin-like molecules in the tissues of the genetically susceptible mice. Although IgM antibodies may not be capable of depositing in tissues, they may be readily accessible to valve endothelial surfaces. Studies will continue to probe for clues to the role and mechanisms of cross-reactive autoantibodies in rheumatic fever as well as other autoimmune diseases and sequelae.

THE MIMICRY MODEL: A STUDY OF THE PATHOGENIC EPITOPES OF STREPTOCOCCAL M PROTEIN AND THEIR MIMICRY WITH CARDIAC MYOSIN

The streptococcal M protein is one of the most well-described virulence factors of an extracellular bacterial pathogen among gram-positive bacteria. Its antiphagocytic property renders the group A streptococcus resistant to phagocytosis (30, 31). However, the infected host produces antibody against the homologous M serotype and eliminates the infection through opsonization and phagocytosis of the group A streptococci. There are more than 80 different M serotypes, and the most efficient protection is by production of homologous M-serotype-specific antibody (28, 30, 45).

The structure of the M protein was determined by amino acid and nucleotide sequence analyses of the M24, M5, and M6 proteins (7, 36, 49, 53). The sequence analysis revealed a seven-amino-acid residue periodicity characteristic of alpha-helical coiled-coil proteins (50–52). Several studies revealed the immunological cross-reactivity of the streptococcal M protein with antibodies that reacted with myosin (25, 26), including the anti-streptococcal and antimyosin MAbs (24, 29). Furthermore, antimyosin antibodies from patients with acute rheumatic fever reacted with specific sequences in the M5 protein (20).

To investigate the pathogenic epitopes of the streptococcal M protein, BALB/c mice were immunized with each of 23 peptides that span the A, B, and C repeat regions of the rheumatogenic streptococcal M5 protein (18). The streptococcal M5 serotype has been associated with rheumatic fever outbreaks for many years (8, 64, 67). The entire set of 23 synthetic peptides is shown in Table 1 and has been reported previously (18). Study of the

Table 1. Overlapping synthetic peptides of streptococcal M5 protein[a]

Peptide	Sequence	Amino acid residues
NT1	AVTRGTINDPQRAKEALD	1–18
NT2	KEALDKYELENHDLKTKN	14–31
NT3	LKTKNEGLKTENEGLKTE	27–44
NT4	GLKTENEGLKTENEGLKTE	40–58
NT5	KKEHEAENDKLKQQRDTL	59–76
NT6	QRDTLSTQKETLEREVQN	72–89
NT7	REVQNTQYNNETLKIKNG	85–102
NT8	KIKNGDLTKELNKTRQEL	98–115
B1A	TRQELANKQQESKENEKAL	111–129
B1B	ENEKALNELLEKTVKDKI	124–141
B1B2	VKDKIAKEQENKETIGTL	137–154
B2	TIGTLKKILDETVKDKIA	150–167
B2B3A	KDKIAKEQENKETIGTLK	163–180
B3A	IGTLKKILDETVKDKLAK	176–193
B2B3B	DKLAKEQKSKQNIGALKQ	189–206
B3B	GALKQELAKKDEANKISD	202–219
C1A	NKISDASRKGLRRDLDAS	215–232
C1B	DLDASREAKKQLEAEHQK	228–245
C1C2	AEHQKLEEQNKISEASRK	241–258
C2A	EASRKGLRRDLDASREAK	254–271
C2B	SREAKKQLEAEQQKLEEQ	267–284
C2C3	KLEEQNKISEASRKGLRR	280–297
C3	KGLRRDLDASREAKKQ	293–308

[a] Sequences from references 49 and 53.

Table 2. M5 protein sequences which produce myocarditis in mice[a]

Peptide	Amino acid sequence	Residues
A repeat region		
NT4	GLKTENEGLKTENEGLKTE	40–58
NT5	KKEHEAENDKLKQQRDTL	59–76
NT6	QRDTLSTQKETLEREVQN	72–89
B repeat region		
B1A	TRQELANKQQESKENEKAL	111–129
B3B	GALKQELAKKDEANKISD	202–219

[a] NT4 produced myocarditis in BALB/c (18) and MRL/++ (37) mice; the other peptides shown produced myocarditis in BALB/c mice. Reproduced from *Effects of Microbes on the Immune System* (13) with permission from Lippincott-Williams & Wilkins.

M5 peptides led to the identification of epitopes which could produce heart disease. Therefore, these undesirable disease-producing epitopes could be eliminated from streptococcal vaccines. Streptococcal M5 peptides which produced lesions in the myocardium were NT4, NT5, NT6, B1A, and B3B (18). None of the other peptides produced myocardial lesions in the mice. Table 2 indicates the protein sequences that produce inflammation in the hearts of BALB/c mice. Peptides of the C repeat region did not produce inflammatory heart lesions. An example of the myocardial cell infiltrate is shown in Fig. 1.

M5 peptides which induced the strongest antibody responses to human cardiac myosin were found in peptides in the A and C repeat regions and one peptide, B2B3B, in the B repeat region (sequences shown in Table 1) (18). The antihuman cardiac myosin responses of M5 peptide-immunized mice are shown in the Western blot in Fig. 2. The titers of the antistreptococcal M5 peptide against human cardiac myosin and skeletal myosin in sera are shown in Fig. 3A and B. The titers of the antipeptide against antihuman cardiac myosin in sera ranged from 25,600 to > 204,800, while the titers against skeletal

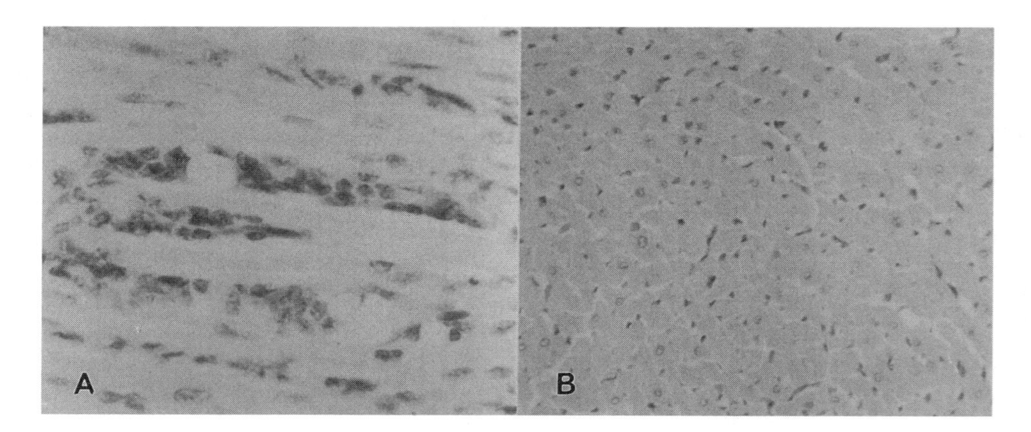

Figure 1. (A) Hematoxylin-eosin-stained heart tissue section from BALB/c mice immunized with NT4 sequence from the streptococcal M5 protein. Similar cellular infiltrates were observed in mice given NT5, NT6, B1A, and B3B. (B) Control myocardium from a mouse given only phosphate-buffered saline and adjuvant. Reprinted from reference 18 with permission from the American Society for Microbiology.

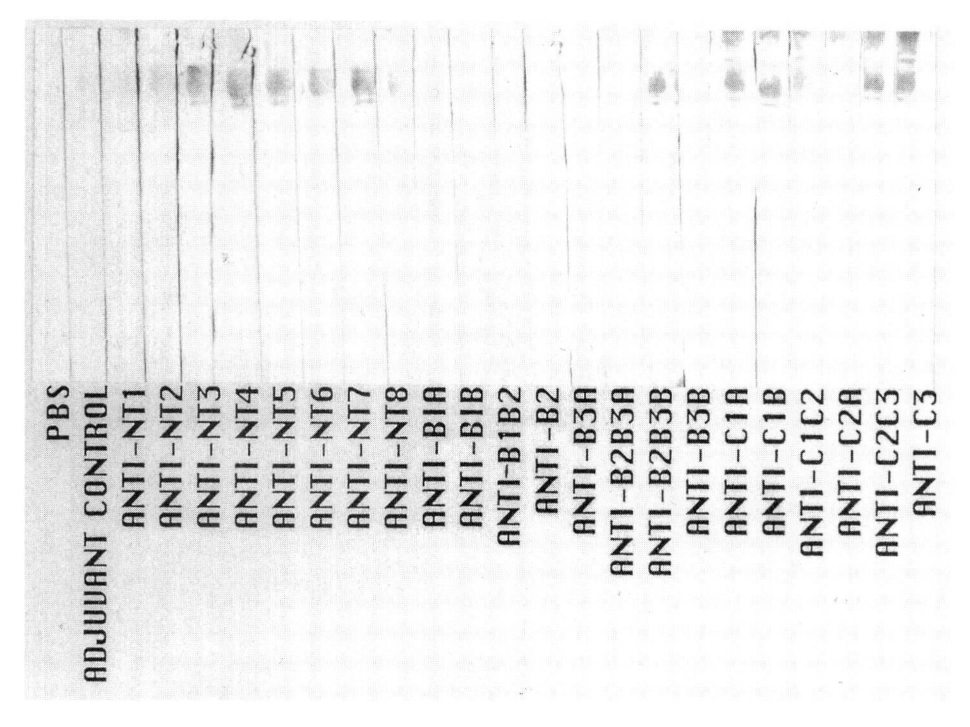

Figure 2. Antisera against streptococcal M5 peptides reacted with human cardiac myosin in the Western immunoblot. Reprinted from reference 18 with permission from the American Society for Microbiology.

myosin ranged from 1,600 to 6,400, as described previously (18). These data suggest that epitopes in type 5 streptococcal M protein mimic cardiac myosin more than skeletal myosin. Sequences within the M proteins which are more like cardiac myosin may be important in breaking tolerance against this self-antigen in heart tissues. Although the role of antimyosin antibodies in acute rheumatic fever is not clear, it is possible that mimicry between streptococcal M protein and myosin-like proteins in the extracellular matrix of the valve and myocardium may be important in producing disease (1, 3, 17, 47). Cross-reactive antistreptococcal and antimyosin mouse and human MAbs have been shown to react strongly with the human endothelium and the basement membrane (33, 35). Cytotoxic cross-reactive antibody has been shown to react with laminin, which shares strong homology with alpha-helical coiled-coil molecules (3, 4).

When human cardiac myosin cross-reactive T-cell epitopes were mapped in the streptococcal M5 protein, several sites appeared to be dominant. The experiments were designed to immunize BALB/c mice with human cardiac myosin and recover the lymph node cells and react the human cardiac myosin-sensitized lymphocytes with each of the streptococcal M5 peptides, as described previously (18). The responses of the myosin-sensitized lymphocytes to streptococcal M5 peptides are shown in Fig. 4A to C (18). The human cardiac myosin-cross-reactive T-cell epitopes were found to reside in the NT4-NT5 region, in the B1B2-B2-B3A region, and in the C2A-C3 region (see Table 1 for sequences). In the same

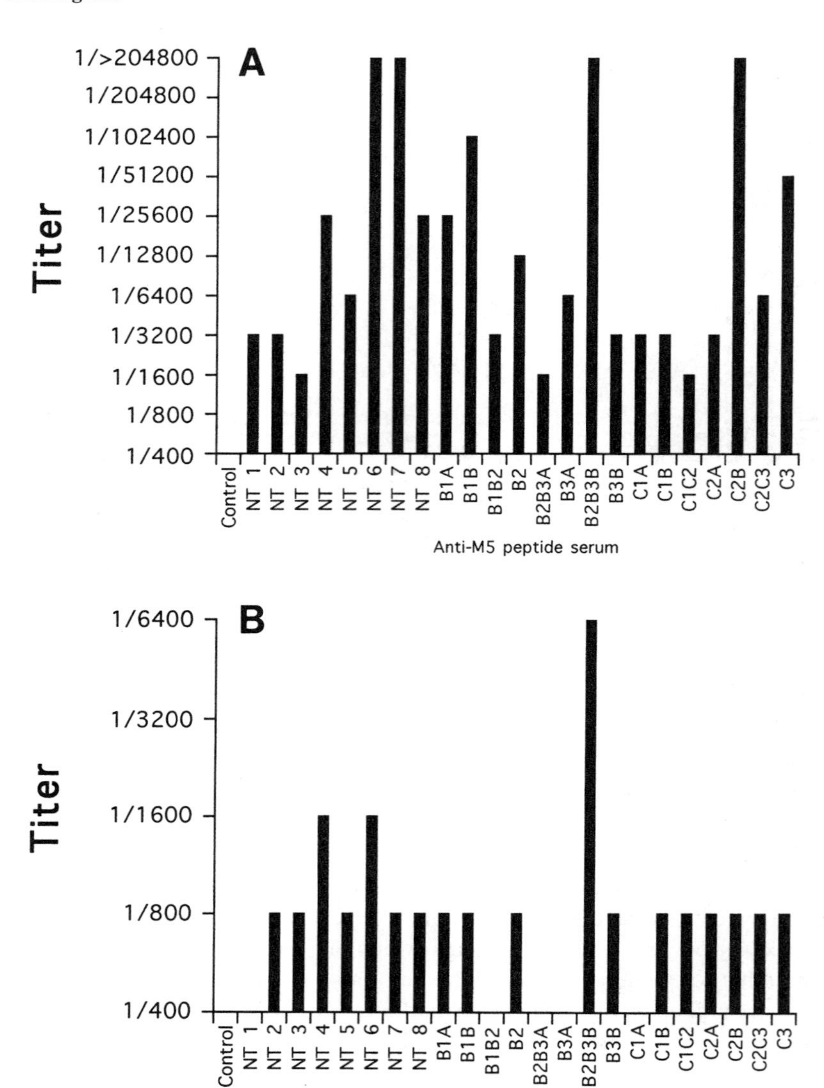

Figure 3. Titers of antistreptococcal M5 peptide against human cardiac myosin in sera determined by reaction of sera with anti-M5 peptide with human cardiac myosin in an enzyme-linked immunosorbent assay. Sera were diluted twofold and were reacted with 10 μg of human cardiac myosin per ml in an enzyme-linked immunosorbent assay. Endpoints were calculated at optical densities of 0.1 and 0.2. The endpoints at an optical density of 0.1 are shown in the figure and were slightly higher but not significantly different from those for the endpoint at an optical density of 0.2. The same relative differences between cardiac and skeletal myosins were observed at both endpoints and in different assays. (A) Titers of the antistreptococcal M5 peptide against human cardiac myosin in sera. (B) Titers of the anti-streptococcal M5 peptide against rabbit skeletal myosin in sera. The titers of the antipeptide against cardiac myosin in sera indicated a greater than fourfold increase in reactivity. Reprinted from reference 18 with permission from the American Society for Microbiology.

study, NT4, NT5, NT6, B1A, and B3B were shown to produce inflammatory heart lesions in mice (Table 2). Both NT4 and NT5 produced inflammatory heart lesions and contained dominant cardiac myosin-cross-reactive T-cell epitopes of M5 protein. However, the three other peptides which produced disease did not appear to contain dominant myosin-cross-reactive T-cell epitopes. It is possible that they were not evident among the myosin-cross-reactive T-cell epitopes because their mimicking sequences in cardiac myosin were not dominant and presented after one immunization of the BALB/c mice with cardiac myosin. It is also possible that another cardiac protein is involved in cross-reactivity with these peptides.

When the amino acid sequences of M5 protein and human cardiac myosin were compared by using the Fast P computer program, 19% identity and 70% homology were observed between the two molecules within a 270-amino-acid overlap in the light meromyosin (LMM) region of the human cardiac myosin β chain. The repeated sequences TIGTL and VKDKIA were found to be overlapping within the streptococcal peptides B1B2 and B2, which produced the strongest proliferative response from myosin-sensitized T lymphocytes (18). The sequences of the A and B repeat regions contained in NT4, NT5, B1B2, and B2, which produced the strongest proliferative responses from myosin-sensitized T cells, share the strongest homology with sites in human cardiac myosin (Fig. 5). Repeated regions of M protein which are homologs of cardiac myosin may break tolerance to this heart protein and produce inflammatory heart disease.

It is well-known that cardiac myosins induce myocardial inflammation, while skeletal myosins do not (43, 54). Amino acid sequences unique to cardiac myosins and mimicked in M proteins may be important in targeting T-cell responses to the heart in acute rheumatic fever. Presentation of mimicking epitopes by major histocompatibility complex (MHC) class II molecules in a susceptible host may lead to enhanced activation of T cells and production of cytokines responsible for disease. The hypothesis that M proteins may break immune tolerance to unique and pathogenic epitopes of human cardiac myosin may be important for defining the immune mechanisms of rheumatic heart disease.

The sequences found within peptides C2A and C3 of the C repeat region did not produce the strongest myosin-cross-reactive T-cell responses, did not produce cardiac inflammation, and were homologous to sites in skeletal as well as cardiac myosins. Previous studies have demonstrated homology and immunological cross-reactivity between the C repeat region, which comprises the class I epitope of M protein, and myosin (59, 70). The RRDL sequence found in both peptides C2A and C3 is present in both skeletal and cardiac myosins within the heavy meromyosin subfragment of myosin. The homology of the C repeat region with skeletal myosin as well as with cardiac myosin may prevent the C repeat region from breaking tolerance and producing cardiac inflammation.

The NT4 peptide, which demonstrates striking amino acid sequence homology with cardiac myosin, was also investigated in studies with MRL/++ mice (37). The sequence repeated in NT4, GLKTENE, is repeated four times in the M5 protein, but the homologous myosin sequence LQTEN is found only once in cardiac myosins. The homology is shown in Fig. 5. NT4 was found to produce cardiac inflammation in the MRL/++ mouse strain (37). Repetitive cardiac myosin-like sequences in the streptococcal M proteins may give them their rheumatogenic potential in a susceptible host. In addition, the study with MRL/++ mice identified $CD4^+$ T cells and expression of the MHC class II IA^k molecules as important in the production of disease since anti-IA^k and anti-CD4 antibodies abrogated

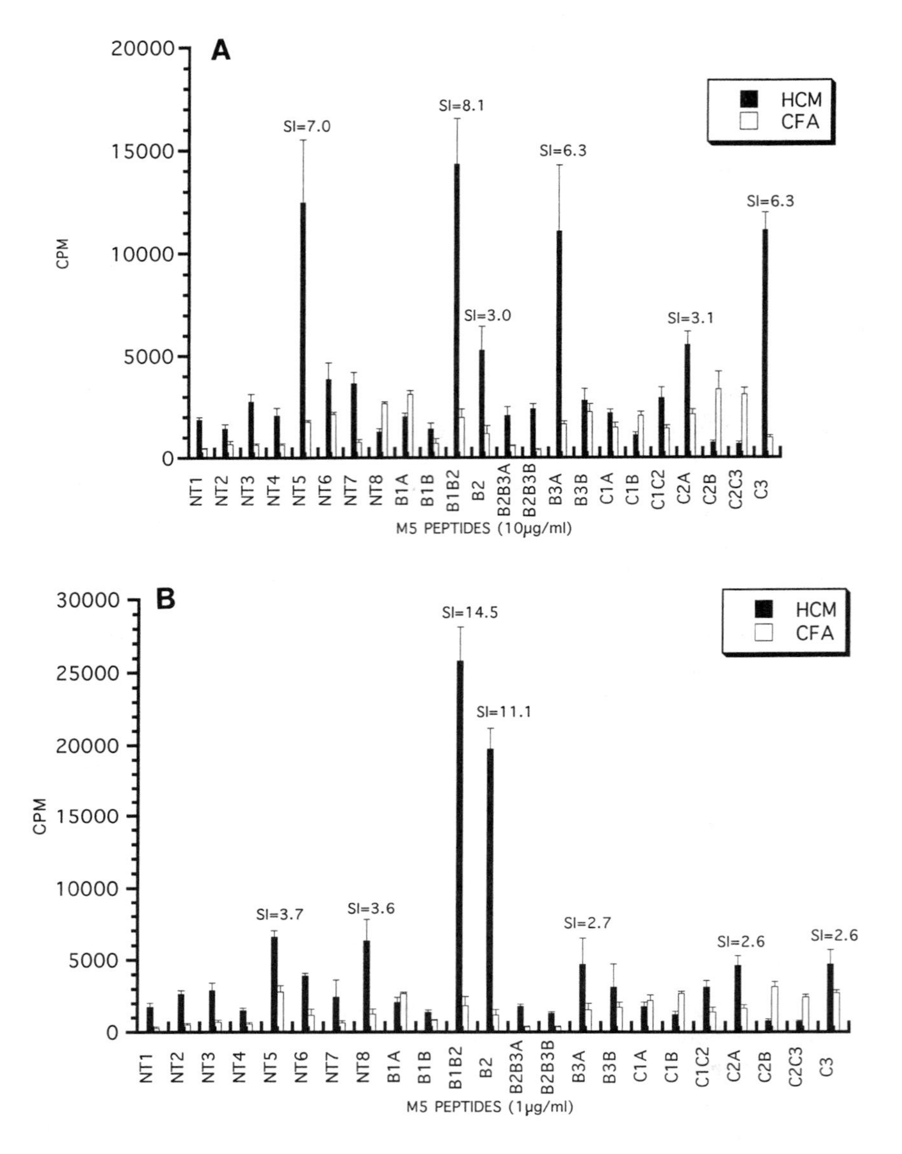

Figure 4. Responses of myosin-sensitized lymphocytes to streptococcal M5 peptides as determined by stimulation of lymphocytes from BALB/c mice immunized with human cardiac myosin (HCM) with each of 23 overlapping M5 peptides in a tritiated thymidine incorporation assay. Panels A, B, and C each contain data from separate experiments. (A and B) Stimulation of lymphocytes by the M5 peptides at concentrations of 10 and 1 µg/ml, respectively. The data are averages from four separate experiments. The no-antigen control was 1,773 cpm. (C) Data from one experiment in which myosin-immune lymphocytes were stimulated with M5 peptides at a concentration of 0.1 µg/ml. The no-antigen control was 3,720 cpm. Standard deviation bars are shown for the data in panel C. Solid bars represent the counts per minute for myosin-immune lymphocytes, and the open bars represent the counts per minute for lymphocytes from mice immunized with complete Freund's adjuvant (CFA) only. Standard error was calculated for the average of four assays, and stimulation indices (SI) are shown for the dominant peptides. The four individual assays demonstrated similar results and are reflected in panels A and B as an average of the data. The most dominant myosin-cross-reactive site was the B1B2-B2 site, which appears in all three panels. Other dominant peptides included NT5, B3A, and C3. Reprinted from reference 18 with permission from the American Society for Microbiology.

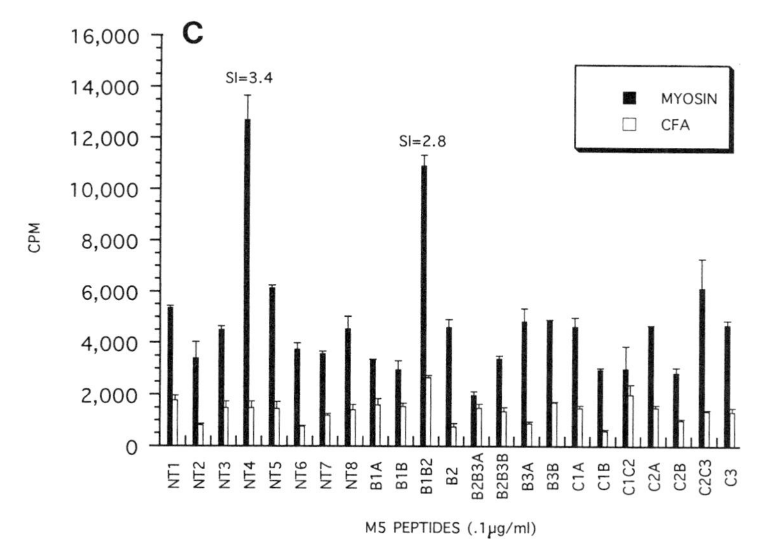

Figure 4. *Continued*

myocarditis (37). Interestingly, the NT4 peptide tolerized the MRL/++ mice against myocarditis produced by coxsackievirus infection (37). Tolerance was observed in dramatically reduced histopathology and reduced T-cell proliferation responses to the peptide. It is possible that the tolerance may be due to a shift in the dominant T-helper-cell subset; however, the mechanism of NT4-induced tolerance has not yet been defined.

Previous studies have presented additional evidence of mimicry between group A streptococcal M6 protein, cardiac myosin, and coxsackieviral capsid proteins VP1, VP2, and VP3 (17). MAbs cross-reactive with group A streptococcal M protein and myosin were shown to neutralize coxsackievirus in vitro in plaque assays. In addition, one of the MAbs was found to be highly cytotoxic for heart cells in culture (17). The MAbs reacted with coxsackieviral capsid proteins in Western blots, and amino acid sequence homology was found to be shared between the capsid protein VP1, human cardiac myosin, and M6 protein (17). Since coxsackievirus is known to produce myocarditis in humans and animals, it is possible that this myocarditic virus shares similar pathogenic mechanisms with the group A streptococcus. Thus, mimicry may be important between humans and microbes as well as between microbes themselves.

In a final note about mimicry between M protein and other microbial proteins, previous studies have shown that streptococcal M proteins and a heat shock protein, hsp65, share immunological cross-reactivity (58). This is most interesting in view of the central role that heat shock proteins may play in the development or prevention of autoimmune diseases (76). Previous studies demonstrated the reactivities of polyclonal and monoclonal anti-hsp65 antibodies with streptococcal M-protein serotypes 5, 6, 19, 24, and 49 but no reactivity with an isogenic mutant strain that lacked the M protein. Anti-hsp65 antibodies also reacted with recombinant M5 and M6 proteins (58). hsp65 has been associated with arthritis in rats and could link M proteins to arthritis symptoms in patients with acute rheumatic fever. Sequence alignments have indicated some homology between M protein and hsp65 (58). Table 3 summarizes host or microbial proteins with known molecular mimicry with M protein.

IMMUNOPATHOGENESIS OF RHEUMATIC FEVER AND RHEUMATIC HEART DISEASE

Acute rheumatic fever is a postinfectious autoimmune sequela that arises from a group A streptococcal infection. It is an important disease that links an infectious agent with autoimmunity because there are few autoimmune diseases for which the infectious etiology has been identified. Antimyosin autoantibodies were found in patients with rheumatic fever (21), and M-protein epitopes recognized by affinity-purified antimyosin anti-

```
A.   streptococcal M5 peptide B2   TIGTLKKILDETVKDKIA
                                    ...::. :.:.:: : :
     human cardiac myosin           LEDLKRQLEEEVKAKNA

B.   streptococcal NT4 peptide     GLKTENEGLKTENEGLKTE
                                    : :::
     human cardiac myosin          KLQTENGE

C.   M5 peptide C2A EASRKGLRRDLDASREAK
                        ::::::::
     M5 Peptide C3      KGLRRDLDASREAKKQ
                         ..::::.
     myosin        NKKREAFQKMRRDLE
```

Figure 5. (A) Examples of homology between human cardiac myosin (residues 1313 to 1329) and streptococcal M5 protein peptide B2 (streptococcal M5 protein; residues 150 to 167). Identity (47%) was observed in a 17-amino-acid overlap. The M5 peptides B2, B1B2, and B3A contain large amounts of overlapping sequence. The colons represent identities, and the periods represent conserved substitutions. Three of the residues shown in the myosin sequence are unique to human cardiac myosin. (B) Amino acid sequence identity (LKTEN) between M5 peptide NT4 and human cardiac myosin (LQTEN). NT4 contains residues 40 to 58 of the streptococcal M5 protein. The cardiac myosin sequence shown is found in residues 1279 to 1286, near the beginning of the LMM tail and the end of the S-2 fragment of myosin. The myosin sequence shown in panel B is conserved among cardiac myosins. Reprinted from reference 18 with permission from the American Society for Microbiology. (C) Alignment of streptococcal M5 peptides C2A and C3 with the heavy chain of human cardiac myosin. Identical residues are indicated with colons, while periods indicate conservative substitutions as determined by the FASTA alignment program (Intelligenetics, Mountain View, Calif.). The sequence identified as homologous between C2A (streptococcal M5 protein; residues 254 to 271), C3 (streptococcal M5 protein; residues 393 to 308), and myosin is highly conserved (identical) between skeletal and cardiac myosins. The identical residues are located in the heavy meromyosin fragment of skeletal and cardiac myosin (residues 1172 to 1186). Reprinted from reference 59 with permission from the American Society for Microbiology.

bodies from rheumatic carditis were identified (20). A specific M5-protein epitope, QKSKQ, from the B repeat region was recognized by the antimyosin antibodies. The MAbs from patients with rheumatic carditis cross-react with the *N*-acetylglucosamine epitope, myosin, and M protein (1, 33).

Although the role of antimyosin antibodies in disease is controversial, cytotoxic antimyosin antibodies from patients with rheumatic carditis have been reported (1, 34). The cross-reactive MAbs from patients with rheumatic carditis were cytolytic for either heart or endothelial cells in the presence of complement. Studies with sera from patients in the rheumatic fever outbreak in Utah showed that approximately 35 to 40% of the serum specimens were cytotoxic for heart cells (M. W. Cunningham, E. E. Adderson, and L. G. Veasy, unpublished data). Furthermore, previous studies suggest that antibodies may play a role in disease. Studies have shown that in patients with rheumatic carditis antibody and complement were bound to both the myocardium and the valve (40). Potentially, the valve could be susceptible to IgM or IgG antibodies with the cross-reactive specificity. If the valvular endothelium is attacked by antibody, then inflammation at the valvular endothelium could promote expression of cell adhesion molecules and recruitment of T cells to the valve. Investigations in my laboratory of heart valve lesions from patients with rheumatic carditis have demonstrated that T cells bind to the valve surface endothelium and extravasate through the endothelium into the valve. The T-cell extravasation event appeared to be associated with expression of vascular cell adhesion molecule type 1 (VCAM-1) on rheumatic valves during disease (S. Roberts, S. Kosanke, S. T. Dunn, D. Jankelow, C. M. G. Duran, and M. W. Cunningham, submitted for publication). Cross-reactive human antistreptococcal and antimyosin MAbs from patients with rheumatic carditis have been shown to react with valve endothelium and recognize laminin, the extracellular matrix molecule found in the laminar basement membrane of the valve and surrounding cardiomyocytes (3, 4, 33).

The mechanism by which the cross-reactive antibodies may react with the valve surface endothelium and/or underlying basement membrane is shown in Fig. 6. Shear stress or damage by cross-reactive antibody and composite fibrin tags may expose the laminar basement membrane and allow binding of cross-reactive antibody between endothelial cells,

Table 3. Host and bacterial antigens with immunological similarity to streptococcal M protein[a]

Host antigens
Cardiac myosin
Skeletal myosin
Tropomyosin
Keratin
Vimentin
Laminin
Retinal S antigen
DNA
Microbial antigens
Coxsackieviral capsid proteins
Mycobacterial heat shock protein Hsp-65
Streptococcal group A carbohydrate *N*-acetylglucosamine

[a] Adapted from reference 12.

which may also be attacked directly on their surfaces by antibody and complement, as shown in Fig. 6. The mechanism could account for the mitral valve being affected more often in patients with rheumatic fever since it receives the largest amount of shear stress from blood flow through the heart. Regions of the valve under the most stress may lose endothelial cell integrity and expose the laminar basement membrane beneath the endothelium to cross-reactive antibodies.

Mimicry between myosin, M protein, and *N*-acetylglucosamine may break tolerance to epitopes of human cardiac myosin in patients with rheumatic heart disease. Once the streptococcal epitope breaks tolerance to a unique epitope of human cardiac myosin, epitope spreading throughout human cardiac myosin may occur. Studies of antimyosin antibodies in patients with rheumatic heart disease indicate that the antibodies do react with a particular group of human cardiac myosin peptides that span the LMM region of the molecule (22). Healthy individuals did not have antibodies against the human cardiac myosin peptides. In this study, healthy individuals demonstrated low titers of antibody against cardiac myosin in the enzyme-linked immunosorbent assay, but the sera did not react with human cardiac myosin in the Western immunoblot, whereas sera from patients with disease reacted with cardiac myosin in the blot and enzyme-linked immunosorbent assay. In cases of streptococcal pharyngitis, antibodies develop against human cardiac myosin most likely due to cross-reactivity with the *N*-acetylglucosamine determinant of the group A carbohydrate and with M protein. Individuals who do not get rheumatic fever likely do not have the risk factors required to induce disease. Risk factors in addition to a group A streptococcal infection may include the presence of susceptibility genes that promote autoimmunity, such as genes that control cytokines or haplotypes of HLA antigens, recognition of unique sequences in human cardiac myosin, and abrogation of tolerance against a pathogenic self-epitope due to mimicry with M protein or *N*-acetylglucosamine, or both. The LMM region of human cardiac myosin has been shown to contain the largest amount of

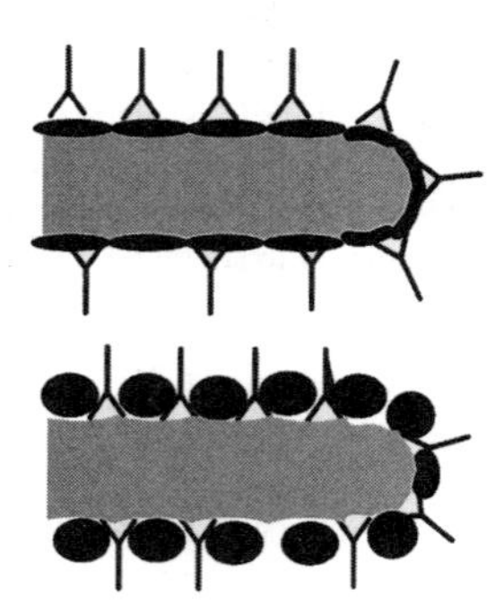

Figure 6. Diagram illustrating the potential mechanism of antibody reactivity in the pathogenesis of rheumatic heart disease. Cross-reactive antibody could bind directly to the endothelium (top diagram) or could bind to the basement membrane of the valve (bottom diagram) if the basement membrane is exposed due to shear stress or damage by antibody and complement. Endothelium is depicted with antibody attached directly in the upper diagram, and then antibody is shown attached to exposed endothelium in the lower diagram.

homology between M protein and myosin (18), although homology exists throughout the molecule, particularly the alpha-helical coiled-coil rod.

The amino acid sequences of M5 protein responsible for inflammatory heart disease have been determined in mice (18, 37), but the pathogenic cross-reactive epitope of human cardiac myosin which may trigger the development of inflammatory heart disease following streptococcal infection is under investigation. It is well-known that cardiac myosins produce autoimmune myocarditis in susceptible strains of mice and rats (43, 54, 63). Pathogenic cardiac myosin epitopes found to cause myocarditis have been studied in mice (27, 46) and were found in the S1 head fragment of the mouse cardiac myosin α-chain and in the region of rat cardiac myosin containing parts of the S1/S2 fragment. Determination of the sequences of human cardiac myosin peptides which are recognized by T cells in patients with rheumatic fever will be important in the future.

T cells have been cloned from the valves of patients with rheumatic heart disease caused by the M5 serotype of group A streptococci (34). The amino acid sequences of the peptides of the M5 protein recognized by the T-cell clones were similar to the amino acid sequences identified as the dominant myosin cross-reactive M5 epitopes in BALB/c mice (18) and recognized by T-cell lines from an M5 protein-induced valvulitis model (M. W. Cunningham and J. E. Galvin, unpublished data). Table 4 compares the peptides recognized by T cells from heart valves from patients with rheumatic disease and the M5 peptide sequences recognized by T cells from mice sensitized to human cardiac myosin (18).

Mimicry between M protein and human cardiac myosin may break tolerance to unique epitopes in cardiac myosin, and specific T cells may become activated and targeted to the heart. Studies in my laboratory have shown that a rat T-cell line with specificity for M protein and cardiac myosin can infiltrate heart tissues in the Lewis rat (A. Quinn, S. Kosanke, S. M. Factor, V. A. Fischetti, and M. W. Cunningham, submitted for publication). The T-cell line expressed a T-helper-1 profile, with secretion of gamma interferon and interleukin 2.

The manifestations observed in patients with acute rheumatic fever are diverse. The arthritis and chorea may result from cross-reactive antibody and complement deposition, with an acute inflammatory response in the joints and brain. Neither manifestation results in permanent damage to tissues, whereas valvular heart disease is a result of scarring of valve tissues which may be a localized chronic granulomatous cell-mediated immune response driven by T-helper-1 cells. Comparative studies of the different manifestations of rheumatic fever and their immune responses may lead to a better understanding of their pathogenesis.

SUMMARY AND CONCLUSIONS

Acute rheumatic fever and rheumatic heart disease are autoimmune sequelae which can follow group A streptococcal infection. Rheumatic fever is an important example of an autoimmune sequela for which the etiological agent is known. No doubt there are several risk factors for acute rheumatic fever, including the streptococcal strain, host susceptibility genes, and immune responses to infection. Molecular mimicry was proposed as a potential mechanism for streptococcal sequela many years ago, before cardiac myosin and streptococcal M protein were discovered as potential key elements in the pathogenesis of the dis-

Table 4. Summary of myosin or heart cross-reactive T-cell epitopes of streptococcal M5 protein[a]

Peptide residues or peptide (residues)	Sequence[b]	Origin of T-cell clone or response (reference)[c]
1–25	TVTRGTISDPQRAKEALDKYELENH[d]	ARF, valve (34)
81–96	DKLKQQRDTLSTQKETLEREVQN[d]	ARF, valve (34)
163–177	ETIGTLKKILDETVK[d]	ARF, valve (34)
337–356	LRRDLDASREAKKQVEKAL	Normal humans, PBL (56)
347–366	AKKQVEKALEEANSKLAALE	Mice (57), normal PBL (56)
397–416	LKEQLAKQAEELAKLRAGKA	ARF, PBL (56)
NT4 (40–58)	GLKTENEGLKTENEGLKTE	BALB/c, lymph node[e]
NT5 (59–76)	KKEHEAENDKLKQQRDTL	
B1B2 (137–154)	VKDKIAKEQENKETIGTL	
B2 (150–167)	TIGTLKKILDETVKDKIA	
C2A (254–271)	EASRKGLRRDLDASREAK	
C3 (293–308)	KGLRRDLDASREAKKQ	

[a] Taken from reference 25 with permission from the American Society for Microbiology.
[b] The amino-terminal TVTRGTIS sequence was taken from the M5 amino acid sequence published by Manjula et al. (49) and deviates from the M5 sequence published by Miller et al. (53) at positions 1 and 8. All other sequences are from the M5 gene sequence reported by Miller et al. (53). Underlined sequences are those shared among the epitopes or sequences compared in the table.
[c] PBL, peripheral blood lymphocytes; ARF, acute rheumatic fever.
[d] Other marked sequences (residues 81 to 96) and (163 to 177) were taken from the PepM5 sequence as reported by Manjula et al. (49). These two sequences are found in the sequence as residues 67 to 89 and 174 to 188, respectively, as reported by Miller et al. (53).
[e] BALB/c mice immunized with purified human cardiac myosin and the lymph node lymphocytes recovered from the mice were stimulated with each of the peptides in tritiated thymidine uptake assays.

ease. Studies have reported the cross-reactive antibody specificities which have led to the identification of host molecules present in the heart which might be related to the disease pathogenesis. Cytotoxic cross-reactive antibodies reacted with extracellular matrix protein laminin which is located in the valve basement membrane. Cross-reactive antibodies may play a role in valve injury by binding to the valve surface endothelium as well as to the basement membrane. Valvular injury would lead to the events of VCAM-1 expression on the endothelium and subsequent translocation of activated T cells from the blood into the valve. Studies of cross-reactive MAbs and T cells have led to a better understanding of how mimicry between the M protein and cardiac myosin could play a role in inflammatory heart disease in acute rheumatic fever.

Acknowledgments. Gratitude is expressed to Jeff Galvin for Fig. 6, Janet Heuser for technical assistance, and the Molecular Biology Resource Center and Kenneth Jackson for production of the synthetic peptides. My work is supported by grants HL35280 and HL56267 from the National Heart, Lung, and Blood Institute.

REFERENCES

1. **Adderson, E. E., A. R. Shikhman, K. E. Ward, and M. W. Cunningham.** 1998. Molecular analysis of polyreactive monoclonal antibodies from rheumatic carditis: human anti-*N*-acetyl-glucosamine/anti-myosin antibody V region genes. *J. Immunol.* **161:**2020–2031.
2. **Agarwal, B. L.** 1981. Rheumatic heart disease unabated in developing countries. *Lancet* **ii:**910–911.
3. **Antone, S. M., E. E. Adderson, N. M. J. Mertens, and M. W. Cunningham.** 1997. Molecular analysis of V gene sequences encoding cytotoxic antistreptococcal/anti-myosin monoclonal antibody 36.2.2 that recognizes the heart cell surface protein laminin. *J. Immunol.* **159:**5422–5430.

4. **Antone, S. M., and M. W. Cunningham.** 1992. Cytotoxicity linked to a streptococcal monoclonal antibody which recognizes laminin. *Int. J. Med. Microbiol.* **277**(Suppl. 22)**:** 189–191.
5. **Ayoub, E. M.** 1992. Resurgence of rheumatic fever in the United States. *Postgrad. Med.* **92:**133–142.
6. **Baird, R. W., M. S. Bronze, W. Kraus, H. R. Hill, L. G. Veasey, and J. B. Dale.** 1991. Epitopes of group A streptococcal M protein shared with antigens of articular cartilage and synovium. *J. Immunol.* **146:**3132–3137.
7. **Beachey, E. H., J. M. Seyer, and A. H. Kang.** 1978. Repeating covalent structure of streptococcal M protein. *Proc. Natl. Acad. Sci. USA* **75:**3163–3167.
8. **Bisno, A. L.** 1995. Non-suppurative poststreptococcal sequelae: rheumatic fever and glomerulonephritis, p. 1799–1810. *In* G. L. Mandell, J. E. Bennett, and R. Dolin (ed.), *Principles and Practice of Infectious Diseases,* vol. 2. Churchill Livingstone, New York, N.Y.
9. **Bisno, A. L.** 1995. Streptococcus pyogenes, p. 1786–1799. *In* G. L. Mandell, R. G. Bennett, and R. Dolin (ed.), *Principles and Practice of Infectious Diseases,* vol. 2. Churchill Livingstone, New York, N.Y.
10. **Cavelti, P. A.** 1945. Autoantibodies in rheumatic fever. *Proc. Soc. Exp. Biol. Med.* **60:**379–381.
11. **Coburn, A. F.** 1931. *The Factor of Infection in the Rheumatic State.* The Williams & Wilkins Co., Baltimore, Md.
12. **Cunningham, M. W.** 1996. Streptococci and rheumatic fever, p. 13–66. *In* N. R. Rose and H. Friedman (ed.), *Microorganisms and Autoimmune Disease.* Plenum Publishing Corp., New York, N.Y.
13. **Cunningham, M. W.** 2000. Cross-reactive antigens of group A streptococci, p. 66–77. *In* V. A. Fischetti, R. P. Novick, J. J. Ferretti, D. A. Portnoy, and J. I. Rood (ed.), *Gram-Positive Pathogens.* ASM Press, Washington, D.C.
14. **Cunningham, M. W.** 2000. Group A streptococcal sequelae and molecular mimicry, p. 123–157. *In* M. W. Cunningham and R. S. Fujinami (ed.), *Effects of Microbes on the Immune System.* Lippincott, Williams & Wilkins, Philadelphia, Pa.
15. **Cunningham, M. W.** 2000. Pathogenesis of acute rheumatic fever, p. 102–132. *In* D. S. Stevens and E. H. Kaplan (ed.), *Streptococcal Infections.* Oxford University Press, New York, N.Y.
16. **Cunningham, M. W., S. M. Antone, J. M. Gulizia, B. A. McManus, and C. J. Gauntt.** 1993. α-Helical coiled-coil molecules: a role in autoimmunity against the heart. *Clin. Immunol. Immunopathol.* **68:** 129–134.
17. **Cunningham, M. W., S. M. Antone, J. M. Gulizia, B. M. McManus, V. A. Fischetti, and C. J. Gauntt.** 1992. Cytotoxic and viral neutralizing antibodies crossreact with streptococcal M protein, enteroviruses, and human cardiac myosin. *Proc. Natl. Acad. Sci. USA* **89:**1320–1324.
18. **Cunningham, M. W., S. M. Antone, M. Smart, R. Liu, and S. Kosanke.** 1997. Molecular analysis of human cardiac myosin-cross-reactive B- and T-cell epitopes of the group A streptococcal M5 protein. *Infect. Immun.* **65:**3913–3923.
19. **Cunningham, M. W., N. K. Hall, K. K. Krisher, and A. M. Spanier.** 1986. A study of anti-group A streptococcal monoclonal antibodies cross-reactive with myosin. *J. Immunol.* **136:**293–298.
20. **Cunningham, M. W., J. M. McCormack, P. G. Fenderson, M. K. Ho, E. H. Beachey, and J. B. Dale.** 1989. Human and murine antibodies cross-reactive with streptococcal M protein and myosin recognize the sequence GLN-LYS-SER-LYS-GLN in M protein. *J. Immunol.* **143:**2677–2683.
21. **Cunningham, M. W., J. M. McCormack, L. R. Talaber, J. B. Harley, E. M. Ayoub, R. S. Muneer, L. T. Chun, and D. V. Reddy.** 1988. Human monoclonal antibodies reactive with antigens of the group A streptococcus and human heart. *J. Immunol.* **141:**2760–2766.
22. **Cunningham, M. W., H. C. Meissner, J. Heuser, and D. Y. M. Leung.** 1999. Anti-human cardiac myosin autoantibodies in Kawasaki syndrome. *J. Immunol.* **163:**1060–1065.
24. **Cunningham, M. W., and R. A. Swerlick.** 1986. Polyspecificity of antistreptococcal murine monoclonal antibodies and their implications in autoimmunity. *J. Exp. Med.* **164:**998–1012.
25. **Dale, J. B., and E. H. Beachey.** 1985. Epitopes of streptococcal M proteins shared with cardiac myosin. *J. Exp. Med.* **162:**583–591.
26. **Dale, J. B., and E. H. Beachey.** 1986. Sequence of myosin-crossreactive epitopes of streptococcal M protein. *J. Exp. Med.* **164:**1785–1790.
27. **Donermeyer, D. L., K. W. Beisel, P. M. Allen, and S. C. Smith.** 1995. Myocarditis-inducing epitope of myosin binds constitutively and stably to I-A^K on antigen presenting cells in the heart. *J. Exp. Med.* **182:**1291–1300.
28. **Facklam, R. R.** 1997. Screening for streptococcal pharyngitis: current technology. *Infect. Med.* **14:**891–898.

29. **Fenderson, P. G., V. A. Fischetti, and M. W. Cunningham.** 1989. Tropomyosin shares immunologic epitopes with group A streptococcal M proteins. *J. Immunol.* **142:**2475–2481.
30. **Fischetti, V. A.** 1991. Streptococcal M protein. *Sci. Am.* **264**(6):48–65.
31. **Fischetti, V. A.** 1989. Streptococcal M protein: molecular design and biological behavior. *Clin. Microbiol. Rev.* **2:**285–314.
32. **Fleming, A.** 1929. On the bacterial action of cultures of a penicillium, with special reference to their use in the isolation of *B. influenza*. *Br. J. Exp. Pathol.* **10:**226–236.
33. **Galvin, J. E., M. E. Hemric, K. Ward, and M. W. Cunningham.** Cytotoxic monoclonal antibody from rheumatic carditis reacts with human endothelium: implications in rheumatic heart disease. Submitted.
34. **Guilherme, L., E. Cunha-Neto, V. Coelho, R. Snitcowsky, P. M. A. Pomerantzeff, R. V. Assis, F. Pedra, J. Neumann, A. Goldberg, M. E. Patarroyo, F. Pileggi, and J. Kalil.** 1995. Human heart-filtrating T cell clones from rheumatic heart disease patients recognize both streptococcal and cardiac proteins. *Circulation* **92:**415–420.
35. **Gulizia, J. M., M. W. Cunningham, and B. M. McManus.** 1991. Immunoreactivity of anti-streptococcal monoclonal antibodies to human heart valves. Evidence for multiple cross-reactive epitopes. *Am. J. Pathol.* **138:**285–301.
36. **Hollingshead, S. K., V. A. Fischetti, and J. R. Scott.** 1986. Complete nucleotide sequence of type 6 M protein of the group A streptococcus: repetitive structure and membrane anchor. *J. Biol. Chem.* **261:**1677–1686.
37. **Huber, S. A., and M. W. Cunningham.** 1996. Streptococcal M protein peptide with similarity to myosin induces $CD4^+$ T cell-dependent myocarditis in MRL/++ mice and induces partial tolerance against coxsakieviral myocarditis. *J. Immunol.* **156:**3528–3534.
38. **Husby, G., I. van de Rijn, J. B. Zabriskie, Z. H. Abdin, and R. C. Williams.** 1976. Antibodies reacting with cytoplasm of subthalamic and caudate nuclei neurons in chorea and acute rheumatic fever. *J. Exp. Med.* **144:**1094–1110.
39. **Kaplan, E. L.** 1991. The resurgence of group A streptococcal infections and their sequelae. *Eur. J. Clin. Microbiol. Infect. Dis.* **10:**55–57. (Editorial.)
40. **Kaplan, M. H., R. Bolande, L. Rakita, and J. Blair.** 1964. Presence of bound immunoglobulins and complement in the myocardium in acute rheumatic fever. Association with cardiac failure. *N. Engl. J. Med.* **271:**637–645.
41. **Kaplan, M. H., and M. Meyeserian.** 1962. An immunological cross reaction between group A streptococcal cells and human heart tissue. *Lancet* **i:**706–710.
42. **Kaplan, M. H., and K. H. Svec.** 1964. Immunologic relation of streptococcal and tissue antigens. III. Presence in human sera of streptococcal antibody cross reactive with heart tissue. Association with streptococcal infection, rheumatic fever, and glomerulonephritis. *J. Exp. Med.* **119:**651–666.
43. **Kodama, M., Y. Matsumoto, M. Fujiwara, M. Massani, T. Izumi, and A. Shibota.** 1991. A novel experimental model of giant cell myocarditis induced in rats by immunization with cardiac myosin fraction. *Clin. Immunol. Immunopathol.* **57:**250–262.
44. **Krisher, K., and M. W. Cunningham.** 1985. Myosin: a link between streptococci and heart. *Science* **227:**413–415.
45. **Lancefield, R. C.** 1962. Current knowledge of type-specific M antigens of group A streptococci. *J. Immunol.* **89:**307–313.
46. **Liao, L., R. Sindhwani, L. Leinwand, B. Diamond, and S. Factor.** 1993. Cardiac α-myosin heavy chains differ in their induction of myocarditis: identification of pathogenic epitopes. *J. Clin. Invest.* **92:**2877–2882.
47. **Liao, L., R. Sindhwani, M. Rojkind, S. Factor, L. Leinwand, and B. Diamond.** 1995. Antibody-mediated autoimmune myocarditis depends on genetically determined target organ sensitivity. *J. Exp. Med.* **187:**1123–1131.
48. **Lyampert, I. M., L. V. Beletskaya, N. A. Borodiyuk, E. V. Gnezditskaya, B. L. Rassokhina, and T. A. Danilova.** 1976. A cross-reactive antigen of thymus and skin epithelial cells common with the polysaccharide of group A streptococci. *Immunology* **31:**47–55.
49. **Manjula, B. N., A. S. Acharya, S. M. Mische, T. Fairwell, and V. A. Fischetti.** 1984. The complete amino acid sequence of a biologically active 197-residue fragment of M protein isolated from type 5 group A streptococci. *J. Biol. Chem.* **259:**3686–3693.

50. **Manjula, B. N., and V. A. Fischetti.** 1986. Sequence homology of group A streptococcal Pep M5 protein with other coiled-coil proteins. *Biochem. Biophys. Res. Commun.* **140:**684–690.
51. **Manjula, B. N., and V. A. Fischetti.** 1980. Tropomyosin-like seven residue periodicity in three immunologically distinct streptococcal M proteins and its implications for the antiphagocytic property of the molecule. *J. Exp. Med.* **151:**695–708.
52. **Manjula, B. N., B. L. Trus, and V. A. Fischetti.** 1985. Presence of two distinct regions in the coiled-coil structure of the streptococcal Pep M5 protein: relationship to mammalian coiled-coil proteins and implications to its biological properties. *Proc. Natl. Acad. Sci. USA* **82:**1064–1068.
53. **Miller, L. C., E. D. Gray, E. H. Beachey, and M. A. Kehoe.** 1988. Antigenic variation among group A streptococcal M proteins: nucleotide sequence of the serotype 5 M protein gene and its relationship with genes encoding types 6 and 24 M proteins. *J. Biol. Chem.* **263:**5668–5673.
54. **Neu, N., N. R. Rose, K. W. Beisel, A. Herskowitz, G. Gurri-Glass, and S. W. Craig.** 1987. Cardiac myosin induces myocarditis in genetically predisposed mice. *J. Immunol.* **139:**3630–3636.
55. **Poynton, E. J., and A. Paine.** 1900. The aetiology of rheumatic fever. *Lancet* **ii:**861–869.
56. **Pruksakorn, S., B. Currie, E. Brandt, C. Phornphutkul, S. Hunsakunachai, A. Manmontri, J. H. Robinson, M. A. Kehoe, A. Galbraith, and M. F. Good.** 1994. Identification of T cell autoepitopes that cross-react with the C-terminal segment of the M protein of group A streptococci. *Int. Immunol.* **6:**1235–1244.
57. **Pruksakorn, S., A. Galbraith, R. A. Houghten, and M. F. Good.** 1992. Conserved T and B cell epitopes on the M protein of group A streptococci. Induction of bactericidal antibodies. *J. Immunol.* **149:**2729–2735.
58. **Quinn, A., T. M. Shinnick, and M. W. Cunningham.** 1996. Anti-Hsp65 antibodies recognize M proteins of group A streptococci. *Infect. Immun.* **64:**818–824.
59. **Quinn, A., K. Ward, V. Fischetti, M. Hemric, and M. W. Cunningham.** 1998. Immunological relationship between the class I epitope of streptococcal M protein and myosin. *Infect. Immun.* **66:**4418–4424.
60. **Shikhman, A. R., and M. W. Cunningham.** 1994. Immunological mimicry between *N*-acetyl-beta-D-glucosamine and cytokeratin peptides. Evidence for a microbially driven anti-keratin antibody response. *J. Immunol.* **152:**4375–4387.
61. **Shikhman, A. R., N. S. Greenspan, and M. W. Cunningham.** 1994. Cytokeratin peptide SFGSGFGGGY mimics *N*-acetyl-beta-D-glucosamine in reaction with antibodies and lectins, and induces in vivo anti-carbohydrate antibody response. *J. Immunol.* **153:**5593–5606.
62. **Shikhman, A. R., N. S. Greenspan, and M. W. Cunningham.** 1993. A subset of mouse monoclonal antibodies cross-reactive with cytoskeletal proteins and group A streptococcal M proteins recognizes *N*-acetyl-beta-D-glucosamine. *J. Immunol.* **151:**3902–3913.
63. **Smith, S. C., and P. M. Allen.** 1991. Myosin-induced acute myocarditis is a T cell mediated disease. *J. Immunol.* **147:**2141–2147.
64. **Stollerman, G. H. (ed.).** 1975. *Rheumatic Fever and Streptococcal Infection.* Grune & Stratton, New York, N.Y.
65. **Stollerman, G. H.** 1993. The global impact of penicillin. *Mt. Sinai J. Med.* **60:**112–119.
66. **Stollerman, G. H.** 1997. Changing streptococci and prospects for the global eradication of rheumatic fever. *Perspect. Biol. Med.* **40:**165–189.
67. **Stollerman, G. H.** 1997. Rheumatic fever. *Lancet* **349:**935–942.
68. **Stollerman, G. H.** 1998. The changing face of rheumatic fever in the 20th century. *J. Med. Microbiol.* **47:**1–3.
69. **van de Rijn, I., J. B. Zabriske, and M. McCarty.** 1977. Group A streptococcal antigens cross-reactive with myocardium: purification of heart reactive antibody and isolation and characterization of the streptococcal antigen. *J. Exp. Med.* **146:**579–599.
70. **Vashishtha, A., and V. A. Fischetti.** 1993. Surface-exposed conserved region of the streptococcal M protein induces antibodies cross-reactive with denatured forms of myosin. *J. Immunol.* **150:**4693–4701.
71. **Veasy, L. G., L. Y. Tani, and H. R. Hill.** 1994. Persistence of acute rheumatic fever in the intermountain area of the United States. *J Pediatr.* **124:**9–16.
72. **Veasy, L. G., S. E. Wiedmeier, and G. S. Orsmond.** 1987. Resurgence of acute rheumatic fever in the intermountain area of the United States. *N. Engl. J. Med.* **316:**421–427.
73. **Zabriskie, J. B.** 1967. Mimetic relationships between group A streptococci and mammalian tissues. *Adv. Immunol.* **7:**147–188.
74. **Zabriskie, J. B., and E. H. Freimer.** 1966. An immunological relationship between the group A streptococcus and mammalian muscle. *J. Exp. Med.* **124:**661–678.

75. **Zabriskie, J. B., K. C. Hsu, and B. C. Seegal.** 1970. Heart-reactive antibody associated with rheumatic fever: characterization and diagnostic significance. *Clin. Exp. Immunol.* **7:**147–159.
76. **Zugel, U., and S. H. E. Kaufmann.** 1999. Role of heat shock proteins in protection from and pathogenesis of infectious diseases. *Clin. Microbiol. Rev.* **12:**19–39.

Molecular Mimicry, Microbes, and Autoimmunity
Edited by M. W. Cunningham and R. S. Fujinami

Chapter 5

Antigenic Mimicry between Self and Coxsackievirus Proteins Leads to Both Humoral and Cellular Autoimmunity to Heart Proteins

Sally A. Huber and Charles J. Gauntt

Picornaviruses are small (diameter, ~30 nm), nonenveloped icosahedral viruses that contain a single-stranded RNA genome of 7,200 to 8,450 nucleotides with a small (7,000 to 8,000 Da) protein (VPg) covalently bound to the 5′ terminus. The surrounding capsid contains three surface polypeptides and one subsurface polypeptide (72). The enterovirus capsid is an extremely stable structure that is impervious to penetration by ions and stable at pH 2 and that can remain viable in an aqueous environment for months, perhaps years. Translation of the genome into at least 11 polypeptides results in production of capsid polypeptides, two proteases, the small VPg protein, and several proteins required for synthesis of complementary and genomic RNA in concert with the final translation product, a core RNA-dependent RNA polymerase (72). Replication of these viruses occurs exclusively in the cytoplasm. Enteroviruses have evolved mechanisms for terminal shutoff of host protein synthesis. These mechanisms involve viral protease 2A degradation of the cellular p220 protein cap-dependent binding complex (72) or degradation of poly(A)-binding protein (36). These mechanisms are not operative in all infected cells that replicate these viruses to the usual titers of 100 to 2,500 virions (25,000 to 100,000 total particles) per cell. Indeed, many types of cells, including cardiac fibroblasts, that are permissive for replication of coxsackievirus group B (CVB) strains exhibit little to no viral cytopathic effects during production of virus for days to weeks (19).

EVIDENCE ASSOCIATING CVB WITH MYOCARDITIS AND DCM

Of the estimated 3 million people who experience a nonpoliovirus enterovirus illness in the United States each year, approximately 150,000 will have an infection of the heart and an unknown proportion will develop myocarditis (67). About half of the clinical cases have been associated with CVB (24, 35, 46, 55, 70). Among patients with known CVB infections, 5 to 12% may have myocardial involvement on the basis of alterations detected by electrocardiographic studies (99). The signs and symptoms of myocarditis may be

Sally A. Huber • Department of Pathology, University of Vermont, 55A South Park Drive, Colchester, VT 05446. ***Charles J. Gauntt*** • Department of Microbiology, The University of Texas Health Science Center at San Antonio, 7703 Floyd Curl Drive, San Antonio, TX 78284-7758.

absent or may be misinterpreted as a severe respiratory disease, but this disease is underestimated. CVBs, at least the serotypes prevalent in the United States (serotypes 1 to 5), may be the etiological agents of at least 30,000 of the approximately 100,000 new cases of dilated cardiomyopathy (DCM) diagnosed in the United States each year (35). Survival rates (10-year rates) for patients with DCM have been less than 40% in some studies (99).

Serological data obtained for patients with a clinical definition of myocarditis provided the initial evidence that CVB is a dominant etiologic factor in this disease (26, 41, 75, 85). In situ hybridization analyses of biopsy samples taken from patients with biopsy-proven myocarditis or DCM found that 18 to 50% of patients, with a consensus estimate of 30%, have evidence of the presence of enterovirus genomic RNA sequences (8, 35, 89). Evidence of an enteroviral infection likely has prognostic significance. One study found that 26% of DCM patients whose biopsy specimens were positive for enterovirus RNA died during a 2-year follow-up, whereas only 3% of patients who were virus negative died (97).

MECHANISMS OF CARDIAC INJURY: EVIDENCE FOR ANTIGENIC MIMICRY

Humoral Immunity

Three major mechanisms probably cause cardiac dysfunction in patients with viral myocarditis and DCM. These are (i) injury directly due to virus infection and replication in the heart, (ii) damage mediated by specific and nonspecific host defense responses to the infection, and (iii) autoimmunity to cardiac antigens triggered by virus infection. A conundrum exists between poor cardiac function in patients with severe chronic myocarditis or DCM and the often minimal pathologic alterations found in their heart tissues upon autopsy (46, 59, 99). Frequently, cardiac dysfunction is global, but only a small number of focal and sporadic inflammatory lesions are observed in the myocardium. Such observations favor submicroscopic factors in pathogenesis. Virus could directly induce death of myocytes, or persistent infections might lead to disrupted contractility without cell loss (5, 28, 36, 38–40, 74, 89, 96).

Widespread binding of immunoglobulin G (IgG) autoantibodies to heart tissues (2, 7, 54, 56, 76) and the release of soluble mediators (cytokines and chemokines) such as interferon, interleukin 1α (IL-1α), and nitric oxide might produce extensive cardiac injury beyond inflammatory foci (6, 18, 43). Several cytokines can depress myocardial function in animal hearts, including IL-1, tumor necrosis factor alpha (TNF-α), and IL-6 (20). CVB3 infection of mice induces production of IL-1 and TNF-α from infiltrating inflammatory cells in the heart during acute disease (35, 44, 45, 71) and significantly upregulates expression of intracellular adhesion molecule-1 (ICAM-1) on murine heart cells (81). The chemokine macrophage inflammatory protein-1 alpha (MIP-1α) is also likely involved in CVB3-induced acute myocarditis, as MIP-1α knockout mice are resistant to CVB3-induced myocarditis (13).

Various lines of evidence implicate immunopathogenic mechanisms of myocardial injury in myocarditis and DCM. Increased susceptibility to development of autoimmune myocardial disease is associated with HLA-DR4/1 and histidine at position 36 of the HLA-DQ beta 1 gene (49). Data from studies with patients with myocarditis or DCM suggest that heart tissue-reactive autoantibodies participate in the pathogenesis of acute and chronic myocarditis (2, 7, 50, 54, 56, 57, 63, 64, 76, 80). A high proportion of patients (80%) with biopsy-proven myocarditis had heart tissue-bound antibody (2, 7, 9–11, 54, 56, 58, 76, 80), whereas the pathognomonic descriptor of myocarditis, a mononuclear cell infiltrate (3), was focal and sporadic (2). Heart impairment severity in patients with chronic myocarditis has been correlated with levels of eluted autoantibody that had previously been bound to heart tissues (64). The majority of DCM patients (75%) had heart tissue-bound IgG and a poor cardiac performance, i.e., an ejection fraction of <35% (7). In another study, a significant proportion of DCM patients had high titers of anti-β_1 adrenergic receptor autoantibodies in their sera, and these patients also exhibited a decreased ejection fraction (94); significantly, patients in remission showed declines in serum antibody titers against the β_1 adrenoreceptor adenine nucleotide translocator (ANT) and branched-chain keto acid dehydrogenase (BCKD) to normal levels (50). About 60% of patients with myocarditis or DCM contain significant titers of anti-ANT serum antibodies (24, 76, 80). Of interest, the ANT molecule shares epitopes with one or more surface Ca^{2+} channel proteins on myocytes (76) and with CVB3 capsid proteins (80). Patients with active or borderline myocarditis have serum antibodies that immunoblot to CVB3 capsid antigens; elution and reblotting of these antibodies to several extracted cardiac antigens detected antigens of the sizes expected for BCKD and ANT (63). Approximately 75% of patients with myocarditis or DCM had serum autoantibodies to laminin, whereas only 5% of control patients had such antibodies (98). Autoantibodies in most myocarditis patients' sera were found to recognize antimyolemmal antigens on isolated adult human myocytes (98). Antibodies in the sera of some patients with myocarditis or DCM can affect myocyte functions (inhibit nucleoside transport, impair contractility, and increase the levels of CA^{2+} uptake), or they can be directly cytotoxic (56). Cross-adsorption of several of the latter sera with purified CVB3 or CVB4 particles removed antibodies capable of lysing normal rat cardiac myocytes (76), suggesting that the viruses and heart cells have shared epitopes.

Murine strains that develop chronic myocarditis also produce high titers of IgG autoantibodies that bind to heart tissues. In CVB3-murine models of myocarditis, the question is "how can one unequivocally prove that the autoantibodies have a role in disease?" In these models, autoantibodies directed against cardiac myosin, ANT, and BCKD have been found to be bound to heart tissues; the specificity of these autoantibodies was demonstrated following elution from cells in heart tissues by immunobloting (65, 66, 70). In one CVB3-murine model of acute myocarditis, the severity of the cardiopathology correlated with the titers of eluted anti-cardiac myosin, anti-ANT, or anti-BCKD antibodies that had been deposited in heart tissues (66). Murine strains that develop CVB3-induced chronic myocarditis transiently produced IgM autoantibodies to cardiac antigens, but only the IgG autoantibodies against cardiac antigens are considered pathologic, since they (and not IgM) bind to heart tissues (71). In contrast, CVB3-challenged strains that resolve the acute myocarditis transiently produce only IgM autoantibodies to cardiac tissues that minimally, if at all, bind to heart tissues (71). The sera of most mice that develop CVB3-induced chronic myocarditis have detectable levels of autoantibodies to cardiac myosin prior to virus challenge, suggesting a predisposition to the development of autoimmune diseases in which CVB3 can serve as the eliciting antigen (22, 61). Healthy mice have no evidence of any cardiopathology, and the presence of the preexisting autoantibodies is not due to subclinical CVB infections, as these viruses have not been isolated from breeding facilities or wild mice (20). Sustained autoantibody production may be one potential indicator of a chronic disease outcome (23).

A role for molecular mimicry via autoantibodies in murine models of CVB3-induced myocarditis was established in the following studies. Several neutralizing monoclonal antibodies (MAbs) against $CVB3_m$, a highly cardiovirulent Nancy prototype strain, were shown to (i) participate in complement-mediated lysis of normal cardiac fibroblasts, (ii) bind to an antigen(s) on the surface of these fibroblasts and induce synthesis of a soluble chemoattractant for macrophages, (iii) bind to epitopes on murine and human cardiac myosins, (iv) induce minimal but reproducible cardiopathologic alterations in normal mice, and (v) significantly exacerbate $CVB3_m$-induced myocarditis when inoculated 3 days after virus challenge (21, 22, 25). Six of the eight cardiopathologic anti-CVB3 MAbs recognize an epitope on mouse or human cardiac myosins (21). MAbs 5 and 1 have been mapped to two different 18-amino-acid synthetic peptides, LMM4 and LMM10, respectively, among 49 peptides that completely span the light meromyosin (LMM) region of human cardiac myosin (HCM). Several MAbs to different areas of the HCM molecule, including MAbs to light chains, will bind to shared epitopes on CVB3 particles (23). MAbs 1 and 13 readily bind to normal murine cardiac fibroblasts, as detected by immunostaining (21). MAb1 binds to pseudopodial projections and broad patches on ≤50% of cardiac fibroblasts cultured from CD-1 mice, whereas binding was more widespread on ~90% of cardiac fibroblasts cultured from C3H/HeJ mice. Since CD-1 mice resolve $CVB3_m$-induced acute viral myocarditis but C3H/HeJ mice transit from acute to chronic myocarditis, these data suggest that the quantity of autoantibody that targets on cells in heart tissues may be one possible mechanistic explanation for genetic differences in disease outcome. Coronal sections of heart tissues from either murine strain show extensive MAb 1 immunostaining of material in cross-striated bands (21, 23) in patterns identical to that published for an anti-CVB4 neutralizing MAb that bound to an epitope in mouse cardiac myosin within myocytes (73). Additionally, MAbs 1, 5, and 18 recognize epitopes on murine laminin (21, 23).

In addition to cardiac myosins (21), CVB3 shares epitopes with ANT (52, 57, 79) and M protein of group A streptococci (14), a virulence factor associated with rheumatic heart disease. In another study, a neutralizing anti-CVB3 MAb was found to bind to an epitope on normal mouse myocytes and on HeLa and HEp-2 cells (95). Conversely, an MAb (MAb 10A1) generated against murine cardiac myocytes (95) could block adsorption of CVB3 to these cells, could partially protect BALB/c mice from CVB3-induced myocarditis and death (52), and also recognizes an epitope on ANT (52, 80), a highly immunogenic inner mitochondrial membrane protein that shares an epitope(s) with the surface calcium channel protein on cardiac myocytes (92). It was also shown that MAb 10A1 could be used to select an attenuated CVB3 escape mutant (93).

Escape mutants of CVB3 were also generated with CVB3-neutralizing antistreptococcal MAbs to M protein. H3-49 virus, the virus variant made with the antibody capable of differentiating between myocarditic and nonmyocarditic CVB3 wild-type variants (MAb 49.8.9) (14), showed dramatically reduced pathogenicity in mouse strains originally susceptible to the myocarditic wild-type CVB3 (33). Thus, the cross-reactive MAb that recognizes shared epitopes between CVB3, cardiac myosin, and the M protein of group A streptococcus clearly must recognize an important epitope, since alteration of this epitope affected viral pathogenicity.

MAb 49.8.9 recognizes a linear epitope in the VP1 capsid protein of CVB3 (14) and has recently been shown to cross-react with *N*-acetyl-β-D-glucosamine (GlcNAc) (83), a carbohydrate moiety that also shares epitopes with cytokeratin peptides (82). Earlier, it was found that MAb 49.8.9, as well as wheat germ agglutinin, a lectin that recognizes Glc-NAc, bound only to CVB3-infected fibroblasts, likely to a peptide sequence (RRKLEFF) found in VP1, the largest capsid polypeptide on $CVB3_m$ particles that emerged from a non-lytic infection of these cells (53). Thus, some autoantibodies may be recognizing carbohydrate moieties on targeted cardiac cells, but the original sensitizing antigen was protein.

Epitope sharing between CVB3 and cardiac myosins has relevance for myocarditis, because inoculation of either murine (71) or human (25) cardiac myosins into healthy mice can induce cardiopathology. Inoculation of C3H/HeJ mice with HCM 1 week prior to or 1 week after $CVB3_m$ challenge significantly increased the severity of myocarditis compared with that for the virus control (25). Among 49 synthetic peptides (18 amino acids) that span the LMM region of HCM, 12 were found to induce cardiopathology in two C3H strains and 11 induced CVB3-specific antibodies detectable by enzyme-linked immunosorbent assay (ELISA) (25). LMM21 induced both cardiopathology and the production of anti-CVB3 antibodies; 14 of the 18 amino acids were located in a 191-Å sequence that spanned all three capsid polypeptides (C. Gauntt, M. Cunningham, and H. Wood, unpublished data) on the surface of a three-dimensional model of CVB3 generated from crystallographic data (60). In addition, hyperimmune antisera to denatured HCM contained antibodies that recognized epitopes on VP1, the largest capsid polypeptide of CVB3 (25). Studies of antigenic targets for the CVB-induced autoantibodies have focused on protein antigens, and one study shows that this narrow focus is incomplete.

Inoculation of hyperimmune or monoclonal anti-CVB3 neutralizing antibodies into CVB3-challenged mice on day 3 postinoculation exacerbated viral myocarditis (20). Inoculation of healthy mice with only one of the anti-CVB3-neutralizing MAbs induced one to three lesions in the myocardiums of all mice (21, 22). These data suggest that autoantibody to shared epitopes does not induce the severe focal myocardial inflammation observed following virus challenge of mice, but one MAb can act as an accessory factor that contributes to virus replication-initiated lesions. However, inoculation of a mixture of three anti-CVB3 MAbs into healthy C3H/HeJ mice can induce >50 myocardial lesions per heart section, suggesting that targeting of multiple sites on a heart tissue cell(s) can induce severe disease (Gauntt et al., unpublished data). Different neutralizing MAbs were found to recognize epitopes on actin, laminin, elastin, vimentin, and phosphorylase *b* by ELISA (22, 23). The titers of cardium-specific antisarcolemma, antimyosin, anti-ANT, or anti-BCKD autoantibodies in sera or deposited in heart tissues of mice with CVB3-induced chronic disease were found to correlate with the severity of cardiomyopathy (66, 90). Susceptibility to CVB3-induced chronic myocarditis in mice was mapped to two regions on chromosome 14 (91).

Cellular Immunity

Although antigenic mimicry that involves autoantibodies has been recognized for many years, antigenic mimicry as a mechanism for T-cell autoimmunity is less well known. Unlike antibodies which normally recognize conformational antigens in their native state, T cells respond to linear epitopes which undergo processing within an antigen-presenting cell and are presented in the groove of a major histocompatibility complex (MHC) antigen (86, 87). The studies that have delineated the characteristics of MHC antigen presentation and T-cell recognition have used protein antigens. These studies have shown that a T-cell epitope contains elements which specifically bind to the MHC molecule (anchor amino acids), while other elements engage the T-cell receptor (TCR). Investigations with synthetic peptides based on a known T-cell epitope indicate that the TCR reacts to the tertiary structure of the MHC molecule and its bound antigenic peptide. Complete concordance of all amino acids within a peptide is not required for TCR signaling. Indeed, even a single amino acid shared between two peptides might be sufficient for activation of the same T-cell clone, provided both peptides bind to the MHC molecule and produce the same conformational structure (12, 15, 68, 101). This means that (i) antigenic mimicry between T-cell epitopes cannot be predicted through amino acid sequence homologies, (ii) multiple peptides are actually capable of stimulating the same T-cell clone, and (iii) most likely, many different pathogens might produce mimicking peptides capable of triggering the same autoimmune T-cell response (88).

More than 25 years ago, Woodruff and Woodruff (100) first demonstrated that experimental CVB3-induced myocarditis depended upon T-lymphocyte responses. However, the T-cell response could be required for either humoral (T-cell-dependent B-cell help) or cellular immunopathogenic mechanisms. The studies described above implicate humoral pathogenesis in viral myocarditis. Yet, not all mouse strains infected with CVB3 develop heart-reactive antibodies (51). While complement depletion of CVB3-infected DBA/2 mice completely prevented myocarditis, suggesting a humoral pathogenesis in this mouse strain, similar treatment of BALB/c animals was not protective (32). Nor do infected BALB/c animals show significant IgG deposition in the myocardium (30). These results strongly indicate that both humoral and cellular immunities can cause cardiac damage during myocarditis and that the genetic constitution of the host will determine whether one or both mechanisms occur.

Antigenic mimicry at the T-cell level has been demonstrated between heart antigens and various pathogens, including group A streptococci (27), *Trypanosoma cruzi* (1, 17), chlamydia (4), cytomegalovirus (16), and CVB (29, 31). Data from the CVB3-myocarditis model further demonstrate that more than one pathogen will contain shared epitopes capable of stimulating the same heart-reactive, autoimmune T-cell clone, as indicated by Theofilopoulos (88). Mice infected with CVB3 generate activated T cells which respond to peptides of the M5 protein of the group A streptococcus (33). T-cell hybridoma clones that are produced from the lymphocytes infiltrating the hearts of CVB3-infected mice and that respond to virus, cardiac myosin, and the NT4 peptide of the M5 protein have been identified (29). This shows shared mimicry between group A streptococci (bacteria which cause autoimmune heart disease) and CVB3 (a virus which causes autoimmune heart disease). Furthermore, the autoimmunity to this shared T-cell epitope represents the major pathogenic mechanism in CVB3-induced myocarditis in MRL/++ mice, since tolerization of animals to the NT4 peptide protects them from myocarditis induced by subsequent CVB3 challenge (31).

While cardiac myosin has been the dominant autoantigen implicated in myocarditis and immunization with this molecule or its derived peptides is sufficient to induce heart disease (see below), other heart-specific proteins may additionally be targeted by the cellular immune response. Lymphocytes derived from the hearts of patients with myocarditis or dilated cardiomyopathy proliferate in vitro in response both to the ANT protein and to CVB particles (78). Adoptive transfer of lymphocytes from ANT-reactive patients also transferred myocarditis into severe combined immunodeficient mice (77). Finally, murine

T cells from CVB3-infected mice respond to ANT (34), again indicating not only that different pathogens may share the same epitope but also that a single pathogen is likely to have multiple distinct mimicking epitopes with diverse heart proteins.

RELATIONSHIP OF CVB3-INDUCED MURINE MYOCARDITIS MODELS TO ANTIGEN-INDUCED NONVIRAL MURINE MODELS OF AUTOIMMUNE MYOCARDITIS

Autoimmune responses to normal cell proteins are sufficient to induce focal myocardial lesions that are histologically similar to those induced by a CVB. Inoculation of normal weanling or adolescent mice with murine (62, 71, 84) or human (23, 25) cardiac myosin, ANT protein (76), or sarcoplasmic reticulum calcium ATPase (37) will induce cardiopathologic alterations. Several murine strains that develop CVB3-induced chronic myocarditis will also develop murine cardiac myosin-induced chronic myocarditis (47, 69, 70). Susceptibility of murine strains to myosin-induced myocarditis apparently depends on the presence of myosin or a myosin-like molecule(s) in the extracellular matrix of cardiac cells (48). Also, anti-murine cardiac myosin MAbs can mediate myocarditis only in susceptible strains (48). Backcross studies between strains susceptible or resistant to cardiac myosin-induced myocarditis identified on two different chromosomes loci that are associated with myocarditis, one associated with the genetic basis of tissue injury and the other associated with the genetic basis of autoreactivity (42). These data are consistent with data from CVB3-murine models of myocarditis in showing that the genetic background of the host is a major factor in the outcome of a disease, whether virus infection and/or autoimmune responses are the etiology. As noted earlier, several but not all CVB3 strains also share epitopes with either cardiac myosin (22, 23, 25) or ANT (52, 57, 79).

SUMMARY AND CONCLUSION

A significant body of data shows that CVB3 can induce chronic myocarditis in mice of certain strains. At issue is whether the sustained cardiac inflammatory response is due to activities that directly or indirectly result from persistence of viral genomes in heart tissues or whether CVB3 merely acts as a trigger to induce autoimmune reactions. If one examines the data on detection of persisting CVB3 genomes by in situ hybridization on reverse transcription-PCR, only 20 to 35% of hearts from mice with virus-induced chronic myocarditis are positive. That suggests that 65 to 80% of mice could have virus-induced autoimmune reactions as a major factor in the chronic disease. Autoimmune responses could be directed against normal cryptic antigens released from cardiac cells during virus replication, and cardiopathologic consequences would arise when other cells in heart tissues share an epitope(s) with the cryptic antigens on their surfaces. The alternative explanation for CVB-induced autoimmune cardiopathology is molecular mimicry between viral proteins and normal host cell proteins. Data presented herein quite clearly establish that some strains of CVB3 and CVB4 share epitopes with cardiac myosin. CVB3 also clearly shares epitopes with laminin and ANT. Several neutralizing MAbs against CVB3 or CVB4 recognize epitopes in cardiac myosin, as determined by immunoblotting, and significantly, some of these MAbs can induce cardiopathologic reactions in healthy mice.

We conclude that CVB3 particles share epitopes with several normal cell proteins and that this molecular mimicry contributes to autoimmune chronic inflammatory heart disease after clearance of virus or viral genomic RNA sequences from heart tissues of most mice. The cardiopathologic autoimmune reactions are directed at the same or additional shared epitopes in antigens on the surfaces of myocardial cells to sustain the inflammation.

REFERENCES

1. **Abel, L., J. Kalil, J. Cunha, and E. Neto.** 1997. Molecular mimicry between cardiac myosin and Trypanosoma cruzi antigen B13: identification of a B13-driven human T cell clone that recognizes cardiac myosin. *Braz. J. Med. Biol. Res.* **30:**1305–1308.
2. **Anderson, J., E. Hammond, and R. Menlove.** 1987. Determining humoral and cellular-immune components in myocarditis: complementary diagnostic role of immunofluorescence microscopy in the evaluation of endomyocardial biopsy specimens, p. 233–244. *In* C. Kawai and W. Abelmann (ed.), *Pathogenesis of Myocarditis and Cardiomyopathy Update I.* University of Tokyo Press; Tokyo, Japan.
3. **Aretz, H.** 1987. Myocarditis: the Dallas criteria. *Hum. Pathol.* **18:**619–624.
4. **Bachmaier, K., N. Neu, L. de la Maza, S. Pal, A. Hessel, and J. Penninger.** 1999. Chlamydia infections and heart disease linked through antigenic mimicry. *Science* **283:**1335–1339.
5. **Badorff, C., G. Lee, B. Lamphear, M. Martone, K. Campbell, R. Rhoads, and K. Knowlton.** 1999. Enteroviral protease 2A cleaves dystrophin: evidence of cytoskeletal disruption in an acquired cardiomyopathy. *Nat. Med.* **5:**320–326.
6. **Bick, R., J. Liao, T. King, A. LeMaistre, J. McMillan, and L. Buja.** 1997. Temporal effects of cytokines on neonatal cardiac myocyte Ca^{2+} transient and adenylate cyclase activity. *Am. J. Physiol.* **272:**H1937–H1944.
7. **Bolte, H.-D., and P. Schultheiss.** 1978. Immunologic results in myocardial diseases. *Postgrad. Med. J.* **54:**500–503.
8. **Bowles, N., P. Richardson, E. Olsen, and L. Archard.** 1986. Detection of coxsackie B virus-specific RNA sequences in myocardial biopsy samples from patients with myocarditis and dilated cardiomyopathy. *Lancet* **i:**1120–1122.
9. **Caforio, A., E. Bonifacio, and J. Stewart.** 1990. Novel organ-specific circulating cardiac autoantibodies in dilated cardiomyopathy. *J. Am. Coll. Cardiol.* **15:**1527–1534.
10. **Caforio, A., J. Goldman, A. Haven, K. Baig, and W. McKenna.** 1996. Evidence for autoimmunity to myosin and other heart-specific autoantigens in patients with cardiomyopathy and their relatives. *Int. J. Cardiol.* **54:**157–163.
11. **Caforio, A., P. Keeling, and E. Zachara.** 1994. Evidence from family studies for autoimmunity in dilated cardiomyopathy. *Lancet* **344:**773–777.
12. **Chen, J., R. Lorenz, J. Goldberg, and P. Allen.** 1991. Identification and characterization of a T cell-inducing epitope of bovine ribonuclease that can be restricted by multiple class II molecules. *J. Immunol.* **147:**3672–3678.
13. **Cook, D., M. Beck, T. Cofman, S. Kirby, J. Sheridan, I. Pragnell, and O. Smithies.** 1995. Requirement of MIP-1a for an inflammatory response to viral infection. *Science* **269:**1583–1585.
14. **Cunningham, M., S. Antone, J. Gulizia, M. McManus, V. Fishchetti, and C. Gauntt.** 1992. Cytotoxic and viral neutralizing antibodies cross react with streptococcal M protein, enteroviruses, and human cardiac myosin. *Proc. Natl. Acad. Sci. USA* **89:**1320.
15. **Donermeyer, D., K. Bersel, P. Allen, and S. Smith.** 1995. Myocarditis-inducing epitope of myosin binds constitutively and stably to I-Ak on antigen-presenting cells in the heart. *J. Exp. Med.* **182:**1291–1299.
16. **Fairweather, D., C. Lawson, A. Chapman, C. Brown, T. Booth, J. Papadimitriou, and G. Shellam.** 1998. Wild isolates of murine cytomegalovirus induce myocarditis and antibodies that cross-react with virus and cardiac myosin. *Immunology* **94:**263–270.
17. **Felix, J., B. von Kreuter, and C. Santos-Bush.** 1993. Mimicry of heart cell surface epitopes in primary anti-*Trypanosoma cruzi* Lyt 2+ T lymphocytes. *Clin. Immunol. Immunopathol.* **68:**141–146.
18. **Freeman, G., J. Colston, M. Zabalgoitia, and B. Chandrasekar.** 1998. Contractile depression and expression of proinflammatory cytokines and iNOS in viral myocarditis. *Am. J. Physiol.* **274**(1 part 2)**:**H249–H258.
19. **Gauntt, C.** 1988. The possible role of viral variants in pathogenesis, p. 159–179. *In* H. Friedman and M. Bendinelli (ed.), *Coxsackievirus—a General Update,* vol. 1. Plenum Publishing Co., New York, N.Y.

20. **Gauntt, C.** 1997. Roles of the humoral response in coxsackievirus B-induced disease. *Curr. Top. Microbiol. Immunol.* **223:**259–282.
21. **Gauntt, C., H. Arizpe, A. Higdon, H. Wood, D. Bowers, M. Rozeck, and R. Crawley.** 1995. Molecular mimicry: anti-coxsackievirus B3 neutralizing antibodies and myocarditis. *J. Immunol.* **154:**2983–2995.
22. **Gauntt, C., A. Higdon, H. Arizpe, M. Tamayo, R. Crawley, R. Henkel, M. Pereira, S. Tracy, and M. Cunningham.** 1993. Epitopes shared between coxsackievirus B3 (CVB3) and normal heart tissue contribute to CVB3-induced murine myocarditis. *Clin. Immunol. Immunopathol.* **68:**129–134.
23. **Gauntt, C., A. Higdon, D. Bowers, E. Maull, J. Wood, and R. Crawley.** 1993. What lessons can be learned from animal model studies in viral heart disease? *Scand. J. Infect. Dis.* **88**(Suppl.)**:**49–65.
24. **Gauntt, C., P. Sakkinen, N. Rose, and S. Huber.** 2000. Picornaviruses: immunopathology and autoimmunity, p. 313–329. *In* M. Cunningham and R. Fujinami (ed.), *Effects of Microbes on the Immune System.* Lippincott-Raven Publishers, Philadelphia, Pa.
25. **Gauntt, C., C. Winfey, H. Wood, A. Karaganis, C. Lee, and M. Cunningham.** 1996. Anti-coxsackievirus group B antibodies and inflammatory heart disease, p. 257–270. *In* G. Pandalai (ed.), *Recent Research Developments in Antimicrobial Agents and Chemotherapy.* Research Signpost, Trivandrum, India.
26. **Grist, N., and E. Bell.** 1973. A six year study of coxsackievirus B infection in heart disease. *J. Hyg. (London)* **73:**165–172.
27. **Guilherme, L., E. Cunha-Neto, V. Coelho, R. Snitcowsky, P. Pomerantzeff, R. Assis, F. Pedra, J. Neumann, A. Goldberg, M. Patarroyo, F. Pileggi, and J. Kalil.** 1995. Human heart-infiltrating T-cell clones from rheumatic heart disease patients recognize both streptococcal and cardiac proteins. *Circulation* **92:**415–420.
28. **Herzum, M., V. Rupert, B. Kuytz, H. Jomau, I. Nakamura, and B. Maisch.** 1994. Coxsackievirus B3 infection leads to cell death of cardiac myoctytes. *J. Mol. Cell. Cardiol.* **26:**907–913.
29. **Huber, S.** 1997. Animal models of human disease—autoimmunity in myocarditis: relevance of animal models. *Clin. Immunol. Immunopathol.* **83:**93–102.
30. **Huber, S.** 1997. Coxsackievirus-induced myocarditis is dependent on distinct immunopathogenic responses in different strains of mice. *Lab. Investig.* **76:**691–701.
31. **Huber, S., and M. Cunningham.** 1996. Streptococcal M protein peptide with similarity to myosin induces $CD4^+$ T cell dependent myocarditis in MRL/++ mice and induces partial tolerance against coxsackieviral myocarditis. *J. Immunol.* **156:**3528–3534.
32. **Huber, S., and P. Lodge.** 1986. Coxsackievirus B3 myocarditis: identification of different mechanisms in DBA/2 and Balb/c mice. *Am. J. Pathol.* **122:**284.
33. **Huber, S., A. Moraska, and M. Cunningham.** 1994. Alterations in major histocompatibility complex association of myocarditis induced by coxsackievirus B3 (CVB3) mutants selected with monoclonal antibodies to group A streptococci. *Proc. Natl. Acad. Sci. USA* **91:**5543–5547.
34. **Huber, S., J. Polgar, P. Schultheiss, and P. Schwimmbeck.** 1994. Augmentation of pathogenesis of coxsackievirus B3 infections in mice by exogenous administration of interleukin-1 and interleukin-2. *J. Virol.* **68:**195–206.
35. **Huber, S. A., C. J. Gauntt, and P. Sakkinen.** 1998. Enteroviruses and myocarditis: viral pathogenesis through replication, cytokine induction and immunopathogenicity. *Adv. Virus Res.* **51:**35–80.
36. **Kerekatte, V., B. Keiper, C. Badorff, A. Cai, K. Knowlton, and R. Rhoads.** 1999. Cleavage of poly(A)-binding protein by coxsackievirus 2A protease in vitro and in vivo: another mechanism for host protein synthesis shutoff? *J. Virol.* **73:**709–711.
37. **Khaw, B., J. Narula, A. Sharaf, P. Nicol, J. Southern, and M. Carles.** 1995. SR-Ca^{2+} ATPase as an autoimmunogen in experimental myocarditis. *Eur. Heart J.* **16**(Suppl. O)**:**92–96.
38. **Klingel, K., C. Hohenadl, A. Canu, M. Albrecht, M. Seemann, G. Mall, and R. Kandolf.** 1992. Ongoing enterovirus-induced myocarditis is associated with persistent heart muscle infection: quantitative analysis of virus replication, tissue damage and inflammation. *Proc. Natl. Acad. Sci. USA* **89:**314–318.
39. **Klingel, K., and R. Kandolf.** 1993. The role of enterovirus replication in the development of acute and chronic heart muscle disease in different immunocompetent mouse strains. *Scand. J. Infect. Dis.* **88**(Suppl.)**:**79–85.
40. **Klingel, K., S. Stephans, M. Sauter, R. Zell, B. McManus, B. Bultmann, and R. Kandolf.** 1996. Pathogenesis of murine enterovirus myocarditis: virus dissemination and immune cell targets. *J. Virol.* **70:**8888–8895.
41. **Koontz, C., and C. Ray.** 1971. The role of coxsackie group B virus infections in sporadic myopericarditis. *Am. Heart J.* **82:**750–758.

42. **Kuan, A., W. Chamberlain, S. Malkiel, H. Lieu, S. Factor, B. Diamond, and B. L. Kotzin.** 1999. Genetic control of autoimmune myocarditis mediated by myosin-specific antibodies. *Immunogen* **49:**79–85.
43. **Kubota, T., C. McTiernan, C. Frye, S. Slawson, B. Lemster, A. Koretsky, A. Demetris, and A. Feldman.** 1997. Dilated cardiomyopathy in transgenic mice with cardiac-specific overexpression of tumor necrosis factor-alpha. *Circ. Res.* **81:**627–635.
44. **Lane, J., D. Neumann, A. Lafond-Walker, A. Herkowitz, and N. Rose.** 1992. Interleukin 1 or tumor necrosis factor can promote coxsackie B3-induced myocarditis in resistant B10.A mice. *J. Exp. Med.* **175:**1123–1129.
45. **Lane, J., D. Neumann, A. Lanfond-Walker, A. Herskowitz, and N. Rose.** 1993. Role of IL-1 and tumor necrosis factor in coxsackievirus-induced autoimmune myocarditis. *J. Immunol.* **151:**1682–1690.
46. **Leslie, K., R. Blay, C. Haisch, P. Lodge, A. Weller, and S. Huber.** 1989. Clinical and experimental aspects of viral myocarditis. *Clin. Microbiol. Rev.* **2:**191–203.
47. **Liao, L., J. Sindhwani, L. Leinwand, B. Diamond, and S. Factor.** 1993. Cardiac a-myosin heavy chains differ in their induction of myocarditis: identification of pathogenic epitopes. *J. Clin. Investig.* **92:**2877–2882.
48. **Liao, L., R. Sindhwani, M. Rojkind, S. Factor, L. Leinwand, and B. Diamond.** 1995. Antibody-mediated autoimmune myocarditis depends on genetically determined target organ sensitivity. *J. Exp. Med.* **187:**1123–1131.
49. **Limas, C.** 1996. Autoimmunity in dilated cardiomyopathy and the major histocompatibility complex. *Int. J. Cardiol.* **54:**113–116.
50. **Limas, C., and C. Limas.** 1993. Immune-mediated modulation of B-adrenoreceptor function in human dilated cardiomyopathy. *Clin. Immunol. Immunopathol.* **68:**204–207.
51. **Lodge, P., M. Herzum, J. Olszewski, and S. Huber.** 1987. Coxsackievirus B3-myocarditis: acute and chronic forms of the disease caused by different immunopathogenic mechanisms. *Am. J. Pathol.* **128:**455–463.
52. **Loudon, R., A. Moraska, S. Huber, P. Schwimmbeck, and H.-P. Schultheiss.** 1991. An attenuated variant of coxsackievirus B3 preferentially induces immunoregulatory T cells in vivo. *J. Virol.* **65:**5813–5819.
53. **Lutton, C., and C. Gauntt.** 1986. Coxsackievirus B3 infection alters plasma membrane of neonatal kin fibroblasts. *J. Virol.* **60:**294–296.
54. **Maisch, B.** 1989. Autoreactivity to the cardiac myocyte, connective tissue and the extracellular matrix in heart disease and postcardiac injury. *Springer Semin. Immunopathol.* **11:**369–396.
55. **Maisch, B.** 1984. Diagnostic relevance of humoral and cell mediated immune reactions in patients with acute myocarditis and congestive cardiomyopathy, p. 1327–1338. *In* E. Chazov, V. Smirnov, and R. Oganov (ed.), *Cardiology.* Plenum Publishing Co., New York, N.Y.
56. **Maisch, B., E. Bauer, M. Cirsi, and K. Kocksiek.** 1993. Cytolytic cross-reactive antibodies directed against the cardiac membrane and viral proteins in coxsackievirus B3 and B4 myocarditis. *Circulation* **87**(Suppl. IV)**:**49–65.
57. **Maisch, B., M. Herzum, G. Hufuagel, C. Bethge, and U. Schonian.** 1995. Immunosuppressive treatment for myocarditis and dilated cardiomyopathy. *Eur. Heart J.* **16**(Suppl. O)**:**153–161.
58. **Maisch, B., R. Trostel-Soder, E. Stechmesser, P. Berg, and K. Kochsiek.** 1982. Diagnostic relevance of humoral and cell-mediated immune reactions in patients with acute viral myocarditis. *Clin. Exp. Immunol.* **48:**533–545.
59. **McManus, B., C. Gauntt, and R. Cassling.** 1987. Immunopathologic basis of myocardial injury. *Cardiovasc. Clin.* **18:**163–184.
60. **Muckelbauer, J., and M. Rossman.** 1997. The structure of coxsackievirus B3. *Curr. Top. Microbiol. Immunol.* **223:**191–208.
61. **Neu, N., K. Beisel, M. Traystman, N. Rose, and S. Craig.** 1987. Autoantibodies specific for the cardiac myosin isoform are found in mice susceptible to coxsackievirus B3-induced myocarditis. *J. Immunol.* **138:**2488–2492.
62. **Neu, N., N. Rose, K. Beisel, A. Herskowitz, G. Gurri-Glass, and S. Craig.** 1987. Cardiac myosin induces myocarditis in genetically predisposed mice. *J. Immunol.* **139:**3630–3636.
63. **Neumann, D., G. Allen, C. Narins, N. Rose, and A. Herskowitz.** 1993. Heart autoantibodies in human myocarditis and cardiomyopathy: virus and skeletal muscle cross-reactivity, p. 325–334. *In* H. Figulla, R. Kandolf, and B. McManus (ed.), *Idiopathic Dilated Cardiomyopathy.* Springer-Verlag, Berlin, Germany.
64. **Neumann, D., C. Burek, K. Baughman, N. Rose, and A. Herskowitz.** 1990. Circulating heart-reactive autoantibodies in patients with myocarditis or dilated cardiomyopathy. *J. Am. Coll. Cardiol.* **16:**839–846.

65. **Neumann, D., J. Lane, S. Wulff, G. Allen, A. Lafond-Walker, A. Herskowitz, and N. Rose.** 1992. In vivo deposition of myosin-specific autoantibodies in the hearts of mice with experimental autoimmune myocarditis. *J. Immunol.* **148:**3806–3811.
66. **Neumann, D., N. Rose, A. Ansari, and A. Herkowitz.** 1994. Induction of multiple heart autoantibodies in mice with coxsackievirus B3- and cardiac myosin-induced autoimmune myocarditis. *J. Immunol.* **152:**343–350.
67. **Pallansch, M.** 1988. *Epidemiology of Group B Coxsackieviruses.* Plenum Publishing Corp., New York, N.Y.
68. **Quarantino, S., C. Thorpe, P. Travers, and M. Londei.** 1995. Similar antigenic surfaces, rather than sequence homology, dictate T-cell epitope molecular mimicry. *Proc. Natl. Acad. Sci. USA* **92:**10398–10402.
69. **Rose, N., A. Herskowitz, D. Newumann, and N. Neu.** 1988. Autoimmune myocarditis: a paradigm of post-infection autoimmune disease. *Immunol. Today* **9:**117–118.
70. **Rose, N., N. Neu, D. Neumann, and A. Herskowitz.** 1988. Myocarditis: a postinfectious autoimmune disease, p. 139–147. *In* H.-P. Schultheiss (ed.), *New Concepts in Viral Heart Disease.* Springer-Verlag, Berlin, Germany.
71. **Rose, N., D. Neumann, and A. Herskowitz.** 1992. Coxsackievirus myocarditis. *Adv. Intern. Med.* **37:**411–429.
72. **Rueckert, R.** 1996. *Picornaviruses,* 3rd ed. Lippincott-Raven, Philadelphia, Pa.
73. **Saegusa, J., B. Prabhakar, K. Essani, P. McClintock, Y. Fukuda, V. Ferrna, and A. Notkins.** 1986. Monoclonal antibody to coxsackievirus B4 reacts with myocardium. *J. Infect. Dis.* **153:**372–373.
74. **Sato, S., R. Tsutsumi, A. Burke, G. Carson, V. Porro, Y. Seko, K. Okumura, R. Kawana, and S. Virmani.** 1994. Persistence of replicating coxsackievirus B3 in the athymic murine heart is associated with development of myocarditic lesions. *J. Gen. Virol.* **75:**2911–2924.
75. **Schmidt, N., R. Magoffin, and E. Lennette.** 1973. Association of group B coxsackie viruses with cases of pericarditis, myocarditis, or pleurodynia by demonstration of immunoglobulin M antibody. *Infect. Immun.* **8:**341–348.
76. **Schultheiss, H.** 1989. The significance of autoantibodies against the ACP/ATP carrier for the pathogenesis of myocarditis and dilated cardiomyopathy—clinical and experimental data. *Springer Semin. Immunopathol.* **11:**15–30.
77. **Schwimmbeck, P., C. Badorff, H. Schultheiss, and B. Strauer.** 1994. Transfer of human myocarditis into severe combined immunodeficient mice. *Circ. Res.* **75:**156–164.
78. **Schwimmbeck, P., C. Badorff, K. Schulze, and H.-P. Schultheiss.** 1997. The significance of T cell responses in human myocarditis, p. 65–76. *In* H.-P. Schultheiss and P. Schwimmbeck (ed.), *The Role of Immune Mechanisms in Cardiovascular Disease.* Springer-Verlag, Berlin, Germany.
79. **Schwimmbeck, P., H.-P. Schultheiss, and B. Strauer.** 1989. Identification of a main auto-immunogenic epitope of the adenine nucleotide translocator which cross reacts with coxsackievirus B3: use in the diagnosis of myocarditis and dilated cardiomyopathy. *Circulation* **80**(Suppl. II)**:**665–669.
80. **Schwimmbeck, P., N. Schwimmbeck, H. Schultheis, and B. Strauer.** 1993. Mapping of antigenic determinants of the adenine-nucleotide translocator and coxsackie B3 virus with synthetic peptides: use for the diagnosis of viral-heart disease. *Clin. Immunol. Immunopathol.* **68:**135–140.
81. **Seko, Y., H. Matsuda, K. Kato, Y. Hashimoto, H. Yagita, K. Okumura, and Y. Yazaki.** 1993. Expression of intercellular adhesion molecule-1 in murine hearts with acute myocarditis caused by coxsackievirus B3. *J. Clin. Invest.* **91:**1327–1336.
82. **Shikman, A., and M. Cunningham.** 1994. Immunological mimicry between *N*-acetyl-β-D-glucosamine and cytokeratin peptides: evidence for a microbially driven antikeratin antibody response. *J. Immunol.* **152:**4375–4387.
83. **Shikman, A., N. Greenspan, and M. Cunningham.** 1993. A subset of mouse monoclonal antibodies cross-react with cytoskeletal proteins and group A streptococcal M proteins recognizes *N*-acetyl-β-D-glucosamine. *J. Immunol.* **151:**3902–3913.
84. **Smith, S., and P. Allen.** 1991. Myosin-induced acute myocarditis is a T cell-mediated disease. *J. Immunol.* **147:**2141–2147.
85. **Smith, W.** 1970. Coxsackie B myopericarditis in adults. *Am. Heart J.* **80:**34–46.
86. **Theofilopoulos, A.** 1995. The basis for autoimmunity. Part I. Mechanisms of aberrant self-recognition. *Immunol. Today* **16:**90–98.
87. **Theofilopoulos, A.** 1995. The basis for autoimmunity. Part II. Genetic predisposition. *Immunol. Today* **16:**150–159.

88. **Theofilopoulos, A.** 1997. *Mechanisms and Genetics of Autoimmunity.* Springer-Verlag, Berlin, Germany.
89. **Tracy, S., V. Wiegand, B. McManus, C. Gauntt, M. Pallansch, M. Beck, and N. Chapman.** 1990. Molecular approaches to enteroviral diagnosis in idiopathic cardiomyopathy and myocarditis. *J. Am. Coll. Cardiol.* **15:**1688–1694.
90. **Traystman, D., and K. Beisel.** 1991. Genetic control of coxsackievirus B3-induced heart-specific autoantibodies associated with chronic myocarditis. *Clin. Exp. Immunopathol.* **86:**291–298.
91. **Traystman, M., L. Chow, B. M. McManus, A. Herskowitz, M. Nesbitt, and K. Beisel.** 1991. Susceptibility to coxsackievirus B3-induced chronic myocarditis maps near the murine Tcr and Myhc alpha loci on chromosome 14. *Am. J. Pathol.* **138:**721–732.
92. **Ulrich, G., U. Kuhl, B. Melzner, I. Janda, B. Schafer, and H.-P. Schultheiss.** 1988. Antibodies against the adenosine di/triphosphate carrier cross-react with the Ca^{++} channel-functional and biochemical data, p. 226–235. *In* H.-P. Schultheiss (ed.), *New Concepts in Viral Heart Disease.* Springer-Verlag, Berlin, Germany.
93. **Van Houten, N., P. Bouchard, A. Moraska, and S. Huber.** 1991. Selection of an attenuated coxsackievirus B3 variant using a monoclonal antibody reactive to myocyte antigen. *J. Virol.* **65:**1286–1290.
94. **Wallukat, G., A. Kayser, and A. Wollenberger.** 1995. The beta 1-adrenoceptor as antigen: functional aspects. *Eur. Heart J.* **16**(Suppl. O)**:**85–88.
95. **Weller, A., K. Simpson, M. Herzum, N. Van Houten, and S. Huber.** 1989. Coxsackievirus B3 induced myocarditis: virus receptor antibodies modulate myocarditis. *J. Immunol.* **143:**1843–1850.
96. **Wessely, R., K. Klingel, L. Santana, N. Dalton, M. Hongo, W. Jonathan Lederer, R. Kandolf, and K. Knowlton.** 1998. Transgenic expression of replication-restricted enteroviral genomes in heart muscle induces defective excitation-contraction coupling and dilated cardiomyopathy. *J. Clin. Investig.* **102:**1444–1453.
97. **Why, H., T. Meany, P. Richardson, E. Olsen, N. Bowles, L. Cunningham, C. Freeke, and L. Archard.** 1994. Clinical and prognostic significance of detection of enteroviral RNA in the myocardium of patients with myocarditis or dilated cardiomyopathy. *Circulation* **89:**2582–2589.
98. **Wolff, P., U. Kuhl, and H. Schultheiss.** 1989. Laminin distribution and autoantibodies to laminin in dilated cardiomyopathy and myocarditis. *Am. Heart J.* **117:**1303–1309.
99. **Woodruff, J.** 1980. Viral myocarditis. *Am. J. Pathol.* **101:**425–483.
100. **Woodruff, J., and J. Woodruff.** 1974. Involvement of T lymphocytes in the pathogenesis of coxsackievirus B3 heart disease. *J. Immunol.* **113:**1726–1734.
101. **Wucherpfennig, K., and J. Strominger.** 1995. Molecular mimicry in T cell-mediated autoimmunity: viral peptides activate human T cell clones specific for myelin basic protein. *Cell* **80:**695–705.

Molecular Mimicry, Microbes, and Autoimmunity
Edited by M. W. Cunningham and R. S. Fujinami

Chapter 6

Molecular Mimicry and Heart Disease

Josef M. Penninger and Kurt Bachmaier

Heart diseases are the most prevalent cause of morbidity and mortality in rich countries. A common cause of progressive heart disease, heart failure, and sudden death is dilated cardiomyopathy (DCM), a group of disorders in which the heart muscle is weakened and cannot pump effectively. The net result is dilation of the cardiac chambers and thinning of the ventricular walls. The poor cardiac function results in congestive heart failure (7). Few treatment options are available for DCM, and DCM is the principal condition that necessitates heart transplantation. Whereas some cases of DCM have been linked to genetic factors, most patients with cardiomyopathy have a history of inflammatory heart disease. Inflammatory heart disease can be caused by a wide variety of pathogens such as viruses, bacteria, and protozoa (8). RNA of the picornavirus coxsackievirus group B3 (CVB3) can be detected in the heart muscle in as many as 40 to 50% of patients with DCM (11). CVB3 infections have been epidemiologically linked to chronic heart disease, implying the CVB3 might be the most prevalent virus associated with DCM. The CVB infection rate in humans is high, and CVB usually induces mild nasopharyngeal disease with symptoms of a common cold. However, CVB3 can infect and replicate in many different cell types within the brain, heart, liver, pancreas, and T or B lymphocytes, and CVB3 infections are principal causes of very severe cases of myocarditis in children (11). After acute viral replication, CVB3 is cleared from the target organs, leading to the development of necrotic areas due to the cytopathic effect of the virus and the antiviral activity of the immune system. Since almost all people have been infected with CVB3 and harbor anti-CVB3 antibodies, it appears that CVB3 and other viruses might induce myocarditis very frequently in the population without ever being detected. Diagnosis of myocarditis and inflammatory heart disease is a big black box of mysteries; e.g., who would do heart muscle biopsies for a patient with a common cold? Therefore, the challenge will be to identify both genetic and environmental factors that determine the progression from acute infections to chronic heart disease and the development of DCM. In addition to virus infections, bacterial infections may be a causative event in the development of heart diseases (8). *Chlamydia* infections, in particular, are epidemiologically linked to human heart disease. *Chlamydia* infections cause pneumonia and conjunctivitis in children and are a primary cause of sexually transmitted diseases and female infertility. The mechanism by which *Chlamydia* causes cardiovascular disease is unknown (1). However, numerous clinical and experimental studies

---

Josef M. Penninger and Kurt Bachmaier • Amgen Institute, Ontario Cancer Institute, Department of Medical Biophysics, University of Toronto, 620 University Avenue, Suite 706, Toronto, Ontario M5G 2C1, Canada.

imply that chronic stages of myocarditis and DCM are mediated by autoimmune responses to cardiac antigens exposed to the immune system after cardiomyocyte damage (11).

EXPERIMENTAL AUTOIMMUNE MYOCARDITIS IN MICE

The development of a murine model of autoimmune myocarditis was based on genetic differences among inbred mouse strains in the immune response to CVB3-induced myocarditis (9). In certain mouse strains CVB3-mediated myocarditis evolves in an early phase characterized by myocyte damage due to CVB3 cytotoxicity during acute viral infection and a late phase that is associated with the production of heart muscle-specific autoantibodies and inflammatory infiltration of T cells, B cells, and macrophages into the myocardium (10). The later phase of CVB3-induced heart disease can be mimicked by immunization of mice with purified murine cardiac myosin in the absence of virus infection, and experimental cardiac myosin-induced myocarditis has immunological and histopathological features that resemble postviral heart disease in mice and of dilated cardiomyopathy in humans (11).

GENETIC MAP OF AUTOIMMUNE MYOCARDITIS IN MICE

The understanding of the induction phase of autoimmunity and the breakdown of immunological tolerance toward autoantigens will be crucial for the successful prevention of disease. In mice, autoimmune diseases either arise spontaneously or can be experimentally induced by injection of the tissue-specific autoantigen. Experimental organ-specific autoimmune diseases are initiated through activation of T cells which then recirculate to the target organ. Thus, two main mechanisms are involved in the pathogenesis of autoimmunity: (i) activation of autoreactive T cells and (ii) susceptibility of the target organ, which allows inflammatory cells to home into the organ. Transgenic mice are ideal tools for dissection of the complex mechanisms involved in heart disease.

Mouse Experimental Autoimmune Myocarditis Is a T-Cell-Mediated Disease

In experimental autoimmune myocarditis, the inflammatory infiltrate consists of $CD4^+$ T cells (~10%), $CD8^+$ T cells (~5 to 7%), a few B cells (~1 to 2%) with local deposition of heart muscle protein-specific autoantibody, and $CD11b^+$ macrophages (70 to 80% of infiltrating cells) (11). Inflammation of the heart muscle can be transferred only into immunodeficient recipients or lipopolysaccharide (LPS)-pretreated healthy mice by purified T cells from mice with active myocarditis. By contrast, transfer of anti-cardiac myosin autoantibodies does not cause myocarditis. In addition, neonatal depletion of T cells but not of B cells does prevent experimental autoimmune myocarditis, but autoimmune heart disease cannot be induced in mice lacking the T-cell-specific tyrosine kinase $p56^{lck}$. The *src*-family protein tyrosine kinase $p56^{lck}$ is crucial for antigen receptor-mediated signal transductions and activation of T lymphocytes, and the T cells of $p56^{lck}$-deficient mice exhibit a severe defect in their development and effector functions. Autoimmune heart disease also does not occur in mice lacking the tyrosine phosphatase CD45, which regulates the enzymatic activity of $p56^{lck}$, and both $p56^{lck}$- and CD45-deficient mice fail to produce immunoglobulin G (IgG) autoantibodies reactive to cardiac myosin (Table 1).

Table 1. Prevalence and severity of experimental autoimmune myocarditis in transgenic mice in comparison to that in wild-type controls

Genotype	Prevalence	Disease severity
$p56^{lck-/-}$	No disease	
$p56^{lck+/-}$	High	Severe
$CD45^{-/-}$	No disease	
$CD45^{+/-}$	High	Severe
$CD4^{+/-}8^{+/-}$	High	Severe
$CD4^{-/-}$	Intermediate	Mild
$CD8^{-/-}$	High	Severe
$CD28^{+/-}$	High	Severe
$CD28^{-/-}$	Low	Mild
TNF-$Rp55^{-/-}$	No disease	
TNF-$Rp55^{+/-}$	High	Severe
IRF-$1^{-/-}$	High	Severe
IRF-$1^{+/-}$	High	Severe
IRF-$2^{-/-}$	High	Severe
IRF-$2^{+/-}$	High	Severe
hCD4TG	Low	Mild
hCD4/DQ6TG	High	Severe

Thus, all genetic and functional data indicate that T cells are required for the induction of autoimmune heart disease and that production of cardiac myosin-specific autoantibodies and IgG class switching strictly depend on prior activation of T cells (11).

Experimental induction of most autoimmune diseases requires $CD4^+$ T-helper cells responsive to the self-antigen presented on major histocompatability complex (MHC) class II molecules, and administration of monoclonal antibodies (MAbs) against CD4 molecules can block disease in experimental and genetic animal models of autoimmunity. Hence, $CD4^+$ T-helper lymphocytes appear to be crucial regulators for the initiation of T-cell-dependent autoimmunity, and clinical trials that target the $CD4^+$ cell lineage are under way in an examination of treatments for various human autoimmune diseases. Besides $CD4^+$ cells, MHC class I-restricted $CD8^+$ lymphocytes can be found in autoimmune lesions, and $CD8^+$ lymphocytes have a role in disease progression and tissue damage (11).

Similar to other experimental autoimmune diseases, depletion of $CD4^+$ or $CD8^+$ T cells by injection of anti-CD4 or anti-CD8 MAbs can prevent induction of cardiac myosin-induced autoimmune heart disease. In later stages of autoimmune heart disease only depletion of $CD8^+$ cells could prevent the disease, whereas depletion of $CD4^+$ T cells no longer had any ameliorating effect, implying that $CD8^+$ T cells contribute to cardiomyocyte damage and disease progression. A study of the role of $CD4^+$ and $CD8^+$ T-cell subsets in autoimmune myocarditis at the genetic level with mice that lack CD4 or CD8 after gene targeting revealed that mice homozygous for the CD8 mutation developed disease significantly more severe than that in their heterozygous littermates, implying that $CD8^+$ lymphocytes not only act as cytotoxic effector cells in autoimmunity but may also regulate disease severity (Table 1) (3, 7). A similar regulatory role of $CD8^+$ T cells has also been described in experimental autoimmune encephelomyelitis (EAE), i.e., increased frequency of relapses in CD8-deficient mice and resistance to a second induction of EAE in mice depleted of $CD8^+$ cells with MAbs (3, 7).

Remarkably, CD4-deficient mice also develop mild autoimmune myocarditis with myocardial inflammation and the appearance of autoantibodies, indicating that cells other than $CD4^+$ lymphocytes can provide help for immunoglobulin class switching and IgG autoantibody production. A high proportion of infiltrating cells in the hearts of CD4-deficient mice had a $TCR\alpha\beta^+$ $CD4^-$ $CD8^-$ phenotype, a cell population that is also probably responsible for initiation of autoimmune heart disease in these mice (Table 1). Approximately 10% of peripheral T cells in CD4 mice have a $TCR\alpha\beta^+$ $CD4^-$ $CD8^-$ phenotype, and these cells can provide MHC class II-restricted help for T-cell-dependent antibody production. The population of $TCR\alpha\beta^+$ $CD4^-$ $CD8^-$ cells is also responsible for development of EAE in CD4-deficient mice (3, 7).

T-Cell Costimulation via CD28

Antigen-specific activation of T lymphocytes requires two signals generated by the T-cell receptor (TCR) and a costimulatory receptor. TCR-mediated activation crucially depends on the function of the *src*-family protein tyrosine kinase $p56^{lck}$. Interactions of CD28 on T cells and B7.1 (CD80) or B7.2 (CD86) molecules on antigen-presenting cells (APCs) provide the crucial second signal for T-cell proliferation and effector functions in multiple human and murine in vivo and in vitro systems, suggesting that CD28 is the most important costimulatory molecule for all T-cell subsets. Since TCR ligation in the absence of CD28 costimulation renders T lymphocytes anergic and B7 overexpression in peripheral organs can induce autoimmune disease, the hypothesis has been put forward that potentially autoreactive T cells remain dormant because tissue-specific interstitial APCs cannot provide the costimulation necessary for T-cell activation. Thus, autoimmunity may occur due to aberrant expression of costimulatory molecules possibly induced by pathological conditions such as viral or bacterial infections.

Since induction of cardiac myosin-induced myocarditis depends on $CD4^+$ T cells and MHC class II molecules, it is an ideal model for study of the role of CD28 costimulation in antigen-induced autoimmune disease. Remarkably, autoimmune myocarditis develops in mice that lack the T-cell costimulatory molecule CD28 (Table 1) (4). However, disease severity and prevalence were significantly lower in CD28-deficient mice than in their control littermates and depended on the autoantigen dose administered. Diseased wild-type mice produce high titers of (Th2-mediated) IgG1 and (Th1-mediated) IgG2a autoantibodies at a ratio of 3:1 (IgG1 to IgG2a), but the ratio of IgG1 to IgG2a autoantibody titers is reversed in CD28-deficient mice (1:1.4), indicating that the induction of Th2-dependent IgG1 responses requires CD28 costimulation during the course of experimental autoimmune heart disease. B7.1 and B7.2 are expressed in the inflammatory infiltrate 21 days after initial immunization. The results for CD28 gene-deficient mice suggest that CD28 is not essential for the induction of autoimmune heart disease, but CD28 costimulation plays an important role in both induction and maintenance of autoimmune heart disease with suboptimal immunization procedures. These data also provided genetic proof that CD28 expression is preferentially required for Th2-mediated IgG1 responses in an autoimmune setting.

Role of Heart APCs

"Professional" APCs located within the heart myocardium appear to be the primary target of autoreactive T cells during the initial phases of autoimmune heart disease. Evidence for this premise was provided by the fact that heart interstitial cells purified from healthy mice could present cardiac myosin in association with MHC class II molecules and stimulate cardiac myosin-specific T cells. APCs from other organs failed to activate the same cardiac myosin-specific cells. Splenic APCs pulsed with exogenous cardiac myosin induced proliferation of these cells. These data indicate that interstitial APCs located within the heart myocardium process and present heart-specific antigens derived from cardiac myosin (14).

Although heart-specific APCs spontaneously present autoantigens to the immune system, the myocardium of healthy mice is not susceptible to an attack of autoagressive T lymphocytes that recognize cardiac myosin. However, disease can be transferred into syngeneic or MHC congenic recipients after injection of LPS prior to adoptive transfer of autoreactive T cells from cardiac myosin-immunized donors. Whereas the myocardium of healthy mice contains a large number of interstitial APCs, very few of these cells express detectable levels of MHC class II molecules on the cell surface. However, after LPS administration or immunization with cardiac myosin, virtually all heart interstitial cells express high levels of MHC class II. Expression of MHC class II molecules and other adhesion molecules such as intercellular adhesion molecule-1 (ICAM-1) and CD31, both of which are predominantly expressed on endothelial cells, precedes cellular infiltration of T lymphocytes into the heart. MHC class I was not upregulated after immunization, although $CD8^+$ cells infiltrate the heart, implying that cardiomyocytes might not be the primary targets of autoreactive T cells but, rather, that $CD8^+$ T cells act during disease progression. Moreover, inflammation is mostly interstitial around blood vessels in autoimmune myocarditis, and myocyte damage is minimal and occurs only in later stages of inflammation (11).

Since disease can be adoptively transferred only into LPS-pretreated immunocompetent mice, upregulation of MHC class II molecules appears to be a crucial prerequisite for the induction of autoimmune myocarditis. This notion is supported by earlier studies that in vivo injection of MAbs against MHC class II molecules or injection of nonimmunogenic, competitor peptides that bind to MHC class II molecules can prevent autoimmune heart disease. In addition, it has been shown that the induction of autoimmune myocarditis is strain dependent and that the MHC class II haplotype (such as $H\text{-}2^{k/k}$ in disease-susceptible A/J mice or $H\text{-}2^{d/d}$ in susceptible BALB/c mice) is the most important genetic factor for conferral of disease susceptibility (11).

Target Organ Susceptibility: Crucial Role for Low-Molecular-Weight TNF-Rp55

LPSs induce MHC class II expression in various organs including the heart via production of cytokines such as gamma interferon (INF-γ) and tumor necrosis factor (TNF) alpha (TNF-α). The TNF–TNF receptor (TNF-R) system has been implicated in the pathogenesis of murine autoimmune myocarditis, experimental allergic encephalomyelitis, arthritis, insulitis and diabetes mellitus, lupus erythematosus, or human multiple sclerosis. TNF-α and TNF-β bind to either TNF-Rp55 (TNF-Rp55) or TNF-Rp75 and induce a plethora of biological responses such as cell proliferation or programmed cell death, or they can act as key mediators of inflammatory immune responses (3).

Mice that lack TNF-Rp55 after homologous recombination have normal T-cell development but succumb to *Listeria monocytogenes* infections in vivo. In addition, these mice are resistant to toxic shock syndrome induced by LPS or bacterial toxins, implying that

TNF-Rp55 mediates the major effects of TNF in vivo. In autoimmune heart disease, interstitial APCs from TNF-Rp55-deficient mice fail to upregulate MHC class II molecules after immunization with cardiac myosin or LPS treatment, and these mice were completely protected from autoimmune myocarditis (Table 1) (3).

Although these data do not exclude a role of the TNF–TNF-R system in the function of autoreactive T lymphocytes and/or macrophages, our results indicate that TNF-Rp55 controls target organ susceptibility during autoimmune myocarditis via upregulation of MHC class II molecules. Upregulation of MHC class II molecules on heart interstitial cells has important functional consequences for the pathogenesis of autoimmunity since induction of autoimmune myocarditis is strictly dependent on $CD4^+$ T cells, and MHC class II and heart interstitial cells from normal mice present cardiac myosin peptides in association with MHC class II (11).

iNOS Expression and Nitrotyrosine Formation in the Myocardium in Response to Inflammation Is Controlled by IRF-1

One of the consequences of inflammation in autoimmune heart disease is injury of cardiomyocytes and subsequent progression to heart failure and DCM. Although a multitude of factors might ultimately contribute to disease pathogenesis, cytotoxicity via NO is a principle candidate for mediation of cardiomyocyte damage. NO is a product of enzymatic conversion of L-arginine to L-citrulline mediated by NO synthases (NOSs) and functions as an intercellular messenger in neurotransmission, vasodilatation, differentiation of hematopoietic cells, and macrophage cytotoxicity. While neuronal and endothelial cells express constitutive NOSs, macrophages and some other cell types including cardiac endothelial cells and cardiac myocytes have a transcriptionally inducible isoform of NOS (iNOS). Cytokines such as IFN-γ, TNF-α, IFN-α, IFN-β, interleukin-1β (IL-1β), and LPS are potent activators of iNOS expression. Interestingly, iNOS expression has been found in the myocardium of patients with DCM and patients with ischemic heart disease or valvular heart disease. NO synthesis may contribute to a variety of myocardial dysfunctions and heart failure, and NO exerts negative inotropic effects on heart muscle contractility and is toxic for cardiomyocytes in vitro. Interferon regulatory transcription factor type 1 (IRF-1) is the crucial transcription factor for iNOS expression in macrophages, and macrophages from IRF-1-deficient mice cannot produce NO after infections.

During the pathogenesis and progression of autoimmune myocarditis, iNOS expression was elicited in inflammatory macrophages and in heart muscle cells. Autoimmune heart disease was accompanied by formation of the NO reaction product nitrotyrosine in inflammatory macrophages as well as in virtually all heart muscle cells. Interestingly, NO production and subsequent nitrosylation of heart muscle proteins was strictly dependent on an inflammatory response within the heart and did not occur in cardiac myosin-immunized CD45 gene-deficient mice that lack functional T cells and that, hence, do not develop any autoimmune heart disease. On the other hand, very strong nitrosylation of virtually all heart muscle cells was evident in CD28-deficient mice, which develop only a very mild form of the disease. These data indicate that nitrosylation of heart muscle proteins is strictly dependent on local inflammation. However, the extent of inflammation does not appear to determine the extent of nitrosylation; i.e., already weak focal inflammation does induce nitrosylation of tyrosine residues on heart muscle proteins of the whole heart. Mice defective for IRF-1 failed to express iNOS protein and did not show any nitrotyrosine formation in the heart muscle or infiltrating macrophages. Thus, IRF-1 is the crucial transcription factor for iNOS expression in macrophages and cardiomyocytes during autoimmune inflammation (Table 1) (2).

"Humanized" Mice Susceptible to Autoimmune Inflammatory Heart Disease

Chronic heart diseases in humans have been linked to certain HLA alleles, such as HLA-DQ6. Using double CD4- and CD8-deficient mice transgenic for human CD4 (hCD4) and human HLA-DQ6 to specifically reconstitute the human CD4/DQ6 arm of the immune system in mice, we provided experimental evidence showing that human MHC class II molecules are involved in the pathogenesis of myocarditis and DCM. Transgenic hCD4 and HLA-DQ6 expression rendered genetically resistant C57BL/6 mice susceptible to autoimmune myocarditis induced by immunization with cardiac myosin (Table 1) (5). The autoimmune inflammatory heart disease induced by the human heart muscle-specific peptide in hCD4 and HLA-DQ6 double transgenic mice shared functional and phenotypic features with the disease occurring in disease-susceptible nontransgenic mice.

HEART SPECIFIC α-MYOSIN HEAVY CHAIN-DERIVED PEPTIDES WITH MAJOR PATHOGENICITY

T cells recognize peptides presented by MHC class I or MHC class II molecules. Autoimmune attacks in a variety of experimental diseases can be initiated by self-peptides such as the peptide from amino acids 1 to 9 of myelin basic protein which mediates EAE (11). Besides recognition of immunodominant self-peptides, autoimmune attacks induce determinant spreading to other epitopes; i.e., after induction of the autoimmune response by the main antigen, lymphocytes also react against other peptides of the same self-protein. The identification of immunodominant epitopes might allow the design of vaccination protocols by use of the peptide responsible for disease, peptide antagonists, or peptide agonists for specific inhibition of the induction and/or progression of autoimmunity. Thus, EAE can be prevented by intravenous injection or induction of oral and nasal tolerance by using the immunodominant peptide or the whole protein. In addition, one could design for human peptides which specifically tolerize autoreactive T cells in predisposed children or which are presented to the immune system by peptide-loaded dendritic cells.

Immunodominant peptides derived from the cardiac α-myosin heavy-chain isoform that can induce autoimmune myocarditis in BALB/c (2) and A/J mice (5) have been identified. Mapping of the autoantigenic peptide in BALB/c mice was based on the finding that myosin isoforms with similar amino acid sequences, i.e., α-myosin purified from hearts or β-myosin purified from the soleus muscle, significantly differ in their abilities to induce myocarditis (13). α-Myosin is the immunodominant isoform and induces myocarditis at a high prevalence and severity, whereas β-myosin induces little disease. Thus, the immunodominant epitopes had to reside within regions of amino acid sequences that differ between α- and β-myosin. By this approach three pathogenic epitopes of α-myosin were mapped (13). One peptide is located in the head portion of the molecule (amino acids 614 to 643) and induces severe myocarditis, whereas two others that reside in the rod portion of α-myosin possess only minor pathogenicity. The α-myosin isoform is also the immunodominant isoform in A/J mice, and in this mouse strain induction of autoimmune myocarditis as a result of immunization with synthesized peptides is restricted to a peptide derived from the myosin heavy-chain α isoform and not the β isoform (6). This peptide was located in a region on the α-myosin molecule (amino acids 334 to 352) different from the regions of the immunodominant epitopes in BALB/c mice. Importantly, the same group also showed that this peptide binds strongly to I-A^k molecules and forms a stable complex with MHC class II molecules (6). It should also be noted that two immunodominant cardiac myosin peptides (CM1 [TRGKLSYTQQMEDLKRQ] and CM2 [KLELQSALEEAEASLEH]) which can induce experimental myocarditis in rats have been identified (11).

The finding that the α- and β-myosin isoforms differ so dramatically in their abilities to induce myocarditis suggests that the ability to induce immunological tolerance varies between these two myosin isoforms. Interestingly, the expression of α- and β-cardiac myosin isoforms is developmentally regulated; i.e., the α-myosin isoform is expressed much earlier in embryonic development than the β-myosin isoform, which is expressed only after birth (6). Thus, during the fetal period the developing immune system can encounter only the β-isoform of myosin but not the α-myosin. The implications of developmental myosin isoform switching for the immunodominance of the α-myosin isoform and immunological tolerance are not known. However, immunological tolerance to heart muscle proteins is maintained under normal physiological conditions in mice and humans.

ANTIGENIC MIMICRY

Experimental in vivo and in vitro data provide evidence of molecular mimicry between bacterial antigens and heart-specific proteins and indicate that bacterial peptides can trigger tissue-specific inflammation of the heart. The possibility that antigenic mimicry exists between *Chlamydia*-derived peptides and endogenous heart-specific peptides was tested in the murine model of antigen-induced inflammatory heart disease. Immunization with a 30-amino-acid peptide (amino acids 614 to 643) of the cardium-specific α-myosin heavy-chain molecule (αmhc; amino acids 614 to 643) induces severe inflammatory heart disease in BALB/c mice. The first 16 amino acids of αmhc amino acids 614 to 629 [SLKLMATLF-STYASAD] constitute a dominant autoaggressive epitope that was designated M7Aα (Table 2; Fig. 1A). In contrast, the homologous region of the β-myosin heavy-chain isoform, designated M7Aβ, does not induce disease (Table 2; Fig. 1B). The introduction of single amino acid substitutions into M7Aα further revealed the importance of the xxxMAxxxSTxxx motif for the pathogenicity of M7Aα in vivo. These immunogenic amino acids are conserved between murine and human α-myosin heavy chains, and injection of the human M7Aα homologue into BALB/c mice also induced inflammatory heart disease (1).

Peptide sequences from the 60-kDa cysteine-rich outer membrane protein (CRP) from different serovars of *Chlamydia trachomatis* matched the M7Aα motif and were designated ChTR1 (serovar E), ChTR2 (serovar C), and ChTR3 (serovars L1, L2, and L3). The homologous peptides from the 60-kDa CRPs of *Chlamydia pneumoniae,* designated ChPN, and *Chlamydia psittaci,* designated ChPS, shared sequence identities with the M7Aα motif, albeit to a lesser extent (Table 2). Apart from identity at the MAxxxST motif, there were no other conserved regions in the primary sequences of the murine M7Aα peptide and all three *Chlamydia* 60-kDa CRP peptides.

Table 2. Prevalence and severity of autoimmune myocarditis induced by *Chlamydia*-derived peptides that mimic heart-specific α-myosin heavy-chain-derived peptides

Peptide	Amino acid sequence[a]	Prevalence[b]	Severity
M7Aα	SLKL**MA**TLF**ST**YASAD	High	High
ChTR1	VLETS**MA**EFT**ST**NVIS	High	Mild
ChTR2	VLETS**MA**ESL**ST**NVIS	High	Mild
ChTR3	VLETS**MA**EFI**ST**NVIS	High	Mild
ChPN	GIEAAV**A**ESLI**T**KIVA	High	Mild
ChPS	KIEAAA**A**ESLA**T**RFIA	High	Mild
M7Aβ	SLKLLSNLFANYASAD	No disease	

[a] Boldface indicates amino acids constituting the pathogenic motif.
[b] Prevalence is diseased animals/immunized animals.

All of the *Chlamydia*-derived peptides induce inflammatory heart disease at similar frequencies, albeit at significantly lower severity, compared to that after immunization of mice with M7Aα (Table 2). Like the inflammation that follows immunization with the endogenous autoantigen M7Aα, the disease induced by all the *Chlamydia*-derived peptides was characterized by perivascular and pericardial infiltration of mononuclear cells and fibrotic changes (Fig. 1A, C, and D). Immunohistochemical characterization revealed that the inflammatory infiltrate in ChTR1 peptide-induced heart disease is similar to that in cardiac myosin- and cardiac myosin-derived peptide-induced myocarditis and consists of about 11% $CD4^+$ and 12% $CD8^+$ T cells, 16% $B220^+$ B cells, and 61% $CD11b^+$

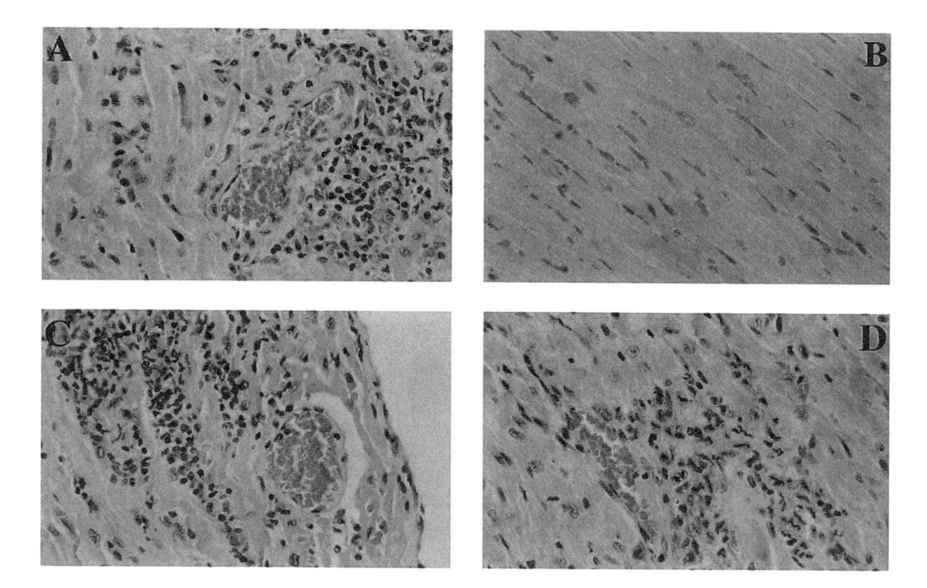

Figure 1. Inflammatory heart disease in BALB/c mice that were immunized with the endogenous mouse M7Aα peptide from the α myosin heavy chain (A), the control endogenous M7Aβ peptide from the homologous region of the β-myosin heavy chain (B), the 60-kDa CRP-derived peptide from *C. trachomatis* (ChTR1) (C), and the 60-kDa CRP-derived peptide from *C. pneumoniae* (ChPN) (D) (1). Hearts were analyzed 21 days after the initial immunization. Hematoxylin-eosin staining was used. Magnifications, ×320. (For color version of figure, see Color Plates, p. 275.)

macrophages. As in the disease induced with the endogenous peptide, inflammation is restricted to the heart and is not observed in the skeletal muscle, lung, liver, pancreas, kidney, intestine, or uterus of peptide-immunized mice. In contrast to the peptides that match the M7Aα motif, injection of mice with human immunodeficiency virus type 2 (HIV-2)- or parainfluenza virus type 1-derived peptides that share homology with other immunogenic regions of the mouse αmhc molecule does not cause inflammatory heart disease (Table 3).

C. pneumoniae has been linked to atherosclerosis and the clogging of blood vessels. Experimental *C. pneumoniae* infections in rabbits and mice accelerate atherosclerosis and lead to focal periarteritis, and *C. trachomatis* infections have been shown to lead directly to myocarditis. Mice immunized with *Chlamydia* peptides developed perivascular fibrosis (Fig. 2A and B), fibrinous occlusions of cardiac blood vessels, and thickening of the arterial walls (Fig. 2C and D). Fibrinous occlusions that originated from blood vessel endothelium, a minimum of one per individual heart, occurred in 19 of 32 (60%) hearts analyzed from mice immunized with *Chlamydia*-derived peptides. Similarly, fibrinous occlusions that originated from blood vessel endothelium occurred in 14 of 21 (67%) hearts analyzed from mice immunized with M7Aα. No fibrinous occlusions were detected in hearts from control mice (1).

Splenic T cells from mice immunized with the endogenous peptide M7Aα proliferate when they are incubated with splenocytes pulsed with the M7Aα peptide. Splenic T cells from these mice also showed a strong proliferative response to the *C. trachomatis*-derived peptide ChTR1. Splenic T cells from mice immunized with ChTR1 proliferated in response to ChTR1 and to the endogenous M7Aα peptide. Thus, ChTR1 peptide immunizations can cross-prime for T-cell reactivity against endogenous M7Aα.

Murine autoimmune myocarditis is accompanied by the T-cell-dependent production of autoantibodies to cardiac epitopes (10). Immunization with endogenous M7Aα peptide leads to the production of serum antibodies to the M7Aα peptide and to the ChTR1 peptide (Fig. 3). Likewise, immunization with the *C. trachomatis*-derived peptide ChTR1 induces the production of serum antibodies to the ChTR1 peptide and to the endogenous M7Aα peptide. Mice immunized with either M7Aα or ChTR1 also had antibodies to the kkα peptide, an unrelated heart-specific peptide (Fig. 3), suggesting that M7Aα- and ChTR1-induced heart disease leads to epitope spreading at the B-cell level.

Because activation of autoaggressive T and B cells occurred in the absence of an overt bacterial infection, we then determined whether actual *Chlamydia* infections would lead to the activation of autoaggressive lymphocytes reactive to heart-specific antigens. Inflammation of either the respiratory tract or the reproductive organs leads to the production of IgG

Table 3. Peptides derived from parainfluenza virus 1- or HIV-2 with homology to heart-specific α-myosin heavy-chain-derived peptides do not cause autoimmune myocarditis

Peptide (amino acid positions)	Amino acid sequence	Prevalence[a]	Severity
αmhc (314–332)	DSAFDVLSFTAEEKAGVYK	High	High
Parainfluenza virus type 1 (HT83b) (291–309)	DLVFDILDLKGKTKSPRYK	No disease	
αmhc (735–747)	GQFIDSGKGAEKL	Low	Intermediate
HIV-2 (gp160)	INFIGPGKGSDPE	No disease	

[a] Prevalence is diseased animals/immunized animals.

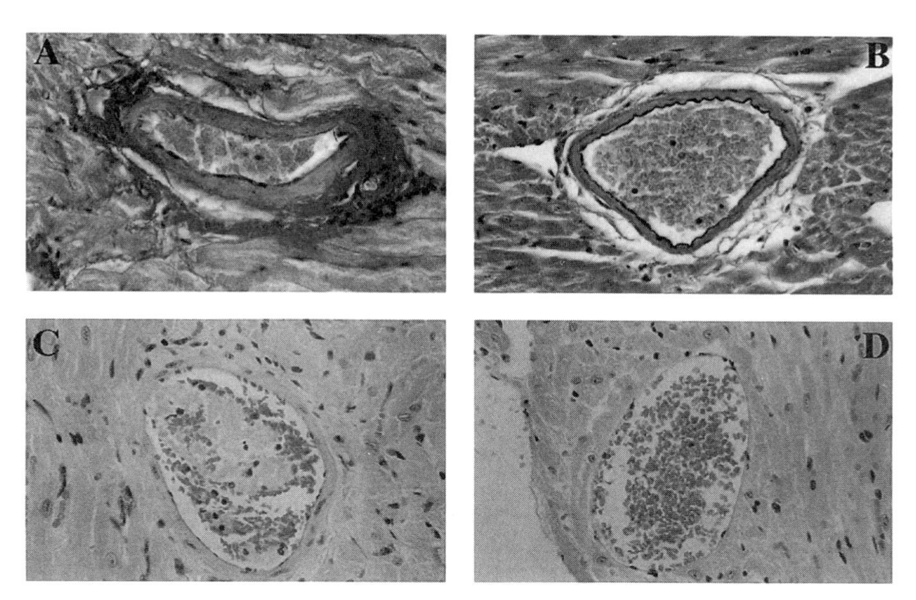

Figure 2. Blood vessels in mice immunized with *C. trachomatis* 60-kDa CRP-derived peptide. (A) Thickening of the arterial wall and perivascular fibrotic changes in mice immunized with ChTR1. Note the perivascular mononuclear inflammatory cells. (B) Normal morphology of the cardiac artery in mice immunized with Freud's complete adjuvant (FCA) alone. (C) Occlusion of cardiac blood vessels in mice immunized with ChTR1. (D) No occlusions in cardiac blood vessel were seen in control mice immunized with FCA alone. (A and B) Elastica staining for collagen (red) for detection of fibrotic changes. (C and D) Hematoxylin-eosin staining was used. Magnifications, ×320. (For color version of figure, see Color Plates, p. 275.)

antibodies to heart-specific epitopes in BALB/c mice (Fig. 4). Because in the mouse model of autoimmune myocarditis, the production of IgG antibodies to heart-specific epitopes is dependent on the activation of autoaggressive T and B cells, these data show that infection with *C. trachomatis* can activate autoaggressive lymphocytes in BALB/c mice.

In mice, the development of peptide-triggered inflammatory heart disease is related to genetic differences among inbred mouse strains (10). Similarly, genetic and environmental risk factors may determine susceptibility to *Chlamydia*-related heart diseases in humans. *Chlamydia* infections are common, and most people can expect to experience a *Chlamydia* infection at least once during their lifetimes. Our data suggest that antigenic mimicry of autoaggressive myosin epitopes by peptides present not only in *C. pneumoniae* but also in *C. trachomatis* and *C. psittaci* may be linked to inflammatory heart disease. Molecular mimicry between bacterial and viral proteins and endogenous molecules has been implicated in various autoimmune diseases, including insulin-dependent diabetes, multiple sclerosis, and autoimmune herpes stromal keratitis. After initiation of the disease, epitope spreading leads to the maintenance and progression of inflammation. Other mechanisms that could also contribute to the pathogenesis of cardiovascular diseases after *Chlamydia* infection include the production of inflammatory cytokines or bystander activation of lymphocytes, or both (1).

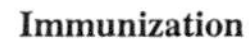

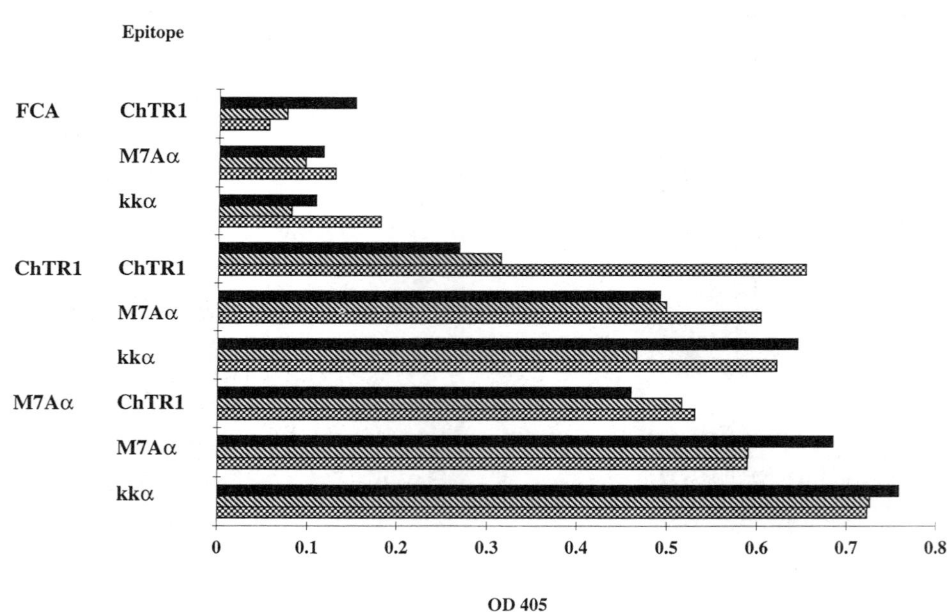

Figure 3. Serum IgG antibodies reactive to cardium-specific epitopes and ChTR1. Specific antibody production was determined by enzyme-linked immunosorbent assay (1). OD 405, optical density at 405 nm.

BLUEPRINT OF AUTOIMMUNE HEART DISEASE

From functional data obtained from studies with transgenic and healthy mice, we propose the following model for the pathogenesis of experimental autoimmune heart disease. Immunization with the autoantigen or mimicking antigens triggers local activation of T cells and systemic activation of the immune system through cytokines such as IL-1 or TNF. Activated autoreactive T cells then circulate through the organism and recognize heart-specific peptide in association with MHC class II molecules expressed on heart interstitial cells. Genetic and functional data indicate that autoimmune myocarditis is mediated by functional $CD4^+$ T cells. CD28 costimulation is not absolutely required for the induction of myocarditis, but CD28 signaling is required for induction of disease with suboptimal doses of the autoantigen. The systemic effects of cytokines may allow presentation of short cardiac myosin heavy-chain (α isoform) peptides in association with MHC class II molecules and the upregulation of adhesion and homing receptors on APCs resident within the target organ. TNF-Rp55 appears to be a crucial receptor that controls target organ susceptibility in autoimmune heart disease via upregulation of MHC class II on heart interstitial cells. After initiation of the inflammatory process by $CD4^+$ T cells, $CD8^+$ cells and macrophages are recruited into the heart muscle and contribute to disease pathogenesis. In addition, activation of B cells and production of autoantibodies may be involved in the progression of heart disease. However, expression of iNOS and nitrotyrosine formation in

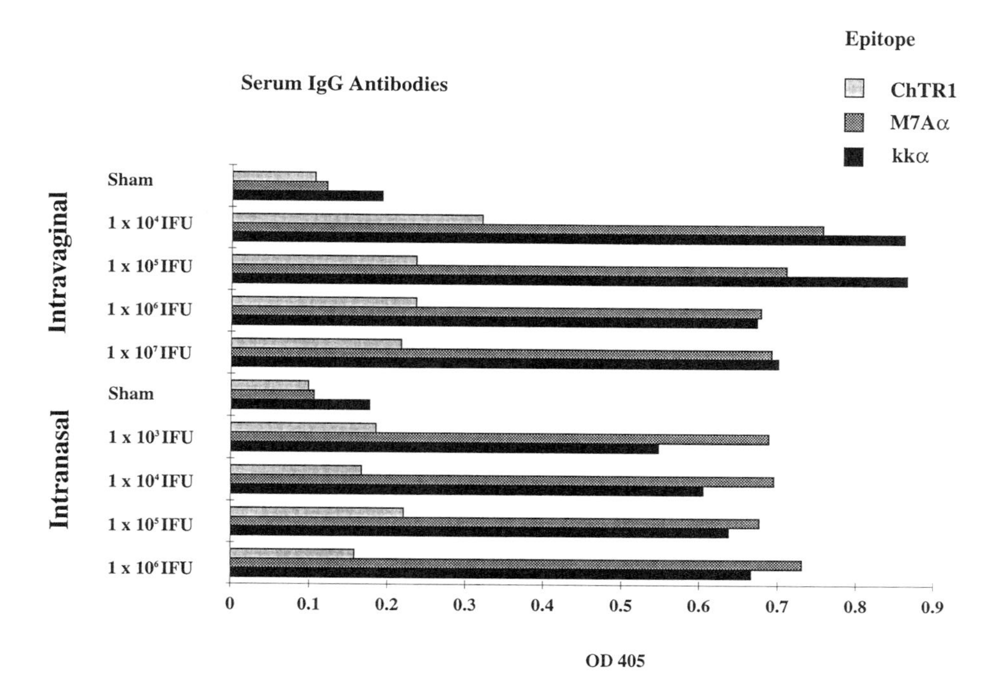

Figure 4. Serum IgG antibody production in *C. trachomatis*-infected mice. Eight-week-old female BALB/c mice were inoculated either intranasally or intravaginally with the indicated doses of *C. trachomatis* MoPn inclusion-forming units (IFU) (1). At 36 days (intranasal infection) or 42 days (intravaginal infection) after the inoculation, serum was collected and specific IgG antibody production was determined by enzyme-linked immunosorbent assay. OD 405, optical density at 405 nm.

infiltrating macrophages and cardiomyocytes might be the crucial factors that ultimately lead to alterations of heart muscle physiology and progression to chronic dilated cardiomyopathy in humans. This molecular level scenario of autoimmunity also implies that induction of disease depends on a fine balance between activation of self-reactive T cells and target organ susceptibility. The challenge for the development of successful prevention and treatment strategies will be to diagnose infection in those few individuals among the many with *Chlamydia* or coxsackievirus infections at risk for development severe heart disease.

REFERENCES

1. **Bachmaier, K., N. Neu, L. M. de la Maza, S. Pal, A. Hessel, and J. M. Penninger.** 1999. Chlamydia infections and heart disease linked through antigenic mimicry. *Science* **283:**1335–1339.
2. **Bachmaier, K., N. Neu, C. Pummerer, J. Ionescu, G. Duncan, T. W. Mak, T. Matsuyama, and J. M. Penninger.** 1997. iNOS expression and nitrotyrosine formation in the myocardium in response to inflammation is controlled by the interferon regulatory transcription factor 1 (IRF-1). *Circulation* **96:**585–591.
3. **Bachmaier, K., C. Pummerer, I. Kozieradzki, K. Pfeffer, T. W. Mak, N. Neu, and J. M. Penninger.** 1997. Low-molecular-weight tumor necrosis factor receptor p55 controls induction of autoimmune heart disease. *Circulation* **95:**655–661.

4. **Bachmaier, K., C. Pummerer, A. Shahinian, J. Ionescu, N. Neu, T. W. Mak, and J. M. Penninger.** 1996. Induction of autoimmunity in the absence of CD28 costimulation. *J. Immunol.* **157:**1752–1757.
5. **Bachmaier, K., N. Neu, R. S. M. Yeung, T. W. Mak, P. Liu, and J. M. Penninger.** 1999. Generation of humanized mice susceptible to peptide-induced inflammatory heart disease. *Circulation* **99:**1885–1891.
6. **Donermeyer, D. L., K. W. Beisel, P. M. Allen, and S. C. Smith.** 1995. Myocarditis-inducing epitope of myosin binds constitutively and stably to I-A(k) on antigen-presenting cells in the heart. *J. Exp. Med.* **182:**1291–1300.
7. **Keating, M. T., and M. C. Sanguinetti.** 1996. Molecular genetic insights into cardiovascular disease. *Science* **272:**681–685.
8. **Mason, J. W.** 1998. Classification of cardiomyopathies, p. 2031–2038. *In* R. W. Alexander, R. C. Schlant, and V. Fuster (ed.), *Hurst's the Heart,* 9th ed. McGraw-Hill, New York, N.Y.
9. **Neu, N., S. W. Craig, N. R. Rose, F. Alvarez, and K. W. Beisel.** 1987. Coxsackievirus induced myocarditis in mice: cardiac myosin autoantibodies do not cross-react with the virus. *Clin. Exp. Immunol.* **69:**566–574.
10. **Neu, N., N. R. Rose, K. W. Beisel, A. Herskowitz, G.-G. Gurri, and S. W. Craig.** 1987. Cardiac myosin induces myocarditis in genetically predisposed mice. *J. Immunol.* **139:**3630–3636.
11. **Penninger, J. M., N. Neu, and K. Bachmaier.** 1996. A genetic map of autoimmune heart disease. *Immunologist* **4:**131–141.
12. **Penninger, J. M., N. Neu, E. Timms, V. A. Wallace, D. R. Koh, K. Kishihara, C. Pummerer, and T. W. Mak.** 1993. The induction of experimental autoimmune myocarditis in mice lacking CD4 or CD8 molecules. *J. Exp. Med.* **178:**1837–1842.
13. **Pummerer, C. L., K. Luze, G. Grassl, K. Bachmaier, F. Offner, S. K. Burrell, D. M. Lenz, T. J. Zamborelli, J. M. Penninger, and N. Neu.** 1996. Identification of cardiac myosin peptides capable of inducing autoimmune myocarditis in BABL/c mice. *J. Clin. Invest.* **97:**2057–2062.
14. **Smith, S. C., and P. M. Allen.** 1992. Expression of myosin-class II major histocompatibility complexes in the normal myocardium occurs before induction of autoimmune myocarditis. *Proc. Natl. Acad. Sci. USA* **89:**9131–9135.

Molecular Mimicry, Microbes, and Autoimmunity
Edited by M. W. Cunningham and R. S. Fujinami

Chapter 7

Peptide Mimicry of Streptococcal Group A Carbohydrate

Neil S. Greenspan, Clemencia Pinilla, and Alexander R. Shikhman

This chapter explores the immunochemical relationships, as defined by antibodies, between the cell wall polysaccharide of *Streptococcus pyogenes* (group A streptococci) and peptides. This section will provide basic information on the structure and immunobiology of the streptococcal cell wall polysaccharide (group A carbohydrate [GAC]). Following comments on the nature of molecular mimicry in general, an overview of prior studies on the cross-reactions between GAC and epitopes on host structures is provided. The next two sections review studies of antibody cross-reactivity between GAC and peptides either that correspond in amino acid sequence to host proteins or that are derived from combinatorial peptide libraries.

S. pyogenes is a gram-positive coccus of 0.5 to 1 μm in diameter that tends to grow in chains. It causes a variety of suppurative human infections (e.g., pharyngitis, scarlet fever, toxic shock-like syndrome, impetigo, erysipelas, wound infections, and burn infections), and it is associated with the nonsuppurative postinfectious sequelae rheumatic fever and post-streptococcal glomerulonephritis. Several of these conditions are potentially life-threatening.

There are numerous streptococcal antigens of microbiological and medical interest. Among the structural antigens of group A streptococci is the capsular polysaccharide, which is composed of hyaluronic acid, a substance also found in human connective tissue. The cell wall contains peptidoglycan (an inducer of inflammation), the group-specific polysaccharide, and several proteins, including the pathogenetically important M protein. Among group A streptococci, all of which produce the same cell wall polysaccharide, there are more than 80 serotypes that vary in M-protein amino acid sequence. Group A streptococci also synthesize soluble products, such as the streptolysins O and S, which are responsible for the characteristic beta-hemolysis associated with group A streptococci grown on blood-agar plates. Other soluble products include erythrogenic toxin (which is associated with scarlet fever), streptokinase, hyaluronidase, nucleases, and C5a peptidase.

The antigen that will be the focus of this chapter is the streptococcal cell wall polysaccharide (GAC). Antibodies specific for the streptococcal cell wall polysaccharides have been used as the basis for classifying streptococcal isolates into the Lancefield groups (A,

Neil S. Greenspan • Institute of Pathology, Biomedical Research Building, Room 927, Case Western Reserve University, 10900 Euclid Avenue, Cleveland, OH 44106-4943. ***Clemencia Pinilla*** • Torrey Pines Institute for Molecular Studies, 3550 General Atomics Court, San Diego, CA 92121. ***Alexander R. Shikhman*** • Division of Rheumatology, Scripps Clinic-MS 113, 10666 North Torrey Pines Road, La Jolla, CA 92037.

B, C, etc.), but antibodies specific for the group A polysaccharide have not been thought to contribute to protection against group A streptococci. Consistent with this notion, Lancefield found that protective immunity to group A streptococci was type specific, not group specific (30). However, a relatively recent report argues, on the basis of epidemiological evidence and new in vitro data, that antibodies specific for GAC may contribute to immunity against group A streptococci (38).

The *S. pyogenes* cell wall polysaccharide is composed of a polyrhamnose backbone, with alternating α-1,2 and α-1,3 linkages (6). Attached to every other rhamnose, via β-1,3 linkages, are *N*-acetyl-β-D-glucosamine (GlcNAc) residues (6). These terminal GlcNAc moieties are the dominant structures recognized by many GAC-specific antibodies (16, 32, 38). This fact suggests the possibility that host carbohydrate structures containing GlcNAc at nonreducing termini might cross-react with some antibodies specific for GAC.

Antibodies specific for GlcNAc occur in sera from healthy adult individuals and, according to one study, can constitute between 1.0 and 13% of total serum immunoglobulin M (IgM) and between 0.5 and 9.5% of total serum IgG (10). If B lymphocytes derived from peripheral blood are stimulated in vitro with a polyclonal activator, such as pokeweed mitogen, between 0.3 and 15% of these cells can secrete antibody with specificity for GlcNAc. Individuals with the IgG heavy-chain allotype Gm(f;n;b) are found to have higher than average levels of the GlcNAc-specific serum antibodies (33). In adults, GlcNAc-specific IgG antibodies are primarily of the IgG2 subclass, while in children, GlcNAc-specific IgG antibodies are found in both the IgG1 and the IgG2 subclasses (40). The clonal diversity of these antibodies tends to increase with age (40).

Elevated levels of antibodies reactive with GlcNAc are seen in patients with acute streptococcal infection (36) and in the contexts of erysipelas (43), rheumatic fever and rheumatic heart disease (9), and post-streptococcal glomerulonephritis (2). The persistence of GAC-specific antibodies has been noted to be particularly extended in patients with rheumatic valvular disease (3). Increased titers of GlcNAc-specific antibodies have also been observed in patients with psoriasis and psoriatic arthritis (35).

PRINCIPLES OF MOLECULAR MIMICRY

There are multiple reasonable interpretations of the term "molecular mimicry" (15). The first and perhaps the most obvious form of mimicry of one molecule (the model) by another (the mimic) is that based on structural resemblance. Two molecules are distinguishable and will have different names as long as they are nonidentical in atomic composition or connectivity, even if the difference reduces to a single atom or covalent bond. Any two such related molecules could share extensive elements of structure. However, even in the case of two molecules that are relatively unrelated with respect to atomic composition, there exists a formal possibility that they will share regions of structure that are similar in their three-dimensional distributions of electron density. Such similarity could arise from the identical atoms residing at the same relative positions or even from nonidentical atoms occupying more or less equivalent relative positions in space over some patch of molecular surface. In this case, the mimicry would likely be limited to one region or face on each of the molecules being compared.

A second conceivable form of mimicry can be termed immunochemical. This form of mimicry is concerned with noncovalent binding. Although there is a tendency to presume that an impressive level of immunochemical mimicry implies a comparable degree of structural mimicry, this presumption should be resisted because although such a correlation does exist, it is not a reliable guide to the extent of structural similarity exhibited by pairs of molecules. Noncovalent interactions between biological macromolecules or between macromolecules and small ligands involve multiple factors, several of which do not derive solely from the structural details of the intermolecular interface.

The ultimate measure of how strongly two biomolecules interact with one another is the free energy change of complex formation. This free energy change reflects both the forces operative at the interface and the effects of binding that are distributed throughout the chemical system that includes the interacting species. An important principle to consider in attempting to explain the contributions of structural subunits to affinity and specificity of binding is that the free energy change for the binding of molecules *A* and *B* reflects the net change in free energy between the unbound reactants and the complex. Therefore, a mutation that destabilizes *A* or *B* can contribute to increasing the affinity characterizing the *A*-*B* complex, just as a mutation that directly stabilizes the *A*-*B* complex can increase the affinity of *A* for *B*.

Studies of an insulin mutant (20) exemplify the cautionary statements about the relationship between structural and immunochemical mimicry. The investigators describe prior work on a mutant form of insulin whose sequence differs from the wild-type amino acid sequence by a single amino acid substitution. This mutant form of insulin exhibits a crystallographic structure that is, with the exception of the single altered side chain, virtually identical to that of the wild-type form. Nevertheless, this molecule does not exhibit the insulin function. In contrast, Hua et al. (20) describe a point mutant of insulin that retains the insulin function, despite having a structure significantly different from that of the wild-type molecule by nuclear magnetic resonance spectroscopy. Thus, insulin and these insulin mutants provide a compelling example of the potential for divergence between structural similarity and functional similarity.

Another factor that may contribute to degeneracy in the relationship between structural and immunochemical mimicry is flexibility in the paratope. Consider the crystallographic study of a monoclonal antibody Fab fragment with specificity for a peptide derived from the influenza virus hemagglutinin. The method of X-ray diffraction was used to determine structures of both the unbound and the peptide-bound Fab fragments, revealing that there was a large conformational change in the backbone of the third hypervariable region in the heavy chain associated with binding of the peptide antigen (37). An important implication of this study and a number of other reports (for a review, see reference 51) is that an antibody paratope potentially possesses not a single structure but a family of structures. Therefore, on this basis alone, one could imagine ligands of modest structural resemblance all having the property of interacting noncovalently with a single pair of antibody variable domains.

The immunochemical form of mimicry can also be extended to include recognition by the antigen-specific receptors of T lymphocytes as well as those of B lymphocytes. In the case of protein antigens being recognized by conventional αβ T-cell receptors, what is being recognized is generally not the native antigen but peptides derived from the antigen in noncovalent complex with a self class I or class II major histocompatibility complex molecule. Consequently, the degree of similarity between model and mimic can be quite different for antibodies (or B cells) versus T cells.

A third form of mimicry, and the form that may be of the greatest interest to biomedical investigators, is functional mimicry. In this form of mimicry, a model and a mimic both elicit a biological response following noncovalent interaction with, most typically, a macromolecular receptor. As in the case of immunochemical mimicry, this form of mimicry will be evaluated in light of one or more appropriate operational measures. It is wise not to presume that a given degree of functional similarity necessarily reflects any given magnitude of structural or immunochemical mimicry.

For example, one can envision molecules that elicit similar functional responses from different receptors on the same or different cells or even different responses from the same receptor under different cellular or environmental circumstances. For example, binding to T-cell receptors can have distinct consequences for immature T-lineage cells in the thymus in comparison to those for mature T lymphocytes in the periphery. Another interesting example is provided by a study (47) of antibodies to a cell surface integrin that mediates uptake of bound particles, including the bacterial pathogen *Yersinia pseudotuberculosis.* The authors of this report found that the protein invasin, expressed by *Y. pseudotuberculosis,* as well as different monoclonal antibodies specific for the host cell integrin, exhibited the ability to be endocytosed, even though they bound to the receptor at distinct sites and, therefore, presumably lacked a high degree of structural similarity. Presumably, in this case, the ability to bind to and cross-link the cell surface integrin, regardless of the location of the contacted site on the integrin, is the basis for the similar functional activities of the molecules active in internalization. Such degeneracy in the relationships among different forms of molecular similarity is also observed in cases in which the end point is simply binding and not cellular function.

With respect to structure, binding, and the elicitation of higher-order biological function, one can ask not only if mimicry of a model molecule by a mimic is present or absent, but one can inquire as to the extent of any such mimicry. Any effort to quantitate mimicry is likely to reveal that there are multiple ways of precisely defining the extent of mimicry. Consider functional mimicry. Binding to a cell surface receptor, such as an antigen-specific B- or T-lymphocyte receptor, can have multiple consequences, such as proliferation or cytokine secretion. Each end point that can be envisioned can be used as the basis for the quantitation of the similarity in functional effects associated with either of the two molecules. There is reason to believe that these end points will not always correspond exactly. For example, Evavold and Allen (11) demonstrated that changing a single amino acid in a peptide presented to a $CD4^{+}$ T cell by a class II major histocompatibility complex molecule can result in activation of the T cell with respect to one function (interleukin 4 [IL-4] secretion) but not another (proliferation). In other words, the modified peptide is a better mimic of the original peptide for the elicitation of IL-4 secretion than for the elicitation of a proliferative response.

MOLECULAR MIMICRY AND GROUP A STREPTOCOCCUS

In the annals of molecular mimicry, group A streptococcus holds a special place of honor (for a review, see reference 50). Twenty-five years ago rheumatic fever was demonstrated to follow pharyngitis with *S. pyogenes* (5). Attempts to explain the clinical manifestations of rheumatic fever on the basis of direct bacterial infection of affected organs or on the basis of toxic streptococcal products proved unfruitful (4). Then several groups described the first of numerous immunological cross-reactions between streptococcal and host antigens, leading to the conceptualization of rheumatic fever as an autoimmune disease, perhaps the first to have a well-defined microbe-associated etiology. These examples of molecular mimicry involved several different streptococcal components and provided tantalizing potential clues to the pathogenesis of rheumatic fever. However, the serum of individuals with streptococcal infections that do not lead to rheumatic fever can have substantial concentrations of antistreptococcal antibodies that cross-react with host antigens, and there is no direct proof that passive transfer of antistreptococcal antibodies can cause rheumatic fever-like condition in animals. Therefore, it is not possible at present to conclude that cross-reactive antibodies actually cause the clinical manifestations of this condition (4).

In the 1960s, Kaplan (21) and Kaplan and Meyeserian (23) presented evidence for cross-reactivity between streptococcal membrane protein antigens and antigens in human heart tissue. Others followed with reports that documented the ability of antistreptococcal antibodies to bind to epitopes in human heart tissue or the ability of antibodies against heart antigens to react with streptococcal antigens (7, 28, 49, 52). Investigators also showed that rheumatic fever patients produced antibodies that reacted with epitopes on heart antigens (22, 24, 25, 53). It has also been shown that the M protein, one of the critical virulence factors of the group A streptococcus, is cross-reactive with host molecules found in the heart (8).

In 1967, Goldstein and colleagues (13) demonstrated that antibodies specific for GAC bound to glycoproteins isolated from human and bovine heart valves. Conversely, antibodies elicited by immunization with glycoproteins extracted from cardiac valves cross-reacted with the group A streptococcal cell wall polysaccharide but not with the polysaccharides that define other streptococcal groups (13). In a later study by this group, it was found that immunization of rabbits with GAC could, after 5 or 6 months, result in rheumatic fever-like carditis (14). Damaged areas of cardiac tissue were found to contain deposits of IgG and the C3 component of complement.

Additional cross-reactivities between GAC and host epitopes were subsequently identified. Lyampert and colleagues (31) found by indirect immunofluorescence that antibodies specific for GAC can bind to a cross-reactive antigen expressed by thymic and skin epithelial cells from humans and other animals. Rook and colleagues (41) produced a monoclonal antibody following immunization with the peptidoglycan-polysaccharide complex of group A streptococci and showed that this antibody can react with agalactosyl IgG, which lacks the galactose residues typically located at the termini of the conserved N-linked oligosaccharides attached to the C_H2 domain. With this monoclonal antibody, they also demonstrated staining of normal epithelium (from tonsil, skin, small intestine, and salivary gland) and of macrophages and extracellular material from patients with rheumatoid arthritis (42). Monoclonal antibodies specific for GAC were also demonstrated by indirect immunofluorescence to react with intracellular epitopes in rat fibroblasts (48). These epitopes were shown to be O-linked GlcNAc residues, which were found throughout the cytoplasm and nucleus, with the most intense staining occurring at the nuclear membrane.

IMMUNOLOGICAL CROSS-REACTIVITY BETWEEN GAC AND CYTOKERATIN PEPTIDES

While a number of cross-reactions between GAC and host molecules have been attributed to terminal GlcNAc residues expressed by host glycans, another possibility is that anticarbohydrate antibodies can bind directly to proteins or peptides. A number of murine GAC-specific monoclonal antibodies were found to bind to a variety of cytoplasmic proteins, including, actin, keratin, vimentin, and myosin, each with a unique reactivity profile (44). The ability of GlcNAc conjugated to bovine serum albumin (BSA) to completely inhibit this binding suggested that this binding was specific. Evidence that the antibodies were actually recognizing protein epitopes was provided by the failure of pretreatment with either sodium periodate or *N*-acetylglucosaminidase to inhibit binding of the antibodies to the proteins. Interestingly, wheat germ agglutinin, a lectin with specificity for terminal GlcNAc residues, exhibited a reactivity profile different from those exhibited by the monoclonal antibodies.

Monoclonal antibodies secreted by eight human hybridomas, generated from mice immunized with group A streptococcal peptidoglycan-polysaccharide complex and pokeweed mitogen, also reacted with keratins from human skin by both enzyme-linked immunosorbent assay and Western blotting (45). This reactivity was also exhibited by affinity-purified GlcNAc-specific polyclonal serum antibodies.

The majority of the anti-GlcNAc human monoclonal antibodies reacted in the Western immunoblot with three of four predominant skin cytokeratin types, including types 1, 5, and 10. Two of the monoclonal antibodies recognized cytokeratin type 14 as well as types 1, 5, and 10, and binding to cytokeratin 14 correlated with binding to human epidermis by indirect immunofluorescence. The reactivities of these monoclonal antibodies with synthetic peptides corresponding in amino acid sequence to cytokeratin 14 confirmed that these antibodies were specific for the protein itself. One decapeptide from the cytokeratin 14 amino acid sequence, SFGSGFGGGY, was found to react with GlcNAc-specific lectins as well as with GlcNAc-specific monoclonal antibodies.

When BALB/c mice were immunized with GlcNAc-BSA they produced antibodies reactive with some of the synthetic cytokeratin 14 peptides. Such antibodies were not produced by BALB/c mice immunized with unconjugated BSA. Reciprocally, one of the cytokeratin 14 peptides was able to elicit antibodies reactive with GAC when used as an immunogen (46).

Molecular analysis of human monoclonal antibodies cross-reactive with GlcNAc and peptides revealed that they used a variety of variable-region genes and exhibited little evidence of antigen-driven somatic mutations (1). It remains unknown if GlcNAc-specific antibodies are pathogenic and, if so, what structural features predispose the antibodies to pathogenicity.

IMMUNOLOGICAL CROSS-REACTIVITY BETWEEN GAC AND PEPTIDES DERIVED FROM COMBINATORIAL LIBRARIES

Since it cannot be assumed that antigens cross-reactive with a given monoclonal antibody must share some structural feature that is recognized by that monoclonal antibody, it would be useful to have a method for determining antibody reactivities in an unbiased manner. Peptide combinatorial libraries composed of millions of peptides with different amino acid sequences prepared chemically (12, 19, 29) or by phage display (39) offer just such a strategy for the identification of molecules capable of immunochemical mimicry of a model molecule. Phage-display libraries have been used to identify L-amino acid peptides that mimic carbohydrate antigens in interacting with antibodies (e.g., see reference 18).

Harris et al. (17) used the phage-display approach to investigate peptide mimicry of GAC with respect to five GAC-specific monoclonal antibodies, including the murine monoclonal antibody HGAC 39.G3 (17). This antibody (IgG3, kappa chain) was generated with the group A vaccine (GAV; heat-killed, pepsin-treated *S. pyogenes*) as the immunogen and has specificity for the GlcNAc residues of group A carbohydrate. However, it does not bind detectably to rhamnose monosaccharide (16).

For four of the five monoclonal antibodies used to isolate peptide-bearing phage particles, there were unique peptide amino acid sequence consensus groups. No phage clone, each of which displays a unique peptide insert, bound well to more than two of the antibodies. The phage clones isolated by the same four monoclonal antibodies all bound best to the antibody used in its isolation. Similar results were obtained with polyclonal antibody preparations elicited by immunization with GAC-related synthetic oligosaccharides conjugated to BSA. Thus, even though all of the GAC-specific monoclonal and polyclonal antibodies bind to the same carbohydrate antigen, they have clearly different profiles of reactivity with the phage-displayed peptides. For HGAC 39.G3 and one other monoclonal antibody, a relatively high concentration (1.1 mM) of a synthetic oligosaccharide corresponding to the structure of GAC was able to compete directly with the cognate phage clones. This result suggests that, at least for these two monoclonal antibodies, the carbohydrate and peptide antigens bind in overlapping sites.

The reactivity of HGAC 39.G3 has also been examined with synthetic positional scanning peptide libraries composed of all-L or all-D hexapeptides (34). Peptides from these libraries were screened for inhibition of binding by monoclonal antibody HGAC 39.G3 to a synthetic antigen that displayed multiple GlcNAc residues per molecule. Inhibitory activity by mixtures from the all-D hexapeptide library was greater than the activity from the all-L hexapeptide libraries. The D-amino acids defined in the most active mixtures in each of the six positions of the library were selected to prepare 27 different individual hexapeptides. The sequence Ac-yryygl-NH_2 (Fig. 1) was bound most strongly by monoclonal antibody HGAC 39.G3; a peptide concentration of 0.3 μM was required to inhibit binding of HGAC 39.G3 to solid-phase GlcNAc-BSA by 50%. In contrast, a concentration of 430 μM GlcNAc was required to achieve 50% inhibition of the binding of HGAC 39.G3 to solid-phase GlcNAc-BSA.

The contributions of the residues of the all-D peptide (Ac-yryygl-NH_2) to the affinity of binding to monoclonal antibody HGAC 39.G3 were examined with a series of truncations, L- and D-amino acid substitutions, and retro analogs. A set of D-alanine substitution analogs for each residue of Ac-yryygl-NH_2 found that the contributions to affinity, as assessed by the effect of alanine substitution, were greatest for the first three positions. Alanine substitutions for the second-position arginine and the third-position tyrosine resulted in peptides with binding activities that were decreased by more than 4,000-fold. The fourth and fifth positions were less contributory, with alanine substitutions at these positions decreasing binding activity by fivefold or less. Also, the stereospecificity of each residue of the all-D peptide was examined with L-amino acid substitution analogs. Overall, the activities of these analogs were similar to those of the D-alanine substitution analogs. The relative affinities for additional analogs are shown in Table 1. Interestingly, the all-L enantiomer (Ac-YRYYGL-NH_2) did not inhibit monoclonal antibody HGAC 39.G3 at the highest concentration tested, suggesting that a specific chiral arrangement of the amino acid side chains of Ac-yryygl-NH_2 is required for high-affinity recognition. Other retro-inverso analogs were also prepared, but they also were not recognized by monoclonal antibody HGAC 39.G3.

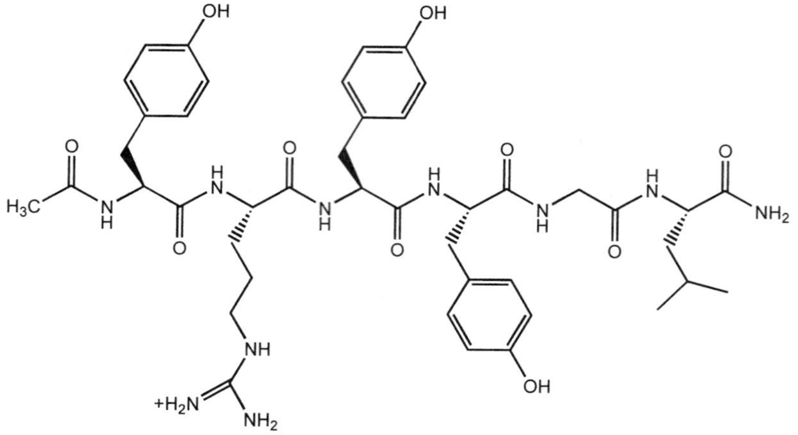

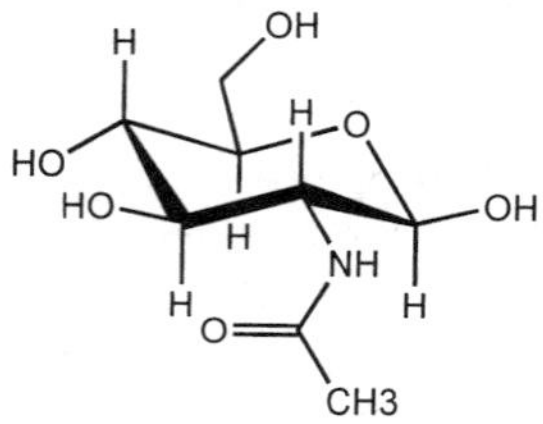

Figure 1. Molecular structures of Ac-yryygl-NH_2 (molecular weight, 876; top) and GlcNAc (molecular weight, 221; bottom).

The identification of all-D peptides that are recognized by monoclonal HGAC 39.G3 with relatively higher affinities than the carbohydrate analogs clearly illustrates an advantage of using peptide libraries to identify antigenic mimics: no prior information regarding the specificity of the antibody is needed. This study also supports the concept that at least some monoclonal antibodies can be defined as functionally polyspecific, given their ability to recognize multiple antigens with different chemical characteristics. This conclusion was most impressively verified for a monoclonal antibody specific for a peptide from human immunodeficiency virus type 1 p24, which was demonstrated to react with several unrelated peptide sequences and with native proteins containing some of those amino acid sequences (26, 27).

SUMMARY AND CONCLUSIONS

In rheumatic fever, the pathogenic role of antibodies cross-reactive with GAC (or other streptococcal antigens) and host tissue antigens remains incompletely understood. What is beyond dispute is that there are many cross-reactivities between streptococcal antigens and host antigens. In some cases these cross-reactions involve molecules of distinct biochemical classes, such as a carbohydrate (e.g., GAC) and a peptide or protein (e.g., a

Table 1. Relative affinities for analogs of Ac-yryygl-NH_2 recognized by monoclonal antibody HGAC 39.G3

Analog[a]	IC_{50} (μM)[b]
Ac-yryygl-NH_2	0.3
Truncation	
Ac-yry-NH_2	195
Ac-yryy-NH_2	9.4
Ac-yryyg-NH_2	12
f and y substitution	
Ac-fyryygl-NH_2	0.49
Ac-yrfygl-NH_2	40
Ac-yryfgl-NH_2	0.16
w and y substitution	
Ac-wryygl-NH_2	1.1
Ac-yrwygl-NH_2	>1,250
Ac-yrywgl-NH_2	11
Ac-YRYYGL-NH_2	>1,250

[a] Lowercase letters represent D-amino acids, and uppercase letters represent L-amino acids in the single-letter code.
[b] IC_{50}, concentration of peptide that inhibited 50% of the binding of monoclonal antibody HGAC 39.G3 to solid-phase GlcNAc-BSA. Bound antibody was detected with horseradish peroxidase-labeled anti-mouse kappa-chain antibody.

cytokeratin or a peptide from a combinatorial peptide library). Such mimicry across molecular categories suggests that what we have termed "immunochemical mimicry" (binding comparably to a receptor molecule, such as an antibody) can occur without any particular degree of structural mimicry (especially in the unbound state). The available data on the mimics of GAC do not yet permit us to draw a certain conclusion regarding this question, as there is only very limited information on the three-dimensional structures of the model (GAC) and the mimics (peptides and proteins). Nevertheless, the finding of many different peptides that mimic GAC with respect to only one or a few of the available GAC-specific monoclonal antibodies is consistent with the hypothesis that mimicry of GAC by peptides is at least partly a consequence of the multiple specificities exhibited by these (perhaps any) antibodies. It will be the task of future investigations to determine the range of molecular and physical mechanisms used by peptides and proteins in mimicking GAC in interactions with antibodies. Deeper understanding of the role of such molecular mimicry in rheumatic fever, or in streptococcal pathogenesis more generally, must also await the results of future experiments.

REFERENCES

1. **Adderson, E. E., A. R. Shikhman, K. E. Ward, and M. W. Cunningham.** 1998. Molecular analysis of polyreactive monoclonal antibodies from rheumatic carditis: human anti-*N*-acetylglucosamine/anti-myosin antibody V region genes. *J. Immunol.* **161:**2020–2031.
2. **Appleton, R. S., B. E. Victorica, D. Tamer, and E. M. Ayoub.** 1985. Specificity of persistence of antibody to the streptococcal group A carbohydrate in rheumatic valvular heart disease. *J. Lab. Clin. Med.* **105:**114–119.

3. **Araj, G. F., H. A. Majeed, A. M. Yousof, and E. M. Ayoub.** 1988. Immunoglobulin isotype response to the group-A streptococcal carbohydrate in humans. *Acta Pathol. Microbiol. Immunol. Scand. Suppl.* **3:**8–12.
4. **Ayoub, E. M.** 1996. Rheumatic fever, p. 1134–1144. *In* R. R. Rich, T. A. Fleisher, B. D. Schwartz, W. T. Shearer, and W. Strober (ed.), *Clinical Immunology: Principles and Practice.* Mosby-Year Book, Inc., St. Louis, Mo.
5. **Catanzaro, F. J., C. A. Stetson, A. J. Morris, R. Chamovitz, C. H. Rammelkamp, Jr., B. L. Stolzer, and W. D. Perry.** 1954. The role of the streptococcus in the pathogenesis of rheumatic fever. *Am. J. Med.* **17:**749–756.
6. **Coligan, J. E., T. J. Kindt, and R. M. Krause.** 1978. Structure of the streptococcal groups A, A-variant and C carbohydrates. *Immunochemistry* **15:**755–760.
7. **Cunningham, M. W., and S. M. Russell.** 1983. Study of heart-reactive antibody in anti-sera and hybridoma culture fluids against group A streptococci. *Infect. Immun.* **42:**531–538.
8. **Dale, J. B., and E. H. Beachey.** 1982. Protective antigenic determinant of streptococcal M protein shared with sarcolemmal membrane protein of human heart. *J. Exp. Med.* **156:**1165–1176.
9. **Dudding, B. A., and E. M. Ayoub.** 1968. Persistence of streptococcal group A antibody in patients with rheumatic valvular disease. *J. Exp. Med.* **128:**1081–1098.
10. **Emmrich, F., B. Schilling, and K. Eichmann.** 1985. Human immune response to group A streptococcal carbohydrate (A-CHO). I. Quantitative and qualitative analysis of the A-CHO-specific B cell population responding in vitro to polyclonal and specific activation. *J. Exp. Med.* **161:**547–562.
11. **Evavold, B. D., and P. M. Allen.** 1991. Separation of IL-4 production from Th cell proliferation by an altered T cell receptor ligand. *Science* **252:**1308–1310.
12. **Geysen, H. M., S. J. Rodda, and T. J. Mason.** 1986. A priori delineation of a peptide which mimics a discontinuous antigenic determinant. *Mol. Immunol.* **23:**709–715.
13. **Goldstein, I., B. Halpern, and L. Robert.** 1967. Immunological relationship between streptococcus A polysaccharide and the structural glycoproteins of heart valve. *Nature* **213:**44–47.
14. **Goldstein, I., L. Scebat, J. Renais, P. Hadjinsky, and J. Dutartre.** 1983. Rheumatic-like carditis induced in rabbits by cross-reacting antigens: streptococcus A polysaccharide and rabbit aortic glycoproteins. *Isr. J. Med. Sci.* **19:**483–490.
15. **Greenspan, N. S.** 1992. Antigen mimicry with anti-idiotypic antibodies, p. 55–79. *In.* M. H. V. Van Regenmortel (ed.), *Structure of Antigens,* vol. I. CRC Press, Inc., Boca Raton, Fla.
16. **Greenspan, N. S., W. J. Monafo, and J. M. Davie.** 1987. Interaction of IgG3 anti-streptococcal group A carbohydrate (GAC) antibody with streptococcal group A vaccine: enhancing and inhibiting effects of anti-GAC, anti-isotypic, and anti-idiotypic antibodies. *J. Immunol.* **138:**285–292.
17. **Harris, S. L., L. Craig, J. S. Mehroke, M. Rashed, M. B. Zwick, K. Kenar, E. J. Toone, N. Greenspan N., F. I. Auzanneau, J. R. Marino-Albernas, B. M. Pinto, and J. K. Scott.** 1997. Exploring the basis of peptide-carbohydrate crossreactivity: evidence for discrimination by peptides between closely related anti-carbohydrate antibodies. *Proc. Natl. Acad. Sci. USA* **94:**2454–2459.
18. **Hoess, R., U. Brinkmann, T. Handel, and I. Pastan.** 1993. Identification of a peptide which binds to the carbohydrate-specific monoclonal antibody B3. *Gene* **128:**43–49.
19. **Houghten, R. A., C. Pinilla, S. E. Blondelle, J. R. Appel, C. T. Dooley, and J. H. Cuervo.** 1991. Generation and use of synthetic peptide combinatorial libraries for basic research and drug discovery. *Nature* **354:**84–86.
20. **Hua, Q. X., S. E. Shoelson, M. Kochoyan, and M. A. Weiss.** 1991. Receptor binding redefined by a structural switch in a mutant human insulin. *Nature* **354:**238–241.
21. **Kaplan, M. H.** 1963. Immunologic relation of streptococcal and tissue antigens. I. Properties of an antigen in certain strains of group A streptococci exhibiting an immunologic cross-reaction with human heart tissue. *J. Immunol.* **90:**595–606.
22. **Kaplan, M. H.** 1964. Immunologic relations of streptococcal and tissue antigens. III. Presence in human sera of streptococcal antibody cross-reactive with heart tissue. Association with streptococcal infection, rheumatic fever, and glomerulonephritis. *J. Exp. Med.* **119:**651–666.
23. **Kaplan, M. H., and H. Meyersian.** 1962. An immunologic cross-reaction between group A streptococcal cells and human heart. *Lancet* **ii:**706–710.
24. **Kaplan, M. H., and K. H. Svec.** 1964. Immunologic relation of streptococcal and tissue antigens. III. Presence in human sera of streptococcal antibody cross-reactive with heart tissue. Association with stretococcal infection, rheumatic fever, and glomerulonephritis. *J. Exp. Med.* **119:**651–666.

25. **Kaplan, M. H., R. Bolande, L. Rakita, and J. Blair.** 1964. Presence of bound immunoglobulins and complement in the myocardium in acute rheumatic fever. *N. Engl. J. Med.* **271:**637–645.
26. **Keitel, T., A. Kramer, H. Wessner, C. Scholz, J. Schneider-Mergener, and W. Hohne.** 1997. Crystallographic analysis of anti-p24 (HIV-1) monoclonal antibody cross-reactivity and polyspecificity. *Cell* **91:**811–820.
27. **Kramer, A., T. Keitel, K. Winkler, W. Stocklein, W. Hohne, and J. Schneider-Mergener.** 1997. Molecular basis for the binding promiscuity of an anti-p24 (HIV-1) monoclonal antibody. *Cell* **91:**799–809.
28. **Krisher, K., and M. W. Cunningham.** 1985. Myosin: a link between streptococci and heart. *Science* **227:**413–415.
29. **Lam K. S., S. E. Salmon, E. M. Hersh, V. J. Hruby, W. M. Kazmierski, and R. J. Knapp.** 1991. A new type of synthetic peptide library for identifying ligand-binding activity. *Nature* **354:**82–84.
30. **Lancefield, R. C.** 1962. Current knowledge of type-specific M antigens of group A streptococci. *J. Immunol.* **89:**307–313.
31. **Lyampert, L. M., L. V. Beletskaya, N. A. Borodiyuk, E. V. Gnezditskaya, L. L. Rassokhina, and T. A. Danilov.** 1976. A cross-reactive antigen of thymus and skin epithelial cells common with the polysaccharide of group A streptococci. *Immunology* **31:**47–55.
32. **McCarty, M.** 1956. Variation in the group-specific carbohydrate of group A streptococci. II. Studies on the chemical basis for serological specificity of carbohydrates. *J. Exp. Med.* **104:**629–643.
33. **Morell, A., G. Vassalli, G. G. DeLange, F. Skvaril, D. M. Ambrosino, and G. R. Siber.** 1989. Ig allotype-linked regulation of class and subclass composition of natural antibodies to group A streptococcal carbohydrate. *J. Immunol.* **142:**2495–2500.
34. **Pinilla, C., J. R. Appel, G. D. Campbell, J. Buencamino, N. Benkirane, S. Muller, and N. S. Greenspan.** 1998. All-D peptides recognized by an anti-carbohydrate antibody identified from a positional scanning library. *J. Mol. Biol.* **283:**1013–1025.
35. **Rantakokko, K., M. Rimpilainen, J. Uksila, C. Jansen, R. Luukkainen, and P. Toivanen.** 1997. Antibodies to streptococcal cell wall in psoriatic arthritis and cutaneous psoriasis. *Clin. Exp. Rheumatol.* **15:**399–404.
36. **Razon Veronesi, S.** 1983. Antipolysaccharide group-specific antibodies of *Streptococcus pyogenes* in children. *Boll. Ist. Sieroter Milan* **62:**433–444.
37. **Rini, J. M., U. Schulze-Gahmen, and I. A. Wilson.** 1992. Structural evidence for induced fit as a mechanism for antibody-antigen recognition. *Science* **255:**959–965.
38. **Salvadori, L. G., M. S. Blake, M. McCarthy, J. Y. Tai, and J. B. Zabriskie.** 1995. Group A streptococcus-liposome ELISA antibody titers to group A polysaccharide and opsonophagocytic capabilities of the antibodies. *J. Infect. Dis.* **171:**593–600.
39. **Scott, J. K., and G. P. Smith.** 1990. Searching for peptide ligands with an epitope library. *Science* **249:**386–390.
40. **Shackelford, P. G., S. J. Nelson, A. T. Palma, and M. H. Nahm.** 1988. Human antibodies to group A streptococcal carbohydrate. Ontogeny, subclass restriction, and clonal diversity. *J. Immunol.* **140:**3200–3205.
41. **Sharif, M., L. S. Wilkinson, J. Edwards, and G. A. Rook.** 1989. Membrane *N*-acetylglucosamine: expression by cells in rheumatoid synovial fluid, and by pre-cultured monocytes. *Br. J. Exp. Pathol.* **70:**567–577.
42. **Sharif, M., G. Rook, L. S. Wilkinson, J. G. Worrall, and J. C. W. Edwards.** 1990. Terminal *N*-acetylglucosamine in chronic synovitis. *Br. J. Rheumatol.* **29:**25–31.
43. **Shikhman, A. R., E. P. Savl'ev, L. V. Nikolaeva, V. L. Cherkasov, and G. I. Anokhina.** 1986. Determination of antibodies against ribosomes and cell wall components, detection of circulating antigens of group A streptococcus in patients with erysipelas. *Zh. Microbiol. Epidemiol. Immunobiol.* **December:**91–95.
44. **Shikhman, A. R., N. S. Greenspan, and M. W. Cunningham.** 1993. A subset of mouse monoclonal antibodies crossreactive with cytoskeletal proteins and group A streptococcal M proteins recognizes *N*-acetyl-β-D-glucosamine. *J. Immunol.* **151:**3902–3913.
45. **Shikhman, A. R., and M. W. Cunningham.** 1994. Immunological mimicry between *N*-acetyl-beta-D-glucosamine and cytokeratin peptides. Evidence for a microbially driven antikeratin antibody response. *J. Immunol.* **152:**4375–4387.
46. **Shikhman, A. R., N. S. Greenspan, and M. W. Cunningham.** 1994. Cytokeratin peptide SFGSFGGGY mimics *N*-acetyl-β-D-glucosamine in reaction with antibodies and lectins, and induces in vivo anti-carbohydrate antibody response. *J. Immunol.* **153:**5593–5606.
47. **Tran Van Nhieu, G., and R. R. Isberg.** 1993. Bacterial internalization mediated by β_1 chain integrins is determined by ligand affinity and receptor density. *EMBO J.* **12:**1887–1895.

48. **Turner, J. R., A. M. Tartakoff, and N. S. Greenspan.** 1990. Cytologic assessment of nuclear and cytoplasmic O-linked *N*-acetyl-glucosamine distribution by using anti-streptococcal monoclonal antibodies. *Proc. Natl. Acad. Sci. USA* **87:**5608–5612.
49. **van de Rijn, I., J. B. Zabriskie, and M. McCarty.** 1977. Group A streptococcal antigens cross-reactive with myocardium. Purification of heart-reactive antibody and isolation and characterization of the streptococcal antigen. *J. Exp. Med.* **146:**579–599.
50. **Williams, R. C., Jr.** 1985. Molecular mimicry and rheumatic fever. *Clin. Rheum. Dis.* **11:**573–590.
51. **Wilson, I. A., and R. L. Stanfield.** 1994. Antibody-antigen interactions: new structures and new conformational changes. *Curr. Opin. Struct. Biol.* **4:**857–867.
52. **Zabriskie, J. B., and E. H. Freimer.** 1966. An immunological relationship between the group A streptococcus and mammalian muscle. *J. Exp. Med.* **124:**661–678.
53. **Zabriskie, J. B., K. C. Hsu, and B. C. Seegal.** 1970. Heart-reactive antibody associated with rheumatic fever: characterization and diagnostic significance. *Clin. Exp. Immunol.* **7:**147–159.

Molecular Mimicry, Microbes, and Autoimmunity
Edited by M. W. Cunningham and R. S. Fujinami

Chapter 8

Role of Superantigens in Molecular Mimicry and Autoimmunity

Malak Kotb

Many autoimmune diseases are believed to have been triggered or exacerbated by infection with pathogenic organisms (for reviews, see references 1, 2, 21, 33, 51, and 57). Pathogenic microorganisms produce virulence factors that, in the susceptible host, can lead to the breach of immunological tolerance and the development of pathogenic autoimmunity. The mechanism by which this occurs may vary considerably for different pathogens, and it is likely that more than one virulence factor, perhaps even more than one pathogen, act in synergy to evoke self-reactivity. Many of these mechanisms are discussed in this book, and therefore, this chapter focuses on the role of superantigens in triggering autoimmunity. Given the unique and diversified mechanisms by which superantigens stimulate the immune system and elicit cytokine responses, there is little doubt that these molecules contribute in some way to autoimmunity. Superantigens can trigger self-reactivity either by directly activating autoreactive T and B cells, by causing tissue damage and aberrant presentation of self-antigens, or by acting as adjuvants to enhance immune responses to microbial antigens that mimic the host. This chapter summarizes the various interactive mechanisms by which superantigens may contribute to the development or exacerbation of pathogenic autoimmunity (for reviews, see references 1, 5, 8, 20, 35, 36, 43, 60, and 62).

SUPERANTIGENS: DEFINITION AND MODE OF INTERACTION WITH IMMUNE CELLS

Superantigens are proteins produced by a number of microbial pathogens, including bacteria and viruses (for reviews, see references 34, 36, and 38). The name *superantigens* was used to reflect the potent immune responses elicited by these molecules (46). The mechanism by which superantigens stimulate T cells is different from conventional antigenic stimulation in many ways (31, 68). Conventional antigens are taken up by the antigen-presenting cells (APCs) and are processed into small peptides in lysozomal compartments. These digested peptides combine with major histocompatibility complex (MHC) class II molecules in the class II compartments, and then they are transported to the cell surface for presentation to T cells. The T-cell receptor (TCR) $\alpha\beta$ heterodimer recognizes nominal antigen in the context of self-MHC molecules, and this recognition is determined

Malak Kotb • Department of Surgery, The University of Tennessee, Memphis, 956 Court Ave., A202, Memphis, TN 38163.

by the five variable components of the receptor, Vα, Jα, Vβ, Dβ, and Jβ. Together these variable components create a very specific structure that allows precise recognition of the target antigenic epitope. By contrast, superantigens interact with the side of the Vβ chain of the TCR by recognizing elements shared by subsets of CD4 and CD8 T cells (27). These elements are responsible for the classification of the TCR β chain into families, and in humans at least 25 major Vβ families exist.

In addition, superantigens bind to MHC class II molecules expressed on APCs (for a review, see reference 36). Binding of superantigen either to the TCR or to MHC class II molecules occurs outside the area where conventional processed antigenic peptides normally bind. This results in the formation of trimolecular complexes that bridge T cells and APCs, which enhances intracellular interaction and the exchange of costimulatory signals that lead to cell activation and the release of inflammatory cytokines (Fig. 1). Cytokines elicited by superantigens are primarily of the Th1 type, some of which have been implicated in the pathogenesis of autoimmunity.

SUPERANTIGENS CAN DIRECTLY STIMULATE AUTOREACTIVE T CELLS

One of the most important features of superantigens is their ability to interact with specific Vβ elements regardless of the antigenic specificity of the TCR (4, 68). Importantly, binding of superantigen to the TCR does not exclude the binding of nominal antigen because they each interact with the receptor at distinct sites (45). Therefore, the activation of T cells by superantigens solely on the basis of their Vβ type can potentially cause the expansion of naturally occurring autoreactive T-cell clones. Once expanded, these cells can potentially mediate autoimmunity. However, it should be noted that, depending on the context of the superantigen-T cell interaction, T cells that express Vβ elements that are recog-

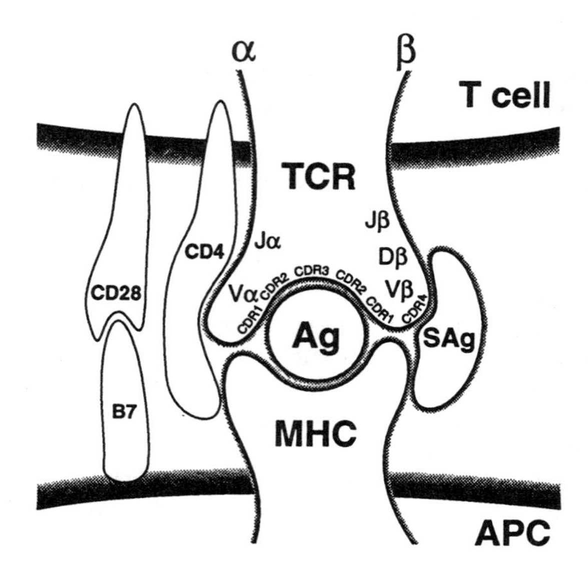

Figure 1. Bridging of T cells and APC: a schematic model of superantigen (SAg) interaction with TCR and class II molecules.

nized by a particular superantigen can be either expanded, deleted, or anergized (36). Only those clones that are expanded may have the potential of eliciting autoimmunity.

THE ENVIRONMENT IN WHICH SUPERANTIGEN AND T CELLS INTERACT AFFECTS THE FATE OF T CELLS AND MAY INFLUENCE EXPANSION OF AUTOREACTIVE T-CELL CLONES

Depending on the context and environment in which the interaction of superantigen with T cells occurs, T cells bearing superantigen-reactive Vβ elements can undergo either clonal expansion, deletion, or anergy. Activation and clonal expansion of T cells require two signals; the first (signal 1) is delivered by TCR engagement, and the second (signal 2) is delivered via the interaction of APC-associated costimulatory molecules with their respective ligands on the T cell (32, 44). Through their ability to bridge T cells and the APCs, superantigens may bring costimulatory molecules such as CD40L, B7, and ICAM-1 in closer proximity to their respective ligands, CD40, CD28, and LFA-1, allowing a better interaction and the more efficient transduction of the signals required to program the activation and proliferation of T cells (52). In the absence of relevant costimulatory signals, the engagement of the TCR by superantigen induces T-cell anergy. On the other hand, in the presence of elevated levels of cytokines such as tumor necrosis factor alpha (TNF-α) and gamma interferon (IFN-γ), reengagement of the TCR by superantigen in preactivated T cells can lead to a process of programmed cell death known as apoptosis, which can be followed by selective deletion of superantigen-specific T cells (66, 67). Therefore, the balance between clonal expansion and clonal deletion will depend on the context of the superantigen-T cell interaction. Conditions that allow expansion rather than deletion of the autoreactive T-cell clones may provide the proper environment for the development of autoimmunity.

It has been hypothesized that during severe infection, superantigen-specific T cells may be deleted by apoptosis due to the elevated inflammatory response and are thus less likely to participate in direct autoreactivity. Similarly, interactions between T cells and superantigens that lead to anergy will not be expected to react to self-antigens, or to any other foreign antigen for that matter. Therefore, clonal expansion of superantigen-specific T cells requires that both signals 1 and 2 be received and that the interaction occur in the absence of strong apoptotic signals. If autoreactive T-cell clones are among those expanded by superantigenic stimulation, these clones may potentially cause tissue destruction. However, even after surviving the superantigenic stimulation, self-reactive T cells must still overcome many barriers before they are capable of actually causing pathogenic autoimmunity. These barriers may include the cytokine environment and the unique presentation of self-antigens by the host MHC allotype in a way that propagates the survival of autoreactive T-cell clones and leads to tissue injury. As will be discussed below, superantigens can induce the cytokine production and the aberrant presentation of sequestered self-antigens, and can therefore propagate the survival of autoreactive clones.

SUPERANTIGENS CAN ACT AS POLYCLONAL MITOGENS FOR B CELLS

The ability of superantigens to activate cells that express class II molecules on their surfaces has led to the suggestion that this may lead to polyclonal activation of B cells and the generation of autoantibodies (18–20). The conventional T-cell superantigen staphylococcal enterotoxin (SE) A (SEA) was also shown to activate B cells in T-cell-dependent manner (13). SEA induces the survival of B cells that express VH3-containing immunoglobulin M (IgM). The SEA-mediated activation of VH3-expressing B cells indicated that SEA functions as a B-cell superantigen.

Superantigens specific to B cells only have also been described. B-cell superantigens are proteins that bind to the Fab regions of immunoglobulin molecules outside their complementarity-determining regions (14). This can lead to polyclonal activation of B cells and their differentiation to antibody-secreting cells. Staphylococcal protein A (SpA) is a prototype of the B-cell superantigens (61). SpA is highly restricted to interactions with products of the *VH3* gene family on IgM, IgA, and IgG.

Another B-cell superantigen, the human immunodeficiency virus type 1 gp120 protein, has been shown to have activity similar to that of SpA. Interestingly, the microbial B-cell superantigens SpA and gp120 mimic an endogenous B-cell superantigen which is a sialoprotein, pFv, that mediates cellular activation. The polyclonal activation and differentiation of B cells that express B-cell receptor specificities that recognize microbial antigens that mimic host proteins can lead to the secretion of autoreactive antibodies that can potentially elicit autoimmunity.

In addition, recent data by Kozlowski et al. (39) provided evidence that B-cell superantigens can react with a substantial amount of host serum immunoglobulins that form complexes. These immune complexes can mediate inflammatory responses in vivo, cause tissue damage, and expose sequestered self-proteins to the immune system, also leading to the development of autoimmunity.

ENHANCEMENT OF T-CELL RECOGNITION OF MIMICRY EPITOPES AND INDIRECT ACTIVATION OF AUTOREACTIVITY BY SUPERANTIGENS: THE ADJUVANT-LIKE EFFECT

It is well established that many pathogens express antigens that may mimic host proteins. The presence of mimicry epitopes shared between the host and pathogen and the ability of virulence components released by the pathogen, such as superantigens, to elicit inflammatory reactions can cause abnormal processing and presentation of sequestered self-antigens through the upregulation of chaperones, costimulatory molecules, heat shock proteins, cell adhesion molecules, and MHC molecules or by causing an imbalance in the cytokine regulatory network (for reviews, see references 35, 40, and 50). Potent immune responses elicited by superantigens may have an adjuvant-like effect, enhancing the ability of autoreactive T cells to respond to microbial epitopes and indirectly causing these clones to expand beyond a threshold that allows them to breach tolerance and to recognize self-epitopes that mimic microbial proteins. Some of these mechanisms are discussed below.

Inflammatory Responses Elicited by Superantigens Can Lead to Tissue Destruction and Aberrant Presentation of Sequestered Self-Antigens

The direct interaction of superantigens with T cells is not the only mechanism by which these molecules can induce inflammation and possibly autoimmunity. The multifunctional nature of superantigens and their ability to interact with and activate a variety of cell types may allow them to cause an imbalance in cytokine regulatory networks, thereby furnishing the appropriate environment for the development of autoimmunity. By interacting with and cross-linking the MHC class II molecules, superantigens can activate B cells, macrophages, and dendritic cells to secrete excessive amounts of inflammatory cytokines, release nitric oxide, express adhesion molecules, upregulate the expression of costimulatory molecules, and enhance tissue damage and the presentation of self-proteins (47, 59).

Inflammatory cytokines have been detected in autoimmune-inflicted sites (12, 16, 30, 40). Cytokines produced locally at the site of disease can maintain the survival and activation of autoreactive T cells and propagate the expansion of pathogenic clones. Furthermore, excessive expression of certain cytokines can upregulate adhesion molecules, increase the level of leukocyte aggregation, enhance the production of matrix proteases, augment the release of oxygen radicals, and induce the production of nitric oxide (17, 42, 55, 64). Some of the inflammatory cytokines elicited by superantigen including interleukin 12 (IL-12) activate natural killer (NK) cells, which can also contribute to the local destructive inflammatory response. Tissue damage can also be caused by the direct cytotoxic effects of some cytokines (54). Together, the various effects of cytokines can lead to tissue destruction and enhancement of the presentation of sequestered self-proteins. It can also alter the processing, trafficking, and presentation of autoimmunogenic epitopes, including those that mimic the pathogen proteins.

In certain cases, the presence of inflammatory cytokines at the site of autoimmune infliction was found to be accompanied by abnormal expression of MHC molecules (7, 29). Under certain inflammatory conditions, cells that normally do not express MHC class II molecules, including endothelial cells, epithelial skin Langerhans cells, Kupffer's cells, and even brain glial cells, can be induced to express these molecules (3, 56). Superantigens, through the action of cytokines, have been shown to induce MHC class II expression on a variety of cells and to activate MHC class II-positive endothelial cells. Expression of MHC class II molecules on these nonprofessional APCs can lead to aberrant presentation of sequestered self-antigens and the activation of autoreactive cells.

In addition to the local actions of cytokines, systemic release of certain cytokines may influence the homing or localization of pathogenic T cells, thereby enhancing their ability to cause damage to target organs. Therefore, in the susceptible host, one would expect potent cytokine responses elicited by superantigens during an infection to play a crucial role in the development of autoimmunity.

Studies with rodents have shown that administration of superantigen by itself may not trigger autoimmunity, but it can reactivate and/or exacerbate disease triggered by other autoimmunogenic agents such as streptococcal peptidoglycans, type II collagen, or myelin basic protein (3, 5, 60). It is believed that the ability of superantigens to cause exacerbation of autoimmune diseases is related primarily to their ability to elicit potent cytokine responses. Recently, Constantinescu et al. (9) reported that in a relapsing model of experimental autoimmune encephalomyelitis (EAE), the administration of SEA or SEB induces immediate relapses. They hypothesized that the actions of superantigen may be related to their ability to elicit IL-12, which drives a Th1 response and activates NK cells. Support for their hypothesis came from studies in which the in vivo effects of superantigens on disease exacerbation were mimicked by the direct administration of IL-12 into mice with EAE. Importantly, spontaneous relapses and enhancement of the severity and frequency of spontaneous relapses induced by superantigen were blocked by anti-IL-12 neutralizing antibodies.

Several studies have been initiated to test the therapeutic potential of blocking inflammatory cytokines in autoimmune diseases. Specifically, clinical trials aimed at blocking TNF-α have shown evidence of its efficacy and promise for the treatment of rheumatoid arthritis and Crohn's disease (48). These encouraging results provide additional support for the role of cytokines in these diseases, as well as for the hypothesis that the contribution of superantigens to the pathogenesis of autoimmunity is, in part, related to the high levels of inflammatory cytokine production in response to these microbial toxins. More trials that target other cytokines, combinations of cytokines, or cytokines and cell-cell interactions should be anticipated.

Superantigen Can Induce Autoimmune Destruction Independent of T-Cell Participation

In most cases, the ability of superantigens to induce the production of cytokines such as TNF-α and IL-1 requires the participation of T cells or the presence of IFN-γ (22, 23, 37). However, under certain conditions, macrophages and even astrocytes can synthesize IFN-γ, and therefore, superantigen activation of these cells can take place entirely independently of the presence or participation of T cells. Interestingly, recent data from Ting and colleagues (26, 28, 49) suggest that this mechanism may contribute to the development of autoimmune destruction in a number of experimental neurological diseases. The signif-

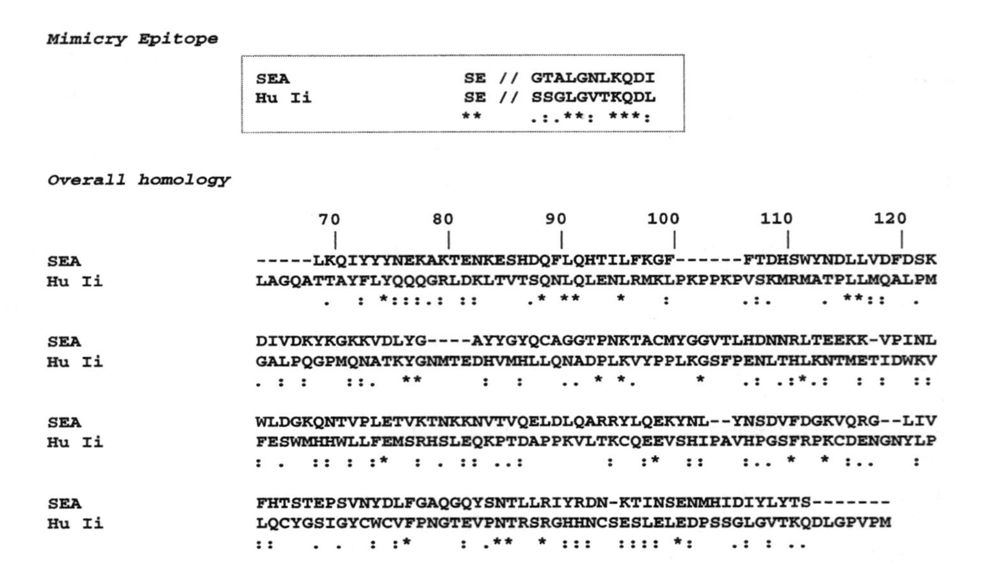

Figure 2. Mimicry epitopes and sequence homology between SEA and human (Hu) Ii. Alignment data between SEA (UNK_28339310) and the human invariant chain (E27011) show 45% overall homology, with 10% of residues being identical (asterisks), 23.3% being strongly similar (colons), and 13% being weakly similar (periods).

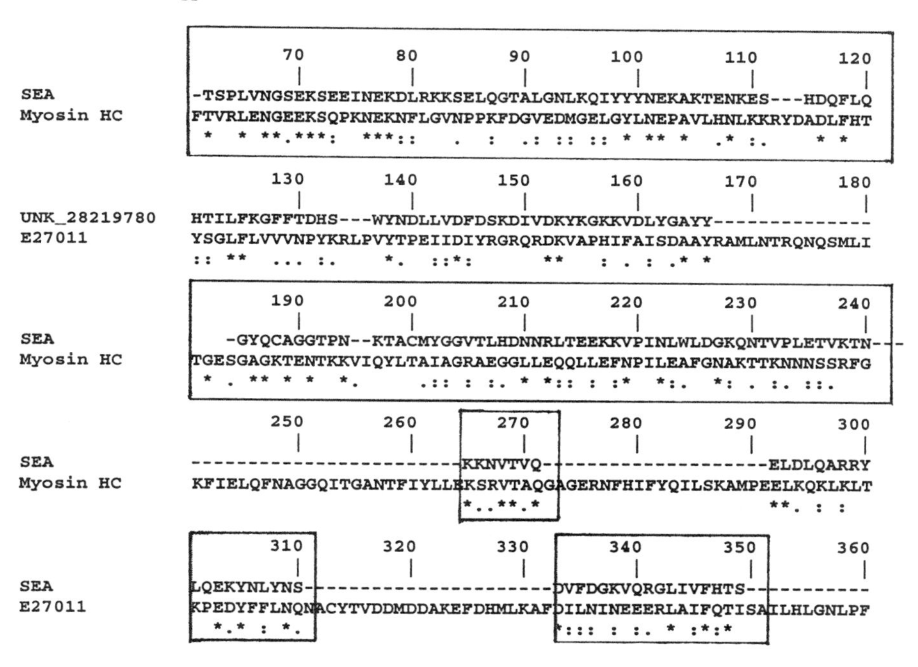

Figure 3. Mimicry epitopes and sequence homology between SE and non-muscle myosin heavy-chain (HC) gene from *Acanthamoeba.* Alignment data between SEA (UNK_28339310) and the myosin heavy chain from *Acanthamoeba* (Q10061).

icance of these observations is that superantigens can elicit autoreactivity at sites not accessible to T cells. It is anticipated that this will be an area of active investigation in the next several years.

DIRECT MIMICRY BETWEEN SUPERANTIGENS AND HOST PROTEINS

Besides the indirect effects of superantigens on autoreactive cells, many of these toxins harbor epitopes that mimic host proteins including mysosin, *S*-crystallin (63), human

Mimicry Epitope

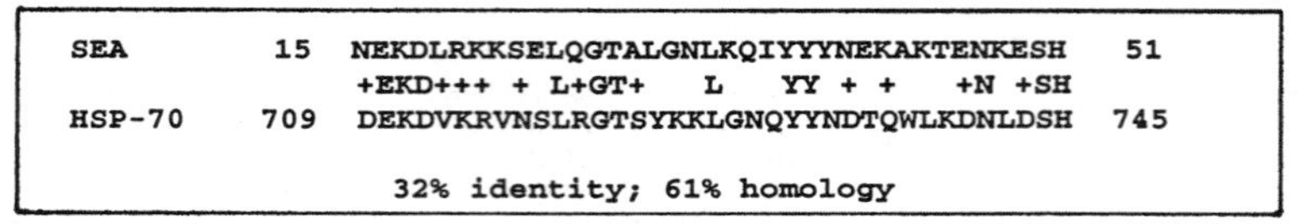

Overall homology

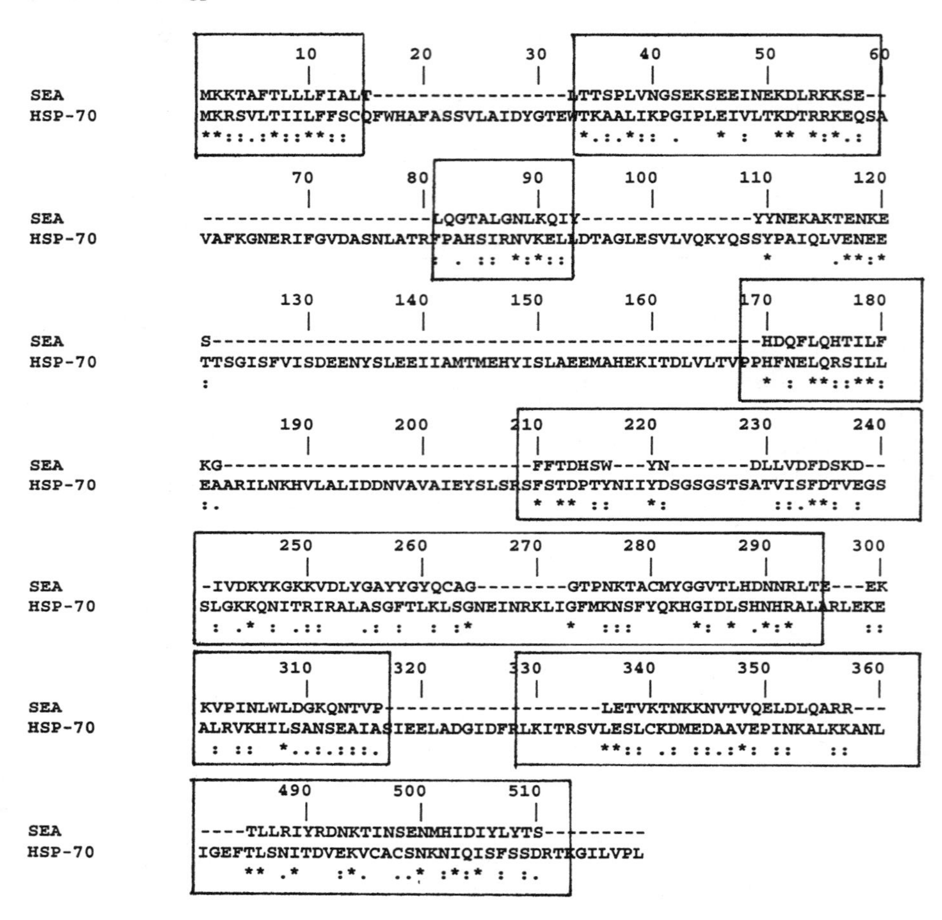

Figure 4. Mimicry epitopes and sequence homology between SEA and hsp70. Alignment data between SEA (UNK_28339310) and HSP-70 (PF00012) (6).

gamma-aminobutyric acid receptors [GABA(A) receptors] Rho-1 and Rho-2 (11), heat shock protein (6), and the MHC class II-associated invariant chain (Ii). In 1993 colleagues and I reported that the motif KSKQXXGAXKQEL, which is found in several superantigens, mimicked the sequences found in the human invariant chain (residues 192 to 211) and mouse invariant chain (residues 145 to 164) (41, 65). This motif is found at the COOH-terminal half of SEA (residues 147 to 155), SEB (residues 152 to 160), SEC1 and

A

```
SEA :          40  NEKAKTENKESHDQFLQ---HTILFKGFFTDHSWYNDLLVDF 78
                   N+   KT+ + S  QFL     HT      F++    WYN L ++F
GABA(A) R:    228  NDSLKTDERISLSQFLIQEFHTTTKLAFYSSTGWYNRLYINF 269

                      35% identity; 86% homology
```

B

```
SEA:        15   NEKDLRKKSELQGTALGNLKQIYYYNEKAKTENKESHDQFLQHTILFKGFFTDHSWYNDL 74
                 NEKD  KK+ELQ          ++  Y EK    NK     F+   IL      T  +  N +
S-Cryst. 108   NEKDAAKKTELQKRFQNTCLRVLPYMEKTLEANKGGAGWFIGDQILLCDMMTHAALENPI 167
                              31% identity; 72% homology
```

Figure 5. Homology epitopes shared data between SEA (UNK_28339310) and GABA(A) receptor (R) RHO-1 (P24046) (11) (A) and *S*-crystallin (S-Cryst.) (P18426) (63) (B).

SEC3 (residues 152 to 160), and streptococcal pyrogenic toxin A (SPE A) (residues 141 to 149) (58). However, this motif is also found in immunologically important areas in the first 27 amino acids of SEA (53) and in the mitogenic 21-mer of toxic shock syndrome toxin 1 residues 58 to 78 (15) and type 5 M-protein residues 84 to 116 (65). Of particular interest to this discussion is the finding that this motif contains a myosin cross-reactive epitope (10) and was reported (24) to stimulate the proliferation of T cells isolated from valvular tissues of rheumatic heart disease patients. Furthermore, I have reported (33) that superantigen stimulation of T cells enhances their ability to recognize cardiac myosin but not muscle myosin.

In addition to the similarities described above, remarkable sequence homology is found between several superantigens and the human invariant chain (Fig. 2), non-muscle myosin heavy chain (Fig. 3), heat shock protein 70 (hsp70) (Fig. 4), GABA(A) receptor (Fig. 5A), and *S*-crystallin (Fig. 5B). Specifically, a highly conserved region shared by many superantigens and represented by residues 15 to 51 of the mature SEA protein bears 61% homology to residues 709 to 745 of hsp70 (6), whereas residues 15 to 74 bear 31% identity and 72% homology to residues 108 to 167 of *S*-crystallin, a lens protein (Fig. 5B). Furthermore, residues 1 to 52 of the mature SEA protein bear 51% homology to residues 62 to 114 of non-muscle myosin heavy chain (25), whereas residues 40 to 78 bear 86% homology to the GABA(A) receptor Rho (Fig. 5A). The implication of these findings to the potential role of superantigens in autoimmunity remains to be elucidated.

SUMMARY AND CONCLUSIONS

Superantigens are unique molecules that stimulate the immune system in a variety of ways that can conceivably break tolerance and cause activation and expansion of autoreactive T cells. Although superantigens alone may not trigger disease, they can enhance its development or exacerbate its symptoms. The role of superantigens in autoimmunity can be either direct or indirect. Indirectly, superantigens can activate autoreactive T cells that bear Vβ elements recognized by the superantigen, or they can elicit the production of inflammatory cytokines that can exert an adjuvant-like effect, thereby enhancing the recognition of mimicry epitopes shared between microbial and host proteins. The inflammatory response can also change the homing patterns of lymphocytes and cause tissue damage and/or abnormal presentation of sequestered self-antigens. Directly, several superantigens harbor epitopes that mimic sequences found in a number of proteins believed to be potential target autoantigens. It is tempting to speculate that the superantigens may act as their own adjuvant, enhancing responses to their own epitopes that mimic host proteins.

However, it should be emphasized that superantigens can initiate very different effects in different individuals. Strong evidence has emerged that suggests that host immunogenetic factors play a crucial role in regulating immune responses to superantigens and are therefore also likely to influence the clinical outcome of superantigenic stimulation. Ongoing work in several laboratories including my own is designed to further investigate the role of superantigens in autoimmunity.

REFERENCES

1. **Acha-Orbea, H.** 1993. Bacterial and viral superantigens: roles in autoimmunity? *Ann. Rheum. Dis.* **52:**S6–16.
2. **Barnett, L. A., and R. S. Fujinami.** 1992. Molecular mimicry: a mechanism for autoimmune injury. *FASEB J.* **6:**840–844.
3. **Baum, D., R. Yaron, and M. J. Yellin.** 1998. TNF-alpha, not CD154 (CD40L), plays a major role in SEB-dependent, CD4(+) T cell-induced endothelial cell activation in vitro. *Cell. Immunol.* **190:**12–22.
4. **Blackman, M. A., and D. L. Woodland.** 1995. In vivo effects of superantigens. *Life Sci.* **57:**1717–1735.
5. **Brocke, S., A. Gaur, C. Piercy, A. Gautman, K. Gijbels, C. G. Fathman, and L. Steinman.** 1993. Induction of relapsing paralysis in experimental autoimmune encephalomyelitis by bacterial superantigen. *Nature* **365:**642–644.
6. **Bukau, B., and A. Horwich.** 1998. The hsp70 and hsp60 chaperone machines. *Cell* **92:**351–366.
7. **Caforio, A. L., J. T. Stewart, E. Bonifacio, M. Burke, M. J. Davies, W. J. McKenna, and G. F. Bottazzo.** 1990. Inappropriate major histocompatibility complex expression on cardiac tissue in dilated cardiomyopathy. Relevance for autoimmunity? *J. Autoimmun.* **3:**187–200.

7a. **Claesson, L., D. Larhammar, L. Rask, and P. A. Peterson.** 1983. cDNA clone for the human invariant r chain of class II histocompatibility antigens and its implications for the protein structure. *Proc. Natl. Acad. Sci. USA* **80:**7395–7399.

8. **Cole, B. C., and M. M. Griffiths.** 1993. Triggering and exacerbation of autoimmune arthritis by the Mycoplasma arthritidis superantigen MAM. *Arthritis Rheum.* **36:**994–1002.
9. **Constantinescu, C. S., M. Wysocka, B. Hilliard, E. S. Ventura, E. Lavi, G. Trinchieri, and A. Rostami.** 1998. Antibodies against IL-12 prevent superantigen-induced and spontaneous relapses of experimental autoimmune encephalomyelitis. *J. Immunol.* **161:**5097–5104.
10. **Cunningham, M. W., J. M. McCormack, L. R. Talaber, J. B. Harley, E. M. Ayoub, R. S. Muneer, L. T. Chun, and D. V. Reddy.** 1988. Human monoclonal antibodies reactive with antigens of the group A Streptococcus and human heart. *J. Immunol.* **141:**2760–2766.
11. **Cutting, G. R., L. Lu, B. F. O'Hara, L. M. Kasch, C. Montrose-Rafizadeh, D. M. Donovan, S. Shimada, S. E. Antonarakis, W. B. Guggino, G. R. Uhl, and H. H. J. R. Kazazian.** 1991. Cloning of the gamma-aminobutyric acid (GABA) rho 1 cDNA: a GABA receptor subunit highly expressed in the retina. *Proc. Natl. Acad. Sci. USA* **88:**2673–2677.
12. **Dinarello, C. A.** 1992. Interleukin-1 and tumor necrosis factor: effector cytokines in autoimmune diseases. *Semin. Immunol.* **4:**133–145.
13. **Domiati-Saad, R., and P. Lipsky.** 1998. Staphylococcal enterotoxin A induces survival of VH3-expressing human B cells by binding to the VH region with low affinity. *J. Immunol.* **161:**1257–1266.
14. **Domiati-Saad, R., and P. E. Lipsky.** 1997. B cell superantigens: potential modifiers of the normal human B cell repertoire. *Int. Rev. Immunol.* **14:**309–324.

15. **Edwin, C., J. A. Swack, K. Williams, P. F. Bonventre, and E. H. Kass.** 1991. Activation of in vitro proliferation of human T cells by a synthetic peptide of toxic shock syndrome toxin 1. *J. Infect. Dis.* **163:**524–529.
16. **Feldmann, M., F. M. Brennan, D. Chantry, C. Haworth, M. Turner, P. Katsikis, M. Londei, E. Abney, G. Buchan, K. Barrett, et al.** 1991. Cytokine assays: role in evaluation of the pathogenesis of autoimmunity. *Immunol. Rev.* **119:**105–123.
17. **Feldmann, M., F. M. Brennan, and R. Maini.** 1998. Cytokines in autoimmune disorders. *Int. Rev. Immunol.* **17:**217–228.
18. **Friedman, S. M., M. K. Crow, J. R. Tumang, M. Tumang, Y. Xu, A. S. Hodtsev, B. C. Cole, and D. N. Posnett.** 1991. Characterization of human T cells reactive with the *Mycoplasma arthritidis*-derived superantigen (MAM): generation of a monoclonal antibody against Vβ 17, the T cell receptor gene product expressed by a large fraction of MAM-derived human T cells. *J. Exp. Med.* **174:**891–900.
19. **Friedman, S. M., D. N. Posnett, J. R. Tumang, B. C. Cole, and M. K. Crow.** 1991. A potential role for microbial superantigens in the pathogenesis of systemic autoimmune disease. *Arthritis Rheum.* **34:**468–480.
20. **Friedman, S. M., J. R. Tumang, and M. K. Crow.** 1993. Microbial superantigens as etiopathogenic agents in autoimmunity. *Rheum. Dis. Clin. N. Am.* **19:**207–222.
21. **Fujinami, R. S., J. A. Nelson, L. Walker, and M. B. Oldstone.** 1988. Sequence homology and immunologic cross-reactivity of human cytomegalovirus with HLA-DR beta chain: a means for graft rejection and immunosuppression. *J. Virol.* **62:**100–105.
22. **Gjorloff, A., H. Fischer, G. Hedlund, J. Hansson, J. S. Kenny, A. C. Allison, H.-O. Sjogren, and M. Dohlsten.** 1991. Induction of IL-1 in human monocytes by the superantigen staphylococcal entertioxin A requires the participation of T cells. *Cell. Immunol.* **137:**61–71.
23. **Grossman, D., J. G. Lamphear, J. A. Mollick, M. J. Betley, and R. R. Rich.** 1992. Dual roles for class II major histocompatibility complex molecules in staphylococcal enterotoxin-induced cytokine production and in vivo toxicity. *Infect. Immun.* **60:**5190–5196.
24. **Guilherme, L., E. Cunha-Neto, V. Coelho, R. Snitcowsky, P. M. Pomerantzeff, R. V. Assis, F. Pedra, J. Neumann, A. Goldberg, M. E. Patarroyo, et al.** 1995. Human heart-infiltrating T cell clones from rheumatic heart disease patients recognize both streptococcal and cardiac proteins. *Circulation* **92:**415–420.
25. **Hammer, J. A. I., B. Bowers, B. M. Paterson, and E. D. Korn.** 1987. Complete nucleotide sequence and deduced polypeptide sequence of a nonmuscle myosin heavy chain gene from Acanthamoeba: evidence of a hinge in the rodlike tail. *J. Cell Biol.* **105:**913–925.
26. **Hellendall, R. P., and J. P. Ting.** 1997. Differential regulation of cytokine-induced major histocompatibility complex class II expression and nitric oxide release in rat microglia and astrocytes by effectors of tyrosine kinase, protein kinase C, and cAMP. *J. Neuroimmunol.* **74:**19–29.
27. **Herman, A., J. W. Kappler, P. Marrack, and A. M. Pullen.** 1991. Superantigens: mechanism of T-cell stimulation and role in immune responses. *Annu. Rev. Immunol.* **9:**745–772.
28. **Hiremath, M. M., Y. Saito, G. W. Knapp, J. P. Ting, K. Suzuki, and G. K. Matsushima.** 1998. Microglial/macrophage accumulation during cuprizone-induced demyelination in C57BL/6 mice. *J. Neuroimmunol.* **92:**38–49.
29. **Itoh, N., T. Hanafusa, A. Miyazaki, J. Miyagawa, K. Yamagata, K. Yamamoto, M. Waguri, A. Imagawa, S. Tamura, M. Inada, et al.** 1993. Mononuclear cell infiltration and its relation to the expression of major histocompatibility complex antigens and adhesion molecules in pancreas biopsy specimens from newly diagnosed insulin-dependent diabetes mellitus patients. *J. Clin. Investig.* **92:**2313–2322.
30. **Jacob, C. O.** 1992. Genetic variability in tumor necrosis factor production: relevance to predisposition to autoimmune disease. *Reg. Immunol.* **4:**298–304.
31. **Janeway, C. A. J., S. Rath, and J. Yagi.** 1991. V beta selective elements: self and non-self. *Behring Inst. Mitt.* **88:**177–182.
32. **Jenkins, M. K., P. S. Taylor, S. D. Norton, and K. B. Urdahl.** 1991. CD28 delivers a costimulatory signal involved in antigen-specific IL-2 production by human T cells. *J. Immunol.* **147:**2461–2466.
33. **Kotb, M.** 1995. Bacterial pyrogenic exotoxins as superantigens. *Clin. Microbiol. Rev.* **8:**411–426.
34. **Kotb, M.** 1995. Infection and autoimmunity: a story of the host, the pathogen, and the copathogen. *Clin. Immunol. Immunopathol.* **74:**10–22.

35. **Kotb, M.** 1994. Post-streptococcal autoimmune sequelae: a link between infection and autoimmunity. *In* A. G. Dalgleish, A. Albertini, and R. Paoletti (ed.), *The Impact of Biotechnology on Autoimmunity.* Kluwer Academic Publishers, Boston, Mass.
36. **Kotb, M.** 1998. Superantigens of gram-positive bacteria: structure-function analyses and their implications for biological activity. *Curr. Opin. Microbiol.* **1:**56–65.
37. **Kotb, M., H. Ohnishi, G. Majumdar, S. Hackett, A. Bryant, G. Higgins, and D. Stevens.** 1993. Temporal relationship of cytokine release by peripheral blood mononuclear cells stimulated by the streptococcal superantigen pep M5. *Infect. Immun.* **61:**1194–1201.
38. **Kotzin, B. L., D. Y. Leung, J. Kappler, and P. Marrack.** 1993. Superantigens and their potential role in human disease. *Adv. Immunol.* **54:**99–166.
39. **Kozlowski, L. M., W. Li, M. Goldschmidt, and A. I. Levinson.** 1998. In vivo inflammatory response to a prototypic B cell superantigen: elicitation of an Arthus reaction by staphylococcal protein A. *J. Immunol.* **160:**5246–5252.
40. **Kroemer, G., and C. Martinez.** 1991. Cytokines and autoimmune disease. *Clin. Immunol. Immunopathol.* **61:**275–295.
42. **Lalani, I., K. Bhol, and A. R. Ahmed.** 1997. Interleukin-10: biology, role in inflammation and autoimmunity. *Ann. Allergy Asthma Immunol.* **79:**469–483. (Erratum, **80:**A-6, 1998.)
43. **Lavoie, P., J. Thibodeau, F. Erard, and R. Sekaly.** 1999. Understanding the mechanism of action of bacterial superantigens from a decade of research. *Immunol. Rev.* **168:**257–269.
44. **Ledbetter, J. A., J. B. Imboden, G. L. Schieven, L. S. Grosmaire, P. S. Rabinovitch, T. Lindsten, C. B. Thompson, and C. H. June.** 1990. CD28 ligation in T-cell activation: evidence for two signal transduction pathways. *Blood* **75:**1531–1539.
45. **Li, H., A. Llera, E. Malchiodi, and R. Mariuzza.** 1999. The structural basis of T cell activation by superantigens. *Annu. Rev. Immunol.* **17:**435–466.
46. **Marrack, P., and J. Kappler.** 1990. The staphylococcal enterotoxins and their relatives. *Science* **248:**705–711.
47. **Mooney, N. A., L. Ju, C. Brick-Ghannam, and D. J. Charron.** 1994. Bacterial superantigen signaling via HLA class II on human B lymphocytes. *Mol. Immunol.* **31:**675–681.
48. **Moreland, L.** 1999. Inhibitors of tumor necrosis factor for rheumatoid arthritis. *J. Rheumatol.* **57:**7–15.
49. **Moses, H., A. Sasaki, and J. P. Ting.** 1991. Identification of an interferon-gamma-responsive element of a class II major histocompatibility gene in rat type 1 astrocytes. *J. Neuroimmunol.* **31:**273–278.
50. **Mosmann, T. R.** 1991. Cytokine secretion patterns and cross-regulation of T cell subsets. *Immunol. Res.* **10:**183–188.
51. **Nygard, N. R., and B. D. Schwartz.** 1993. Infection and autoimmunity. *Adv. Intern. Med.* **38:**337–359.
52. **Ohnishi, H., J. A. Ledbetter, S. B. Kanner, P. S. Linsley, T. Tanaka, A. M. Geller, and M. Kotb.** 1995. CD28 cross-linking augments TCR-mediated signals and costimulates superantigen responses. *J. Immunol.* **154:**3180–3193.
53. **Pontzer, C. H., and J. K. M. Russell.** 1989. Localization of an immune functional site on staphylococcal enterotoxin A using the synthetic peptide approach. *J. Immunol.* **143:**280–284.
54. **Pukel, C., H. Baquerizo, and A. Rabinovitch.** 1988. Destruction of rat islet cell monolayers by cytokines. Synergistic interactions of interferon-gamma, tumor necrosis factor, lymphotoxin, and interleukin 1. *Diabetes* **37:**133–136.
55. **Reed, J. C.** 1999. Caspases and cytokines: roles in inflammation and autoimmunity. *Adv. Immunol.* **73:**265–299.
56. **Riesbeck, K., A. Billstrom, J. Tordsson, T. Brodin, K. Kristensson, and M. Dohlsten.** 1998. Endothelial cells expressing an inflammatory phenotype are lysed by superantigen-targeted cytotoxic T cells. *Clin. Diagn. Lab. Immunol.* **5:**675–682.
57. **Rose, N. R., D. A. Neumann, and A. Herskowitz.** 1992. Coxsackievirus myocarditis. *Adv. Intern. Med.* **37:**411–429.
58. **Schlievert, P. M., K. N. Shands, B. B. Dan, G. P. Schmid, and R. D. Nishimura.** 1981. Identification and characterization of an exotoxin from Staphylococcus aureus associated with toxic shock syndrome. *J. Infect. Dis.* **143:**509–516.
59. **Scholl, P. R., and R. S. Geha.** 1994. MHC class II signaling in B-cell activation. *Immunol. Today* **15:**418–422.
60. **Schwab, J. H., R. R. Brown, S. K. Anderle, and P. M. Schlievert.** 1993. Superantigen can reactivate bacterial cell wall-induced arthritis. *J. Immunol.* **150:**4151–4159.

61. **Silverman, G.** 1998. B cell superantigens: possible roles in immunodeficiency and autoimmunity. *Semin. Immunol.* **10:**43–55.
62. **Soos, J. M., J. Schiffenbauer, B. A. Torres, and H. M. Johnson.** 1997. Superantigens as virulence factors in autoimmunity and immunodeficiency diseases. *Med. Hypotheses* **48:**253–259.
63. **Tomarev, S. I., and R. D. Zinovieva.** 1988. Squid major lens polypeptides are homologous to glutathione *S*-transferase subunits. *Nature* **336:**86–88.
64. **Truman, J. P., M. L. Ericson, C. J. Choqueux-Seebold, D. J. Charron, and N. A. Mooney.** 1994. Lymphocyte programmed cell death is mediated via HLA class II DR. *Int. Immunol.* **6:**887–896.
65. **Wang, B., P. M. Schlievert, A. O. Gaber, and M. Kotb.** 1993. Localization of an immunologically functional region of the streptococcal superantigen pepsin-extracted fragment of type 5 M protein. *J. Immunol.* **151:**1419–1429.
66. **Watanabe-Ohnishi, R., D. E. Low, A. McGeer, D. L. Stevens, P. M. Schlievert, B. Schwartz, B. Kreiwwirth, and M. Kotb.** 1994. Selective depletion of Vβ-bearing T cells in patients with invasive group A streptococcal disease and streptococcal toxic shock syndrome: implications for a novel superantigen. *J. Infect. Dis.* **171:**74–84.
67. **Webb, S., C. Morris, and J. Sprent.** 1990. Extrathymic tolerance of mature T cells: clonal elimination as a consequence of immunity. *Cell* **63:**1249–1256.
68. **White, J., A. Herman, A. M. Pullen, R. Kubo, J. W. Kappler, and P. Marrack.** 1989. The V beta-specific superantigen staphylococcal enterotoxin B: stimulation of mature T cell and clonal deletion in neonatal mice. *Cell* **56:**27–35.

Molecular Mimicry, Microbes, and Autoimmunity
Edited by M. W. Cunningham and R. S. Fujinami

Chapter 9

Peptide Induction of Systemic Lupus Autoimmunity

John B. Harley, R. Hal Scofield, and Judith A. James

The generation of autoimmune disorders in animals by immunization has a long history dating back almost half a century to studies by Noel Rose and colleagues with thyroglobulin (68). In those studies homologous thyroglobulin induced autoimmune thyroid disease in rabbits. Since those early experiments autoimmune thyroid disease has been induced in several animal species including the rabbit, guinea pig, dog, rat, and mouse by immunization with autologous, homologous, or even heterologous thyroid protein (usually in the presence of Freund's adjuvant) (10, 52). Many insights into the operation of the immune response relevant to human disease have resulted. The extent that autoimmune diseases remain serious, enigmatic, and idiopathic problems is an indictment not only of our inadequate present-day applied capabilities but also of our present lack of basic understanding of autoimmune disease pathogenesis.

Systemic lupus erythematosus often is cited as the classic representative of systemic autoimmunity, in contrast to organ-specific autoimmune disorders such as myasthenia gravis or Graves' disease. In animal models of lupus and in humans, many organs are typically affected. The reasons why, for example, skin disease dominates the clinical course in one person for many years while in another lupus patient the skin is not affected or is only transiently involved are poorly understood. The exacerbations and remissions of diseases that affect particular organs often worsen or improve without apparent relationship to other aspects of the disease.

Autoantibodies are the unifying feature of lupus autoimmunity. Historically, the appreciation that the biologically false-positive Venereal Disease Research Laboratory test was related to lupus preceded the antinuclear antibody test by nearly a half century. The LE cell test became available in the late 1940s, almost a decade before the antinuclear antibody test (the most commonly applied screening test at present). Together, these tests along with Ouchterlony immunodiffusion set the stage for the present molecular biological era in lupus research, which has led to the identification and characterization of many of the autoantigens important in lupus.

Various antinuclear antibodies are the most characteristic for our present conception of this disorder. A positive antinuclear antibody test is sensitive for the detection of lupus (>95% in affected patients and virtually all animals affected) but is not specific. These

John B. Harley, R. Hal Scofield, and Judith A. James • Arthritis/Immunology, Oklahoma Medical Research Foundation, 825 N.E. 13th Street, Oklahoma City, OK 73104.

autoantibodies may be present in healthy individuals (~5 to 10%) and in patients with a number of other disorders (progressive systemic sclerosis or scleroderma [90%], polymyositis [69%], dermatomyositis [78%], polyarticular juvenile and adult-onset rheumatoid arthritis [45% and 10%, respectively], and Sjögren's syndrome [60%]).

While determination of the specificity of the antinuclear antibody test over the last several decades has been one of the most important accomplishments in rheumatic disease immunology, there is enormous variation in our knowledge. Some antigens are only partially characterized, while the structures and fine specificities of others are known in extraordinary detail (9, 16, 18, 34, 44, 48, 56). Indeed, new specificities continue to appear (e.g., such as CD45, apolipoprotein A1, and L-selectin), and at present there must be more than 50 known autoantibody specificities in lupus. These autoantibodies, their associations, their implied consequences, and evidence consistent with their having pathogenic potential constitute the evidence that lupus is an autoimmune disease. While more than one autoantibody specificity is found in a number of autoimmune diseases, no other disorder (with the sole exception of infectious mononucleosis) is associated with such a large collection of autoantibody specificities from which the antibodies found in an individual patient may have been selected.

The variation in the number of antibodies among lupus patients is also extreme (Table 1). Some patients have only an antinuclear antibody for which the antigen is unknown, while others simultaneously have autoantibodies with many different specificities. For example, antinuclear antibodies along with anti-native DNA, anti-denatured DNA, anti-Ro (or SS-A), anti-La (or SS-B), anti-SM, anti-nuclear RNP (anti-nRNF), anti-P, rheumatoid factor, and Coombs' antibodies may all be simultaneously present. No particular antibody is, at present, known to be completely specific for lupus. Among the more commonly encountered autoantibodies, anti-native DNA, anti-Sm, anti-ribosomal P, and antiphospholipid are the most specific. These, along with anti-nRNP, antihistone, anti-Ro, anti-La, and rheumatoid factor, appear to be the most common.

Table 1. The more-common autoantibodies in lupus

Autoantibody	% Positive	Comments
Antinuclear antibody	>95	Lacks specificity but is sensitive for disease; many antigenic specificities
Anti-native DNA	~60	Most specific; associated with renal disease
Anti-single-stranded DNA	~80	Common, but not specific
Anti-Ro	~40	Large quantities of antibody; associations with photosensitivity, skin rashes, lung disease, leukopenia, lymphopenia, anti-La antibody, and rheumatoid factor
Anti-La	~20	Lack of renal disease
Anti-nRNP	~40	Associated with mild disease and myositis
Anti-Sm	~20	Relatively specific for lupus
Anti-ribosomal P	~20	Associated with central nervous system manifestations, especially depression and psychosis; relatively specific for lupus

ANIMAL MODELS OF SYSTEMIC LUPUS

Strategies that can be used to better understand lupus autoimmunity include the recognition of its existence in animals. These are of four types. First, lupus has been observed to occur spontaneously in dogs (22, 32, 41, 42). With antibodies as diverse as those in its human counterpart, canine lupus is also characterized by the same host of various disease manifestations and lupus autoantibodies. Antinuclear and antihistone antibodies are frequent findings, although they are usually found at concentrations lower than those found in humans. Familial forms of canine lupus have also been described (13). Although it is found in a variety of breeds, this canine disease is more commonly found in Shetland sheepdogs and German shepherds (15).

Second, some inbred strains of mice have a genetic propensity for the development of a lupus-like illness, including the MRL-*lpr/lpr,* BXSB, *gld,* Palmerson North, and NZB/NZW strains of mice. Lupus in these strains was usually found serendipitously in a large breeding program. These genetic models have impressive variations, including differences in clinical expression, differences in phenotype by sex, and differences in the particular autoantibody specificities found. The MRL-*lpr/lpr* strain tends to develop anti-DNA and anti-Sm antibodies and lupus nephritis. That this strain also has massive lymphoproliferation is of no surprise when considering that the critical mutation is in the *Fas* gene, a receptor responsible for the initiation of apoptosis. The *gld* mouse, on the other hand, has a defective *Fas* ligand and a phenotype similar to that of the MRL-*lpr/lpr* strain when expressed in the same murine background strain. The BXSB animals have a disease which dominates in males and which is attributed to the *yaa* gene on the Y chromosome.

The NZB/NZW strain has attracted the greatest investigative effort. At present, multiple linkages for the phenotypes of premature death or lupus nephritis, anti-DNA antibodies, antichromatin antibodies, and hypergammaglobulinemia have been established (for a review, see reference 43). Congenic inbred strains have been constructed with the expectation that some of the genes that confer risk will be identified in the next few years. Other spontaneous, inbred lupus models, such as the Palmerston North model, have yet to be genetically characterized.

The third kind of animal model of lupus is that discovered after genetic manipulation. One of the first transgenic models of lupus autoimmunity was developed by establishing the expression of human *bcl-2* in animals. These animals developed an increased number of B cells and increased immunoglobulin titers. In addition, within the first year of life most mice developed antinuclear antibodies and 60% developed lupus nephritis (65).

In addition, transgenic mice that overexpress gamma interferon in the epidermis develop features of systemic lupus. For example, these animals develop anti-double-stranded DNA and antihistone antibodies, with female mice developing higher levels of autoantibodies than males. In addition, immunoglobulin G (lgG) deposits in the glomeruli of female mice and severe proliferative glomerulonephritis have also been found in a proportion of these animals (63). Further evaluation of this transgenic model has suggested that alpha-beta T cells are necessary for this induced lupus model (62).

Finally, lupus may be induced by an environmental exposure. In humans, the classic example is drug-induced lupus, most characteristically caused by procainamide or, usually less impressively, by hydralazine. The list of drugs allegedly capable of inducing lupus is lengthy (50). Several animal models of lupus induced by environmental exposure have been described. Treatment of the BALB/c or SJL/J murine strain with pristane by intraperitoneal injection induces lupus autoimmunity, including anti-nRNP, anti-Su, and anti-Sm antibodies (55). Infection of C57Bl/6 *lpr/lpr* mice with murine cytomegalovirus is reported to cause anti-Ro/La RNP autoimmunity and salivary gland inflammation, similar to the symptoms of Sjögren's syndrome (14).

Immunization with peptides is also capable of generating lupus autoimmunity and a clinical illness that closely resembles some forms of human lupus (6, 12, 13, 26, 27, 30, 31, 33, 35, 38–40, 47, 49, 58, 60, 64). This approach and the existing results are the focus of this chapter and are discussed below.

In aggregate, both in humans and in the animal models of lupus, there are powerful genetic and environmental influences in the genesis of this disease.

IMMUNOCHEMICAL CHARACTERIZATION

The peptide immunization model of lupus grew from the immunochemical description of the fine specificity of the autoantigens. The rationale of this work is simple. First, we assume that lupus autoimmunity arises from a limited set of immune environment contexts. We mean that lupus begins from a normal immune response and that the features of the immune system which have the potential to generate lupus are discoverable. We imagine that knowledge of the structural details of the antigen, the autoantibody, and their interaction in sufficient relevant detail would provide the foundation upon which hypotheses concerning the origin of lupus could be promulgated.

This effort began with the evaluation of the carboxyl third of the Ro (or SS-A) antigen. The role of Ro in the cellular economy of the cell is not known, but it is often physically associated with La (or SS-B). Autoantibodies that bind to Ro are found in about 40% of lupus patients. A higher proportion of patients with Sjögren's syndrome or subacute cutaneous lupus have anti-Ro antibodies. Nearly all mothers of infants with complete congenital heart block or neonatal lupus dermatitis have anti-Ro antibodies.

The Ro autoantigen is molecularly complicated. The specificity was originally defined as a precipitin line by Ouchterlony immunodiffusion and was appreciated to contain both RNA and protein (1, 5). With the application of molecular biological techniques the protein was initially shown to have a molecular mass of about 60 kDa, and in humans the RNA was shown to be one of four Y RNAs: hY1, hY3, hY4, or hY5 (37). The hY2 RNA was recognized to be a truncated version of hY1. The 60-kDa protein sequence contains an extensive RNA recognition motif (4, 7). No evidence for posttranscriptional amino acid modifications is known.

A 52-kDa Ro specificity has also been defined (3). The antigenicity of this protein appears to favor the denatured form, while the antigenicity of the native 60-kDa protein predominates over that of the denatured form (23). Some clinical relationships appear to be more closely associated with the anti-52-kDa Ro specificity than with the anti-60-kDa Ro autoantibody specificity, such as complete congenital heart block (8).

Octapeptides that overlap by seven amino acids were constructed from the carboxyl-terminal 120 amino acids of the 60-kDa Ro sequence. Binding by autoantibodies was assessed by a solid-phase assay. A relatively dominant epitope was found at about amino acid position 480 and contained the sequence EYRKKMDI, which was noted to be similar to a primary sequence from the N protein of vesicular stomatitis virus (59).

Maximally overlapping octapeptides from the entire sequence of the 60-kDa Ro polypeptide were then evaluated, and a consistent pattern of relatively dominant and shared epitopes was revealed. An algorithm was derived which maximized the antigenic regions relative to the sequences in the databases at that time. The alignment of the short sequence identities of N protein of vesicular stomatitis virus to the antigenic epitopes was least likely to have arisen by chance from among the sequences then available (57).

The possibility that the anti-Ro response is related to infection by an incompletely characterized relative of vesicular stomatitis virus, such as the Chandipura strain (46), was supported by other work. Among a cohort of patients with systemic lupus erythematoses, those with anti-Ro were more likely to have antibody that bound to the N protein of vesicular stomatitis virus than those without anti-Ro (17). The patients also had antibodies that bound to other vesicular stomatitis virus proteins, including an association of anti-RNP antibody with binding of the M protein of the virus (17). The data were consistent with the possibility of human infection with an incompletely characterized relative of vesicular stomatitis virus.

We have also immunized animals with the vesicular stomatitis virus N protein. Such animals develop high-titer anti-Ro antibodies such that a Ro precipitin forms in double immunodiffusion assays (21). Analyses showed that this induced anti-Ro was indistinguishable from the anti-Ro that arises in lupus patients. Thus, anti-Ro autoimmunity can be induced by immunization with a foreign protein that shares sequence homology with the 60-kDa Ro autoantigen.

On the other hand, the immunological relationships that we have found between anti-Ro and vesicular stomatitis virus may have no causal origin. Many, perhaps most, critical scientists who have evaluated these data would doubt that a vesicular stomatitis virus is likely to cause anti-Ro autoimmunity. If so, then this would be an example of association and molecular mimicry without causation. Logicians might cite this as an example of a relationship which is true-true and unrelated. Nonetheless, these data demonstrate in principle that, under the experimental conditions used, an encounter with a foreign antigen that shares critical structural features with a self-antigen can lead to autoimmunity.

The spliceosome contains small nuclear RNPs (snRNPs) which are common antigenic targets for lupus autoantibodies, commonly called Sm and nRNP. The anti-Sm antibody is sufficiently specific and sensitive for lupus that its detection is one of the diagnostic criteria, even though only about 20% of lupus patients have this precipitin. The spliceosomal polypeptides B/B′ and D are the common targets in this system, with B/B′ being more frequent. In anti-nRNP, which is found in about 40% of lupus patients, the antigenic polypeptides are the nRNP 70K, A, and C proteins.

The antigenic octapeptides from the various spliceosomal polypeptides have interesting properties. For each protein, sufficient antigenic peptide recognition is shared by tested patient sera to identify octapeptides that are commonly antigenic. The degree of similarity, however, varies greatly. Individual patient sera contain antibodies which bind substantially different octapeptides in the nRNP 70K system, and the common antigenic targets are derived from the heterogeneous behavior of the individual sera. In contrast, the B/B′ octapeptides are bound by the Sm precipitin-positive sera in an almost identical pattern. While this response is not homogeneous, the degree of similarity is astonishing. Those sera with anti-nRNP that bind to the A protein can be divided into two groups. One is identical between patients, with the same two octapeptide epitopes and only these two octapeptide epitopes being bound (25). This is the most homogeneous peptide binding response identified among lupus patients to date.

It is not likely that octapeptides identify all of the antigenicity of these autoantigens. On the other hand, absorption experiments have shown that individual octapeptide-defined epitopes are capable of removing from ~5 to 63% of the autoantibodies with the polypeptide specificity. For example, the Ro epitope AIALREYRKKMDIPA removed ~5% of the anti-Ro from the serum of a patient with a complex, mature anti-Ro response (57). As an example from the other extreme, LVSRSLKMRGQAF (A3) and ERDRKREKRKPKS (A6) are the two peptides that constitute the more homogeneous pattern of octapeptide binding to the nRNP A polypeptide. In some sera as much as 75% of the autoantibodies directed against the nRNP particle is removed by absorption with these two peptides (25). For some specificities and sera, therefore, the octapeptides identify the major part of the antigenicity for the entire antigen-antibody system.

The physical properties were explored to determine whether the antigenic octapeptides could be predicted. Of the surface probability, Jameson-Wolff algorithm, hydrophobicity/hydrophilicity index, and isoelectric point, only an alkaline isoelectric point consistently predicted the antigenicities of octapeptides from the U1 snRNP (18).

Of the anti-Sm antibodies, those that bind to B/B′ are the most sensitive and specific. Thus, we explored this specificity for the changes that might occur over time. Most patients have fully mature patterns of octapeptide binding at presentation. However, a few patients (four) who developed anti-Sm autoantibodies while under observation were identified. These patients initially targeted the repeated PPPGMRPP motif. With time these antigenic regions expand to include neighboring octapeptides and then proceed in a predictable, sequential progression of binding to other regions of the Sm B/B′ protein (1a). Two additional patients who initially presented with anti-Sm antibodies at diagnosis were found; however, they presented with a simpler pattern of epitope binding and only targeted the PPPGMRPP and PPPGMRGP motifs. Several serum samples obtained over a span of 2 years show the progressive accrual of epitopes in this patient (Fig. 1). A schematic representation of how this may occur is presented as Fig. 2. (The very careful reader will note that there is a sequence discrepancy with the sequence PPPGMRGP given above and the sequence PPPGIRGP in the original publication (31). This arises from confusion with the sequence of the very closely related SmN, which is expressed only in the nervous system. Both alternatives at this site, PPPGMRGP and PPPGIRGP, bind to anti-Sm after PPPGMRPP [unpublished observations].)

Many other data consistently support PPPGMRPP as the first epitope in the anti-Sm system. A lupus patient who has anti-Sm precipitins and whose sera bound only to the PPPGMRPP group of octapeptides for a period of 10 years has been found (2). The sera of all of the patients with anti-Sm, now numbering more than 50, bind to PPPGMRPP. Most of these bind to epitopes in addition to this repeated sequence, but all bind to PPPGMRPP. A few consistently bind only to PPPGMRPP.

IMMUNIZATION MODEL

The apparent constancy of the initial epitope of the Sm system, PPPGMRPP, focused our attention upon this structure as being the first point of autoimmunity in a subset of lupus patients. We sought to understand this antigenic epitope better. We therefore immu-

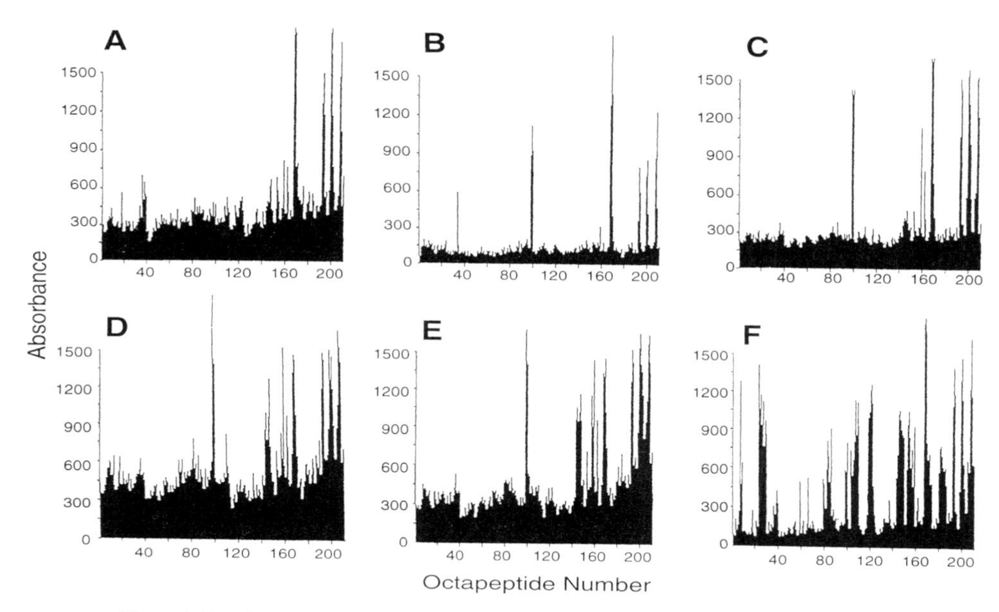

Figure 1. Development of the anti-Sm B/B′ response in one anti-Sm- and anti-nRNP precipitin-positive patient. This anti-Sm- and anti-nRNP precipitin-positive patient shows an increase in the number of antigenic determinants over time: 1 April 1986 (A), 7 July 1987 (B), 13 January 1988 (C), 8 June 1988 (D), 16 October 1988 (E), and 26 December 1988 (F). The data span a 2-year progression of system lupus erythematosus and exhibit an increase from binding of four groups of octapeptides to 15 separate groups (27). This figure is reproduced with the permission of Munksgaard International Publishing Ltd., Malden, Mass.

nized rabbits with this peptide, built as a multimer on a lysine pyramid (MAP). We expected to produce anti-PPPGMRPP antibodies, which occurred. In addition, the binding to Sm was substantial, and titers were found to be $>10^6$ over the background titer. The binding to Sm was not, however, substantially altered by attempted inhibition with the PPPGMRPP peptide or by absorption over an anti-PPPGMRPP column. In fact, virtually all of the anti-PPPGMRPP activity could be removed by absorption, but paradoxically, more than 80% of the anti-Sm activity remained.

We eventually reasoned that one explanation for this unexpected finding could be that autoimmunity had developed against the Sm antigen. If so, then there should be more binding to the Sm antigen than just to the peptide used for immunization. We had already prepared the overlapping octamers from the B/B′ polypeptide and used these to evaluate the human serum. The assay was then modified to evaluate the PPPGMRPP-immunized rabbits. We found that more than 16 apparently distinguishable epitopes of B/B′ were bound. Indeed, these were very similar to the octapeptide-defined epitopes bound by human sera with anti-Sm antibodies. This phenomenon of the progression of the immune response

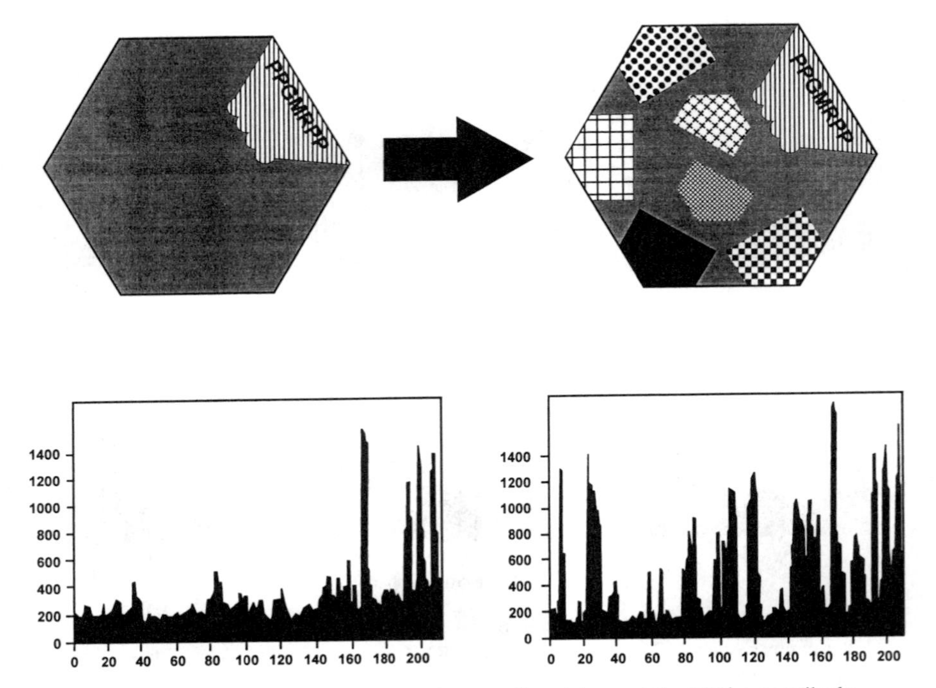

Figure 2. B-cell epitope spreading in the naturally arising anti-Sm B/B′ autoantibody response in human lupus. The cartoon shows the progression from binding of the initial structures, consisting of three repeated PPPGMRPP and PPPGMRGP, to a mature humoral response that involves at least nine other peptide-defined epitopes (none of the others of which have a similar primary sequence) (27). This figure is reproduced with the permission of Munksgaard International Publishing Ltd., Malden, Mass.

from one epitope to many is called "epitope spreading"; in this case it is B-cell (or humoral) epitope spreading.

These peptide-immunized rabbits typically developed anti-Sm precipitins, and many animals had features suggestive of lupus, including positivity for antinuclear antibodies, positivity for anti-native DNA antibodies, thrombocytopenia, rising proteinuria and creatinine levels, seizures, and moderate alopecia. A number of these peptide-immunized rabbits developed sufficiently characteristic features that they satisfied the American College of Rheumatology criteria for the classification of systemic lupus erythematosus (66). A proposed model of how this might occur is presented in Fig. 3.

The difference between the PPPGMRPP monomer and the MAP multimer of PPPGMRPP as immunogens is dramatic. In rabbits the monomer requires repeated boosting over many months for the development of even the smallest evidence of B-cell epitope spreading, and these immunized animals do not have obvious clinical evidence of lupus. This is in contrast to the ease with which B-cell epitope spreading and lupus autoimmunity have been induced in our hands with the MAP multimer of PPPGMRPP. Another laboratory has presented data confirming B-cell epitope spreading by immunization with the MAP multimer of PPPMRPP, but in their hands the response is more muted (39).

In this model of B-cell epitope spreading and induction of lupus autoimmunity we do not know many of the important variables. The improved performance of the MAP multimer of PPPGMRPP relative to the performance of the monomer of PPPGMRPP may be due to minor structural variants in its chemical preparation, to the potential carrier effect of the polylysine pyramid (in the sense of a hapten carrier), to the de novo T-cell (and/or B-cell, for that matter) epitopes now available to the immune system from the lysine pyramid (as well as from combinations of the lysine and PPPGMRPP), to the multivalency of the peptide, or to some other reason.

Subsequent experiments have shown that other peptides taken from the Sm and nRNP antigens are capable of inducing B-cell epitope spreading. The first example of this was immunization with the mistaken PPPGIRGP, mistaken in the sense that this is derived from the sequence of Sm N instead of Sm B/B′. Rabbits immunized with the mistaken PPPGIRGP mount a diversified immune response to the spliceosome in a manner very similar to that for the PPPGMRPP-immunized rabbits. These rabbits also mount a diversified response to the spliceosomal peptides. These PPPGIRGP-immunized rabbits actually developed antibodies to many different regions of the spliceosomal proteins, more than 50

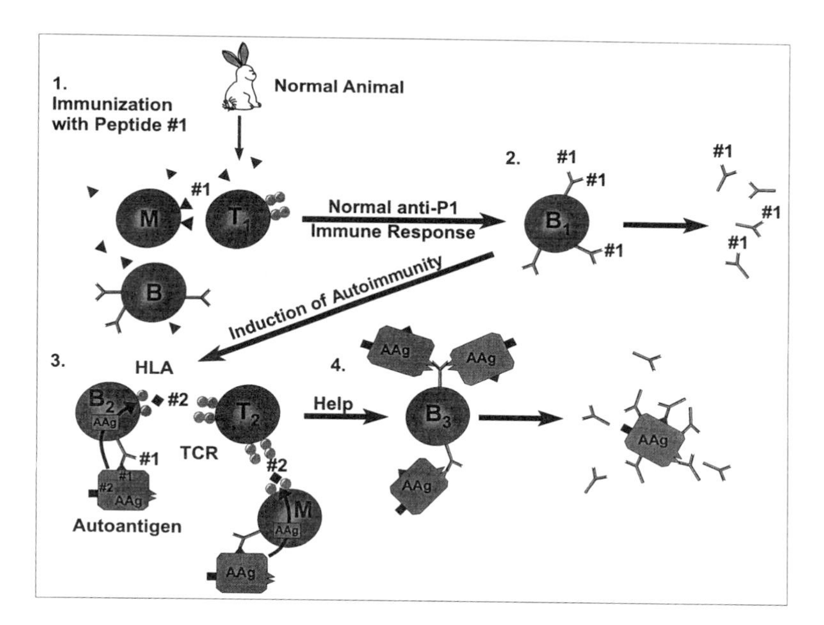

Figure 3. Model for peptide-induced lupus autoimmunity. Step 1, a normal animal is immunized with autoantigenic peptide (peptide #1) from a lupus autoantigen (AAg); step 2, this animal develops a normal antipeptide immune response; step 3, some of the B cells (B_2) which make antipeptide antibody also bind to this peptide on the surface of the lupus autoantigen; alternatively, the component of the antipeptide (Peptide #1) antibody which is also capable of binding to autoantigen binds to the Fc receptors of antigen-presenting cells and monocytes and macrophages bind to the T-cell receptor on T cells (T_2); step 4, these T cells are capable of providing help to other B cells (B_3), which then produce autoantibodies which can bind to structures in addition to the immunization peptide on the surface of the autoantigen. This figure is taken from reference 19. Copyright © 1998 Hogrefe & HuberPublishers. Reproduced with permission.

regions in the example presented (Fig. 4). These rabbits also developed antibodies to double-stranded DNA and to native Sm and/or nRNP proteins. In addition, rabbits have been immunized with the sequence which is commonly targeted after the PPPGMRPP motif in the Sm B/B′ response, amino acids 164 to 171.

Immunization with the major epitope of Sm D, which contains the glycine-arginine repeat, also produces a diversified response directed against Sm D1, D2, and D3, as well as other spliceosomal autoantigens (unpublished observations). The nRNP A protein is also very interesting from this standpoint. Immunization with the two major epitopes, A3 and A6, results in quite different responses. Immunization with A3 constructed on a polylysine backbone produces an autoimmune response directed against various region of the nRNP A, nRNP 70-kDa, nRNP C, Sm B/B′, and Sm D proteins. In addition, these animals develop anti-native DNA antibodies. On the other hand, rabbits immunized with the A6 peptide (again constructed on a branching, polylysine backbone) mount a very limited immune response restricted only to the peptide used for immunization. In addition, immunization with a nonantigenic peptide of Sm B/B′ produces only a limited immune response against the peptide used for immunization.

Not all octapeptides appear to be capable of inducing autoimmunity through B-cell epitope spreading, but many do. The antigenic peptides appear to induce autoimmunity and B-cell epitope spreading in this system more easily than the nonantigenic peptides do, but a few nonantigenic octapeptides of the 60-kDa Ro system clearly induce B-cell epitope spreading.

These observations lead to the following proposition which bears upon the etiology of lupus autoimmunity. Many peptides are capable of inducing lupus autoimmunity upon immunization, but only a very small number (only one PPPGMRPP is defined to date) actually do so in the naturally arising autoimmune response observed in humans. This is consistent with the existence of a similarly restricted number of possible origins of anti-Sm and anti-nRNP autoimmunity and, by extension, the clinical illness lupus associated with these specificities.

Ro IMMUNIZATION

In experiments in which rabbits were immunized with several different short peptides from the sequence of the 60-kDa Ro peptide, we found that use of any of these peptides resulted in epitope spreading in rabbits, including (as mentioned above) peptides not bound by human sera with anti-Ro antibodies. These animals developed antibodies that bound to many epitopes of the 60-kDa Ro peptide as well as anti-La, anti-native DNA, and antinuclear antibodies with different patterns. Absorption studies showed that the epitope spreading was not due to cross-reactive antipeptide antibodies and that the expanded response became progressively complicated over time (60).

Immunization with Ro peptides has also been carried out with mice. We find epitope spreading with development of full-blown anti-Ro and anti-La in some strains of mice such as the DBA/2J, BALB/c, and SJL/J strains but not in others, including the C57BL/6J and PL/J strains, using a peptide that contains amino acid residues 480 to 494 of the 60-kDa Ro peptide. In BALB/c mice, other peptides have also been used for immunization. One peptide, Ro 413, does not induce expansion of the immune response beyond the immunogen in these animals (58). Of interest, immunization of rabbits with this peptide did eventually result in epitope spreading (60). Comparison of sequences across species at

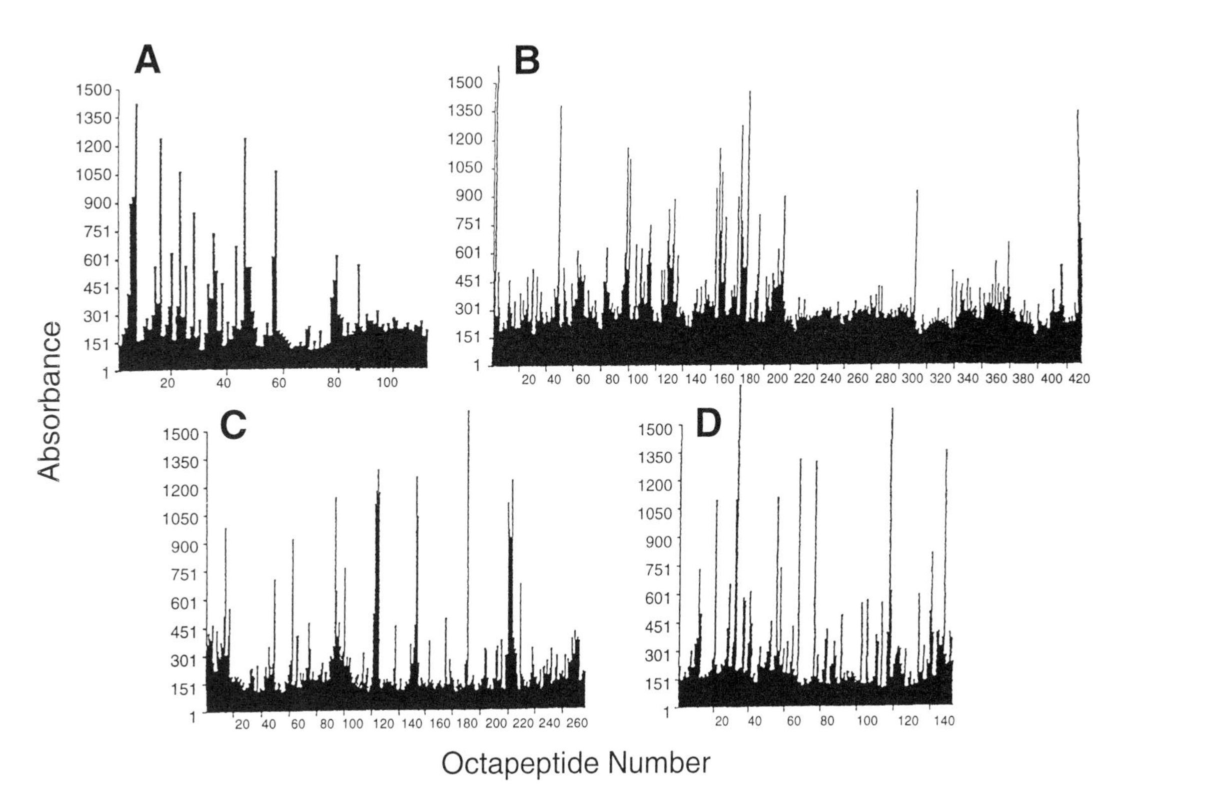

Figure 4. Binding of one PPPGIRGP-immunized rabbit serum sample with overlapping octapeptides of the Sm and nRNP proteins which do not contain the immunization peptide. After 20 weeks into the immunization protocol, serum from this one rabbit binds to many different regions of Sm D (A), nRNP 70K (B), nRNP A (C), and nRNP C (D). Preimmunization sera do not bind to any of these octapeptides at a level above the background binding. The background reactivity for these studies is equal to an optical density of <0.200 (27). A portion of this figure is reproduced with the permission of Munksgaard International Publishing Ltd., Malden, Mass.

these residues of the 60-kDa Ro peptide shows that this peptide is highly conserved between humans and rabbits but is divergent in mice. In addition, in BALB/c mice immunization with Ro 413 may induce immune deviation that favors a TH_2 T-helper-cell phenotype (58). This could account for the lack of B-cell epitope spreading. All these experiments were carried out with monomer peptides, not MAPs.

Recently, we have compared MAP constructs and monomer peptides that represent several known T-cell epitopes of La. These studies with AKR/J mice show that immunization with a MAP results in a higher titer and more rapid epitope spreading when a subdominant T-cell epitope is used. Also, a tolerogenic (dominant) T-cell epitope can be made antigenic by use of the MAP construct (12).

There is no systemic lupus erythematosus-like illness in mice or rabbits immunized with 60-kDa Ro peptides, but preliminary data indicate that SJL/J mice immunized with the Ro 480 peptide have inflammatory infiltrates of the salivary glands similar to those found in humans with Sjögren's syndrome.

Sm PEPTIDE IMMUNIZATION

Findings from immunochemical studies of fine specificity of Sm and nRNP have important implications for the development of human lupus. Our data suggest that it is possible for lupus autoimmunity in the Sm system to begin from an immune response against any one of a number of different structures. Yet, in the naturally arising autoimmune response against Sm in human lupus, anti-Sm appears to always (at least to the time of this writing) begin with PPPGMRPP. This means that the particular route to lupus (when anti-Sm is important) is limited to this structure. This further means that PPPGMRPP is an antigenic structure upon which we should base our understanding of the immune events that precede and lead to Sm autoimmunity.

Most of the initial immunization work with the MAP multimer PPPGMRPP was done with rabbits. This was for the historical reason that we had originally intended to develop serological reagents, not a new model of epitope spreading and lupus autoimmunity. These developments dictated that we establish the model with mice. Thirteen parental strains of available recombinant inbred sets were screened for the ability of lupus autoimmunity and B-cell epitope spreading to be induced in them. All of these strains developed anti-PPPGMRPP antibodies after immunization and boosting with the MAP multimer of PPPGMRPP. About half of the strains underwent B-cell epitope spreading and developed lupus autoimmunity. Only strains AKR/J and A/J developed anti-native DNA antibodies (26). On the basis of the data obtained with the recombinant inbred progenitor strains, the 14 recombinant inbred strains from the set generated from a cross of AKR/J and C57BL/J were obtained and immunized. The determination of the phenotypes of these animals suggests a genetic effect near the telomere of murine chromosome 4 (unpublished data). This genetic effect appears to be autosomal dominant since it is present in the Fl-generation from AKR/J × C57BL/J crosses. In this setting a single gene may be the difference between these two parental strains and may determine whether B-cell epitope spreading and lupus autoimmunity occur (unpublished data).

When we first evaluated the possible analogs with PPPGMRPP in the sequence databases in 1992, the most obvious similarity to PPPGMRPP in a potential etiological agent among the more than 12,000 primary sequences then available was PPPGRRP. This PPP-

GRRP sequence differs by only one amino acid from the amino-terminal seven amino acids of PPPGMRPP. PPPGRRP is found in the Epstein-Barr virus nuclear antigen (EBNA) 1 (EBNA-1; positions 398 to 404). A role for Epstein-Barr virus in lupus has previously been examined, with inconclusive results (11, 36, 53, 61, 67, 69). We reasoned that if this were an important, biologically relevant relationship, then anti-Sm antibody-positive patients should have antibodies to the EBNA-1 structure and we should be able to induce lupus autoimmunity by immunization with this structure. Another interesting parallel to this finding is that the glycine-arginine repeat of Sm D1 (which is also found in EBNA-1) is the major antigenic region of this other spliceosomal protein (29, 51, 54). A truncated form of this sequence is found to be repeated several times in EBNA-1.

If the cross-reactivities of these anti-Epstein-Barr virus responses were in some way to lead to lupus autoantibodies, then patients with anti-Sm antibodies should have antibodies which bind to these EBNA-1 proteins. The sera of patients, controls with rheumatic disease, and healthy controls have been tested for their reactivities with these similar viral and autoantigen sequences. Anti-Sm antibody-positive patients have statistically significant increased titers of antibodies to the PPPGRRP and GRGRGRG of EBNA-1 compared to those of control patients (Fig. 5). In addition, systemic lupus erythematosus patients without anti-Sm antibodies have been shown to have statistically significant increased concentrations of antibodies to the GRGRGRG and GRGRGRGRGRGRRG sequences of EBNA-1 and Sm D1 (29, 54). Antibodies to these peptides have not been found in patients with rheumatoid arthritis (29, 56).

Immunization with the MAP multimer of PPPGRRP generated anti-PPPGMRPP, B-cell epitope spreading throughout the Sm and nRNP polypeptides, and lupus autoimmunity in some animals (30). This result shows that it is possible that an immune response against a structure from the Epstein-Barr virus could generate the autoimmunity of lupus. These findings have served to make Epstein-Barr virus a more serious possible etiology for lupus.

EPSTEIN-BARR VIRUS AS A POTENTIAL ENVIRONMENTAL TRIGGER

If Epstein-Barr virus is a possible etiology, then infection with this virus must be associated with lupus. Finding of an association in adults is difficult because the infection rate is so high in healthy adults (>90%). In healthy children, on the other hand, the rate is much lower.

Consequently, we collected sera from children and teenagers with lupus, along with appropriate controls, and then compared the frequency of Epstein-Barr virus seropositivity in the two groups. More than 99% of the lupus patients showed serological evidence of Epstein-Barr virus infection. Meanwhile, only 70% of the controls were positive, which was the expected rate for average 15-year-old subjects (28). Absorption experiments showed that the anti-Sm and anti-Epstein-Barr virus antibodies used in this assay are separate, apparently not cross-reactive antibody populations. There was no relationship of anti-Epstein-Barr virus seropositivity with immunoglobulin level.

In addition, sera from these patients and controls were tested for their reactivities with cytomegalovirus, herpes simplex virus type 1, herpes simplex virus type 2, and varicella-zoster virus (28). No significant association of lupus with infection with any of the other herpesviruses was detected by use of a linear regression model. (When analyzed alone, the incidence of anti-herpes simplex virus type 2 antibodies after deletion of the Epstein-Barr virus effect is slightly higher than that for matched controls. This finding has not been con-

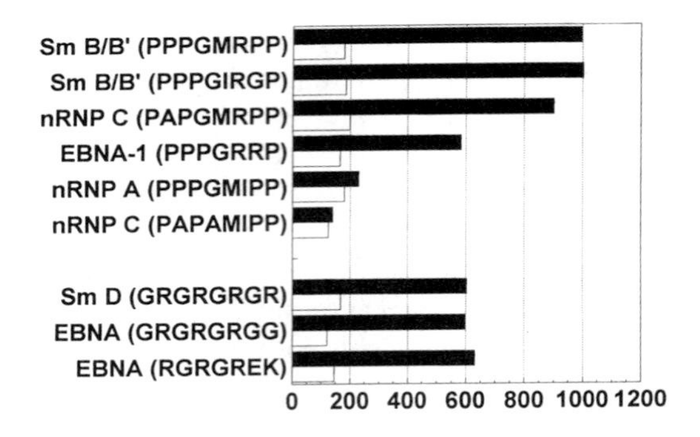

Figure 5. Binding to select peptides of EBNA-1 and spliceosomal autoantigens. Binding is presented for an average of 10 anti-Sm precipitin-positive patient serum samples (black bars). Binding of sera from an average of 10 healthy controls is also presented (white bars). The proline-rich sequences are presented above, and the glycine-arginine repeat sequences are presented below. (Some data are taken from references 18 and 24.)

firmed in a subsequent study of adults with lupus [unpublished observation].) The association of anti-Epstein-Barr virus antibodies and lupus has been confirmed with another group of children and teenagers (20). When analyzed together, the combined results for these three groups of young lupus patients are presented and are quite convincing for a strong association of previous Epstein-Barr virus exposure in systemic lupus erythematosus patients compared to that for controls (Table 2). Again, no strong association with the disease state is found with the other herpesviruses tested (Table 3). The essence of these results has also recently been extended to and confirmed in studies with adults (unpublished data).

Since lupus is a serological disease, possible etiological conclusions supported by only serological data are suspect. Therefore, an assay of Epstein-Barr virus DNA was adapted to the purpose of determining whether the latent Epstein-Barr virus infection of B cells would support the results of the serological assays. A matched case-control design was used in which relatives or acquaintances of the lupus patient were used as controls (28). The study was again restricted to children and teenagers in order to have a sufficiently low rate among the controls that a small study would have the statistical power needed

Table 2. Seroconversion against Epstein-Barr virus viral capsid antigen in sera from young lupus patients and controls[a]

Group	No. positive (total no.)		
	Original study	Confirmatory study	Combined
Lupus patients	116 (117)	26 (26)	142 (143)
Controls	107 (153)	14 (26)	121 (179)
Odds ratio	49.9	≥ 25[b]	68.06
95% CI[c] of odds ratio	9.3, 1,025	4.2, ∞	
Probability	−0.00000000000421	0.00024	0.000197

[a] These data are taken from references 20 and 28.
[b] Using correction for zero cell of 0.5.
[c] CI, confidence interval.

Table 3. Seroconversion frequencies in pediatric lupus and controls for IgG binding to cytomegalovirus, herpes simplex virus type 1, herpes simplex virus type 2, and varicella-zoster virus antigens[a]

Study and group	No. of patients positive (total no.)			
	CMV	HSV-1	HSV-2	VZV
Original study				
Lupus patients	42 (117)	72 (117)	59 (117)	102 (117)
Controls	40 (152)	74 (153)	51 (153)	132 (153)
Confirmatory				
Lupus patients	11 (26)	15 (26)	12 (26)	24 (26)
Controls	8 (26)	14 (26)	7 (26)	23 (26)
Combined				
Lupus patients	53 (143)	87 (143)	71 (143)	126 (143)
Controls	48 (178)	88 (179)	58 (179)	155 (179)
Odds ratio	1.59	1.61	2.06	1.15
Probability	0.11	0.061	0.0056	0.82

[a] These data are taken from references 20 and 28. Abbreviations: CMV, cytomegalovirus; HSV-1, herpes simplex virus type 1; HSV-2, herpes simplex virus type 2; VZV, varicella-zoster virus.

to detect a difference. Epstein-Barr virus DNA was detected in all 32 of the lupus patients and 23 of the controls. Among the Epstein-Barr virus, DNA-positive patients and controls, all except two controls were serologically positive. One of the two controls was available and was serologically positive upon repeat testing, thereby supporting the contention that this teenager had recently been infected with Epstein-Barr virus and that the first specimen had been collected before serological conversion had occurred.

In any case, of the matched case-control pairs, nine pairs contained a positive patient and a negative control subject. Meanwhile, in none of the pairs was a patient negative and a control subject positive for Epstein-Barr virus DNA. The difference here is significant ($P < 0.01$), again powerfully supporting the association of the presence of Epstein-Barr virus infection with lupus (28).

Now we are at the beginning. We have used molecular mimicry to identify a candidate etiological factor. A myriad of additional predictions and molecular relationships must be defined and must be consistent with the possibility that Epstein-Barr virus is importantly involved in the pathogenesis of lupus before it will generally be accepted that Epstein-Barr virus is etiologically relevant to lupus. The animal model of lupus and of B-cell epitope spreading is also likely to teach fundamental lessons about the immune system and of the immunopathology of lupus autoimmunity.

REFERENCES

1. **Anderson, J. R., K. G. Gray, J. S. Beck, W. U. Buchanan, and A. J. McElhinney.** 1962. Precipitating antibodies in the connective tissue diseases. *Ann. Rheum. Dis.* **21:**360–369.

1a **Arbuckle, M. R., M. Reichlin, J. B. Harley, and J. A. James.** 1999. Shared early autoantibody recognition events in the development of anti-Sm B/B′ in human lupus. *Scand. J. Immunol.* **50:** 447–455.

2. **Arbuckle, M. R., A. R. Schilling, J. B. Harley, and J. A. James.** 1998. A limited lupus anti-spliceosomal response targets a cross-reactive proline-rich motif. *J. Autoimmun.* **11:**1–8.

3. **Ben-Chetrit, E., E. K. Chan, K. F. Sullivan, and E. M. Tan.** 1988. A 52-kD protein is a novel component of the SS-A/Ro antigenic particle. *J. Exp. Med.* **1670:**1560–1571.

4. **Ben-Chetrit, E., B. J. Gandy, E. M. Tan, and K. F. Sullivan.** 1989. Isolation and characterization of a cDNA clone encoding the 60-kD component of the human SS-A/Ro ribonucleoprotein autoantigen. *J. Clin. Investig.* **83:**1284–1292.
5. **Clark, G., M. Reichlin, and T. B. Tomasi.** 1969. Characterization of a soluble cytoplasmic antigen reactive with sera from patients with systemic lupus erythematosus. *J. Immunol.* **102:**117–122.
6. **Deshmukh, U. S., J. E. Lewis, F. Gaskin, C. C Kannapell, S. T. Waters, Y. H. Lou, K. S. Tung, and S. M. Fu.** 1999. Immune responses to Ro60 and its peptides in mice. I. The nature of immunogen and endogenous autoantigen determine the specificities of the induced autoantibodies. *J. Exp. Med.* **189:**531–540.
7. **Deutscher, S. L., J. B. Harley, and J. D. Keene.** 1988. Molecular analysis of the 60-kDa human Ro ribonucleoprotein. *Proc. Natl. Acad. Sci. USA* **85:**9479–9483.
8. **Dorner, T., R. Chaoui, E. Feist, B. Goldner, K. Yamamoto, and F. Hiepe.** 1995. Significantly increased maternal and fetal lgG autoantibody levels to 52 kD Ro (SSA) and La (SSB) in complete congenital heart block. *J. Autoimmun.* **8:**675–684.
9. **Elkon, K. B., E. Bonfa, and N. Brot.** 1992. Antiribosomal antibodies in systemic lupus erythematosus. *Rheum. Dis. Clin. N. Am.* **18:**377–390.
10. **Esquivel, P. S., N. R. Rose, and Y. C. M. Kong.** 1976. Induction of autoimmunity in good and poor responder mice with mouse thyroglobulin and lipopolysaccharide. *J. Exp. Med.* **145:**1250–1263.
11. **Evans, A. S., N. F. Rothfield, and J. C. Niederman.** 1971. Raised antibody titers to E. B. virus in systemic lupus erythematosus. *Lancet* **i:**167–168.
12. **Farris, A. D., L. Brown, P. Reynolds, J. B. Harley, J. A. James, R. H. Scofield, J. McCluskey, and T. P. Gordon.** 1999. Induction of autoimmunity by multivalent immunodominant and subdominant T cell determinants of La (SS-B). *J. Immunol.* **162:**3079–3087.
13. **Fatenjad, S., M. J. Mamula, and J. Craft.** 1993. Role of intermolecular/intrastructural B- and T-cell determinants in the diversification of autoantibodies to ribonucleoprotein particles. *Proc. Natl. Acad. Sci. USA* **90:**12010–12014.
14. **Fleck, M., E. R. Kern, T. Zhou, B. Lang, and J. D. Mountz.** 1998. Murine cytomegalovirus induces a Sjogren's syndrome-like disease in C57Bl/6-*lpr/lpr* mice. *Arthritis Rheum.* **41:**2175–2184.15.
15. **Fournel, C., L. Chabanne, C. Caux, J. R. Faure, D. Rigal, J. P. Magnol, and J. C. Monier.** 1992. Canine systemic lupus erythematosus. I. A study of 75 cases. *Lupus* **1:**133–139.
16. **Hahn, B. H.** 1998. Antibodies to DNA. *N. Engl. J. Med.* **338:**1359–1368.
17. **Hardgrave, K. L., B. Neas, R. H. Scofield, and J. B. Harley.** 1993. Antibodies to vesicular stomatitis virus proteins in patients with systemic lupus erythematosus and normals. *Arthritis Rheum.* **36:**962–970.
18. **Harley, J. B., and J. A. James.** 1995. Autoepitopes in lupus. *J. Lab. Clin. Med.* **126:**509–516.
19. **Harley, J. B., and J. A. James.** 1998. Is there a role for Epstein Barr virus in lupus. *Immunologist* **6:**79–83.
20. **Harley, J. B., and J. A. James.** 1999. EBV infection may be an environmental risk factor for SLE in children and teenagers. *Arthritis Rheum.* **42:**1782–1783.
21. **Huang, S. C., Z. Pan, B. T. Kurien, J. A. James, J. B. Harley, and R. H. Scofield.** 1995. Immunization with vesicular stomatitis virus nucleocapsid protein induces autoantibodies to the 60 kD Ro ribonucleoprotein particle. *J. Investig. Med.* **43:**151–158.
22. **Hubert, B., M. Teichner, C. Fournel, and J. C. Monier.** 1988. Spontaneous familial systemic lupus erythematosus in a canine breeding colony. *J. Comp. Pathol.* **98:**81–89.
23. **Itoh, Y., and M. Reichlin.** 1992. Autoantibodies to the Ro/SSA autoantigen are conformation dependent. I. Anti-60 kD antibodies are mainly directed to the native protein; anti-52kD antibodies are mainly directed to the denatured protein. *Autoimmunity* **14:**57–65.
24. **James, J. A., and J. B. Harley.** 1992. Linear epitopes of Sm B/B′. *J. Immunol.* **46:**2073–2077.
25. **James, J. A., and J. B. Harley.** 1996. Human lupus anti-spliceosome A protein autoantibodies bind contiguous surface structures and segregate into two sequential epitope binding patterns. *J. Immunol.* **156:**4018–4026.
26. **James, J. A., and J. B. Harley.** 1998. A model of peptide-induced lupus autoimmune B cell epitope spreading is strain specific and is not H-2 restricted in mice. *J. Immunol.* **160:**502–508.
27. **James, J. A., and J. B. Harley.** 1998. B-cell epitope spreading in autoimmunity. *Immunol. Rev.* **164:**185–200.
28. **James, J. A., K. M. Kaufman, A. D. Farris, E. Taylor-Albert, T. J. A. Lehman, and J. B. Harley.** 1997. An increased prevalence of Epstein-Barr virus infection in young patients suggests a possible etiology for systemic lupus erythematosus. *J. Clin. Invest.* **100:**3019–3026.
29. **James, J. A., M. J. Mamula, and J. B. Harley.** 1994. Sequential autoantigenic determinants of the small

nuclear ribonuclear protein Sm D are shared by human lupus autoantibodies and MRL *lpr/lpr* antibodies. *Clin. Exp. Immunol.* **98:**419–426.

30. **James, J. A., R. H. Scofield, and J. B. Harley.** 1997. Lupus humoral autoimmunity after short peptide immunization. *Ann. N.Y. Acad. Sci.* **815:**124–127.
31. **James, J. A., T. F. Gross, R. H. Scofield, and J. B. Harley.** 1995. Immunoglobulin epitope spreading and autoimmune disease after peptide immunization: Sm B/B′-derived PPPGMRPP and PPPGIRGP induce spliceosome autoimmunity. *J. Exp. Med.* **181:**453–461.
32. **Jones, D. R.** 1993. Canine systemic lupus erythematosus: new insights and their implications. *J. Comp. Pathol.* **108:**215–228.
33. **Kaliyaperumal, A., C. Mohan, W. Wu, and S. K. Datta.** 1995. Nucleosome peptide epitopes for nephritis-inducing T helper cells of murine lupus. *J. Exp. Med.* **183:**2459–2469.
34. **Kaliyaperumal, A., C. Mohan, W. Wu, and S. K. Datta.** 1996. Nucleosomal peptide epitopes for nephritis-inducing T helper cells of murine lupus. *J. Exp. Med.* **183:**2459–2469.
35. **Keech, C. L., T. P. Gordon, and J. McCluskey.** 1996. The immune response to 52-kDa Ro and 60-kDa Ro is linked in experimental autoimmunity. *J. Immunol.* **157:**3694–3699.
36. **Klipple, J. H., J. L. Decker, P. M. Grimley, A. S. Evans, and N. F. Rothfield.** 1973. Epstein-Barr virus antibody and lymphocyte tubuloreticular structures in systemic lupus erythematosus. *Lancet* **i:**1057–1058.
37. **Lerner, M. R., J. A. Boyle, J. A. Hardin, and J. A. Steitz.** 1981. Two novel classes of small ribonucleoproteins detected by antibodies associated with lupus erythematosus. *Science* **211:**400–402.
38. **Mamula, M. J.** 1998. Epitope spreading: the role of the self peptides and autoantigen processing by B lymphocytes. *Immunol. Rev.* **164:**231–239.
39. **Mason, L. J., L. M. Timothy, D. A. Isenberg, and J. K. Kalsi.** 1999. Immunization with a peptide of Sm B/B′ results in limited epitope spreading but not autoimmune disease. *J. Immunol.* **162:**5099–5105.
40. **Miranda Carús, M. E., M. Boutjdir, C. E. Tseng, F. DiDonato, and J. P. Byon.** 1998. Induction of antibodies reactive with SSA/Ro-SSB/La and development of congenital heart block in a murine model. *J. Immunol.* **161:**5886–5892.
41. **Monier, J. C., M. Dardenne, D. Rigal, O. Costa, C. Fournel, and M. Lapras.** 1980. Clinical and laboratory features of canine lupus syndromes. *Arthritis Rheum.* **23:**294–301.
42. **Moore, D. J.** 1976. Canine systemic lupus erythematosus. The disease, clinical manifestations and treatments. *J. S. Afr. Vet. Assoc.* **47:**267–275.
43. **Morel, L., and E. K. Wakeland.** 1998. Susceptibility to lupus nephritis in the NZB/W model system. *Curr. Opin. Immunol.* **10:**718–725.
44. **Peter, J. B., and Y. Shoenfeld.** 1996. *Autoantibodies.* Elsevier, New York, N.Y.
45. **Peterson, J., G. Rhodes, J. Roudier, and J. H. Vaughan.** 1990. Altered immune response to glycine-alanine-rich sequence of Epstein-Barr nuclear antigen-1 in patients with rheumatoid arthritis and systemic lupus erythematosus. *Arthritis Rheum.* **33:**993–1000.
46. **Pringle, C. R.** 1987. Rhabdovirus genetics, p. 167–243. *In The Rhabdoviruses.* Plenum Press, New York, N.Y.
47. **Putterman, C., and B. Diamond.** 1998. Immunization with a peptide surrogate for double-stranded DNA (dsDNA) induces autoantibody production and renal immunoglobulin deposition. *J. Exp. Med.* **188:**29–38.
48. **Reeves, W. H.** 1992. Antibodies to the p70/p80 (Ku) antigens in systemic lupus erythematosus. *Rheum. Dis. Clin. N. Am.* **18:**391–414.
49. **Reynolds, P., T. P. Gordon, A. W. Purcell, D. C. Jackson, and J. McCluskey.** 1996. Hierarchical self-tolerance to T cell determinants within the ubiquitous nuclear self-antigen La (SS-B) permits induction of systemic autoimmunity in normal mice. *J. Exp. Med.* **184:**1857–1870.
50. **Rich, M. W.** 1996. Drug-induced lupus. The list of culprits grows. *Postgrad. Med.* **100:**299–302.
51. **Rivkin, E., M. Vella, and R. G. Lahita.** 1994. A heterogeneous immune response to an Sm D-like epitope by SLE patients. *J. Autoimmun.* **7:**119–132.
52. **Rose, N. R., and E. Talor.** 1991. Antigen-specific immunoregulation and autoimmune thyroiditis. *Ann. N. Y. Acad. Sci.* **636:**306–320.
53. **Rothfield, M. F., A. S. Evans, and J. C. Niederman.** 1973. Clinical and laboratory aspects of raised virus antibody titers in systemic lupus erythematosus. *Ann. Rheum. Dis.* **32:**238–245.
54. **Sabbatini A. S., S. Bombardieri, and P. Migliorini.** 1993. Autoantibodies from patients with systemic lupus erythematosus bind a shared sequence of Sm D and Epstein-Barr nuclear antigen-1. *Eur. J. Immunol.* **23:**1146–1152.
55. **Satoh, M., E. L. Treadwell, and W. H. Reeves.** 1995. Pristane induces high titers of anti-Su and anti-

nRNP/Sm autoantibodies in BALB/c mice. Quantitation by antigen capture ELISAs based on monospecific human autoimmune sera. *J. Immunol. Methods* **182:**51–62.

56. **Scofield, R. H., A. D. Farris, A. C. Horsfall, and J. B. Harley.** 1999. Fine specificity of the autoimmune response to the Ro/SSA and La/SSB ribonucleoproteins. *Arthritis Rheum.* **42:**199–209.
57. **Scofield, R. H., and J. B. Harley.** 1991. Autoantigenicity of Ro/SSA antigen is related to a nucleocapsid protein of vesicular stomatitis virus. *Proc. Natl. Acad. Sci. USA* **88:**3343–3347.
58. **Scofield, R. H., K. M. Kaufman, U. Baber, J. A. James, J. B. Harley, and B. T. Kurien.** 1999. Immunization of mice with human 60-kd Ro peptides results in spreading if the peptides are highly homologous between human and mouse. *Arthritis Rheum.* **42:**1017–1024.
59. **Scofield, R. H., W. D. Dickey, K. W. Jackson, J. A. James, and J. B. Harley.** 1991. A common autoepitope near the carboxyl terminus of the 60-kD Ro ribonucleoprotein: sequence similarity with a viral protein. *J. Clin. Immunol.* **11:**378–388.
60. **Scofield, R. H., W. E. Henry, B. T. Kurien, J. A. James, and J. B. Harley.** 1996. Immunization with short peptides from the sequence of the systemic lupus erythematosus-associated 60-kDa Ro autoantigen results in anti-Ro ribonucleoprotein autoimmunity. *J. Immunol.* **156:**4059–4066.
61. **Sculley, D. G., T. B. Sculley, and J. H. Pope.** 1986. Reactions of sera from patients with rheumatoid arthritis, systemic lupus erythematosus and infectious mononucleosis to Epstein-Barr virus-induced polypeptides. *J. Gen. Virol.* **67:**2253–2258.
62. **Seery, J. P., E. C. Y. Wang, V. Cattell, J. M. Carroll, M. J. Owen, and F. M. Watt.** 1999. A central role for alpha-beta T cells in the pathogenesis of murine lupus. *J. Immunol.* **162:**7241–7248.
63. **Seery, J. P., J. M. Carroll, V. Cattell, and F. M. Watt.** 1997. Antinuclear autoantibodies and lupus nephritis in transgenic mice expressing interferon gamma in the epidermis. *J. Exp. Med.* **186:**1451–1459.
64. **Singh, R. R., V. Kumar, F. M. Ebling, S. Southwood, A. Sette, E. E. Sercarz, and B. H. Hahn.** 1995. T cell determinants from autoantibodies to DNA can upregulate autoimmunity in murine systemic lupus erythematosus. *J. Exp. Med.* **181:**2017–2027.
65. **Strasser, A., S. Whittingham, D. L. Vaux, M. L. Bath, J. M. Adams, S. Cory, and A. W. Harris.** 1991. Enforced BCL2 expression in B-lymphoid cells prolongs antibody responses and elicits autoimmune disease. *Proc. Natl. Acad. Sci. USA* **88:**8661–8665.
66. **Tan, E. M., A. S. Cohen, J. F. Fries, A. T. Masi, D. J. McShane, N. F. Rothfield, J. G. Schaller, N. Talal, and R. J. Winchester.** 1982. The 1982 revised criteria for the classification of systemic lupus erythematosus. *Arthritis Rheum.* **25:**1271–1277.
67. **Tsai Y. T., B. L. Chiang, Y. F. Kao, and K. H. Hsieh.** 1995. Detection of Epstein-Barr virus and cytomegalovirus genome in white blood cells from patients with juvenile rheumatoid arthritis and childhood systemic lupus erythematosus. *Int. Arch. Allergy Immunol.* **106:**235–240.
68. **Witebsky, E., N. R. Rose, K. Terplan, J. R. Paine, and R. W. Egan.** 1957. Chronic thyroiditis and autoimmunization. *JAMA* **164:**1439–1447.
69. **Yokochi T., A. Yanagawa, Y. Kimura, and Y. Mizushima.** 1989. High titer of antibody to the Epstein-Barr virus membrane antigen in sera from patients with rheumatoid arthritis and systemic lupus erythematosus. *J. Rheumatol.* **16:**1029–1032.

Molecular Mimicry, Microbes, and Autoimmunity
Edited by M. W. Cunningham and R. S. Fujinami

Chapter 10

Mimicry between DNA, Carbohydrates, and Peptides: Implications in Systemic Lupus Erythematosus

Czeslawa Kowal and Betty Diamond

Systemic lupus erythematosus (SLE) is an autoimmune disorder that is complex in clinical manifestation and, most likely, in etiology. Aberrations at multiple steps of the normal immune response may lead solely or cumulatively to lupus autoimmunity. Microbial pathogens can influence the immune system of the host by a number of mechanisms, many of which might lead to or augment autoreactivity. These mechanisms include changes in major histocompatibility (MHC) class I and II expression and alterations in antigen processing and cytokine production. The focus of this chapter will be molecular mimicry of host structures by microbial pathogens, one of the possible mechanisms that contributes to autoreactivity in lupus.

That structural similarities between host antigens and antigens present on invading pathogens can lead to autoimmunity has been suggested for many autoimmune disorders and has recently been proved with animal models of some of these diseases (14, 53, 55). Molecular mimicry between microbial and mammalian structures as a cause of cross-reactive immune responses is often considered a consequence of linear sequence identity or homology between molecules that may be functionally unrelated and that are derived from two distinct sources. For the purpose of this chapter, we would like to extend the definition of molecular mimicry to include immunologically defined identity or homology of component parts of self-antigens and microbial antigens without obvious sequence homology. Thus, cross-reactive determinants might be present in distinct molecules, might represent antigenic epitopes with similar chemical and conformational features, and might be recognized by the same antibodies. For example, the repeating phosphate groups linked by phosphodiester bonds present on the sugar backbone of DNA and on phospholipids might explain the cross-reactivity of anti-DNA antibodies with phospholipids and double-stranded DNA (dsDNA) (24).

We will present evidence generated in our laboratory in support of molecular mimicry as a trigger for the autoantibody formation in lupus. We will discuss two antigen classes as potential forces in triggering the anti-self-response in SLE: bacterial polysaccharides

Czeslawa Kowal and Betty Diamond • Departments of Microbiology and Immunology, and Medicine, Albert Einstein College of Medicine, 1300 Morris Park Boulevard, Bronx, NY 10461.

and proteins. We will provide data suggesting that either of these antigens may lead to at least a subset of anti-self-reactivity and will present an animal model for the latter, demonstrating that this cross-reactivity can lead to tissue injury. It is important to remember that the antibody response matures through somatic mutation following B-cell activation. We will demonstrate that molecular mimicry between self-antigens and foreign antigens may be achieved or modified by this process and that antibodies which in their germ line-encoded configuration bind well to microbial antigen but weakly, if at all, to self-antigen may mutate to acquire higher levels of binding to self-antigen and may in the process either retain or lose their antimicrobial activity. Somatic mutation may thus play a crucial role in the induction of the anti-self-response and in shifting protective immune responses to cross-reactive, anti-self-, and antimicrobial responses or strictly anti-self-responses.

SLE AND THE ROLE OF ANTI-dsDNA IN LUPUS

Lupus is a systemic autoimmune disease characterized by the production of antibodies to a number of self-antigens. It is distinguished from other autoimmune disorders by an unprecedented diversity of anti-self-antigen-specificities present in a patient's serum, most of which are directed to nuclear antigens, such as nucleic acids and nucleoproteins, but antibodies that bind to mitochondrial lipid, cardiolipin, some cell surface antigens, and some components of the coagulation pathway may also be present. The most prominent among all autospecificities in SLE is the anti-dsDNA response, which is known to contribute significantly to renal pathology (50). Much effort has been expended, therefore, to understand what triggers the production of these antibodies.

Naked eukaryotic DNA is not immunogenic (30). It has been shown that DNA conjugated with some protein carriers can elicit an anti-dsDNA response, but the response is generally weak and the involvement of such anti-DNA antibodies in the induction of lupus or in pathogenicity is not yet clear (11). Bacterial DNA is immunogenic in mammals, but the resulting anti-DNA antibodies that arise in nonautoimmune strains of mice do not cross-react with mammalian DNA (17, 54). These antibodies are, therefore, unlike lupus anti-DNA antibodies, which bind to both bacterial and eukaryotic DNA. The eliciting antigen of anti-dsDNA antibodies in SLE remains unknown. Understanding of the origin of anti-DNA antibodies and the molecular basis of their interaction with DNA and other self-antigens as well as foreign antigens may lead to a greater understanding of the etiology of lupus and improve our knowledge of lupus pathogenesis.

It is known at present that a great number of anti-DNA antibodies in both patients and mouse models of lupus display two important features of an antigen-driven immune response: somatic hypermutation and heavy-chain class switching (39). Analysis of a large number of anti-DNA antibodies demonstrates that somatic mutation occurs in a manner consistent with an antigen-driven response, since the replacement mutation-to-silent mutation ratio is generally greater at the sites predicted by computer modeling to be DNA contact regions. These regions include all of the complementarity-determining regions (CDRs) and framework 3 (FR3) of the heavy chain (39). The nature of these amino acid replacements indicates an antigen-driven response as well, since the changes often lead to substitutions of amino acids associated with DNA binding (which will be discussed in detail in a later section).

EVIDENCE SUPPORTING MOLECULAR MIMICRY AS A SOURCE OF ANTI-SELF-REACTIVITY IN SLE

Studies of serum from patients with bacterial infections led to the realization that nonautoimmune hosts, during the time of infection, make antibodies with specificities similar to those of autoantibodies in sera from patients with SLE. Titers of anti-dsDNA and anticardiolipin antibodies were shown to increase in a significant percentage of serum samples following *Klebsiella pneumoniae* infection, and the increase was correlated with the presence of a common anti-DNA idiotype, idiotype 16/6 (13). Similar cross-reactivity between antigens of *Mycobacterium tuberculosis* and dsDNA was also reported (45).

Studies from our laboratory demonstrated the presence of two anti-DNA, lupus-associated idiotypes in the serum of patients following vaccination with pneumococcal polysaccharide. In one study, healthy individuals immunized with a polyvalent pneumococcal polysaccharide vaccine, Pneumovax-23, were examined (18). All immunized individuals showed an increased titer of antipolysaccharide antibody. More importantly, these antibacterial antibodies bore 3I, an idiotypic specificity present on anti-DNA antibodies in 70 to 80% of SLE patients. No antitetanus antibodies that bear the 3I idiotype were detected, and no DNA reactivity was detected in that study (18).

In a second study, serum samples from 14 lupus patients and 28 unaffected family members were examined following vaccination with pneumococcal polysaccharide (28). Serum was studied for the presence of antibodies bearing the 8.12 idiotype, a marker present on lambda light chains of anti-DNA antibodies in 50% of lupus patients. A rise in 8.12-positive, antipneumococcal antibodies was observed in 10 individuals, 8 lupus patients and 2 nonaffected family members, and serum samples from all but one lupus patient displayed cross-reactivity with dsDNA (28).

There is indirect evidence in a mouse model of SLE that microbial antigens are involved in the creation of lupus specificities; NZB mice bred in a germfree environment show delayed onset and diminished severity of disease (51). Conversely, (NZB × NZW)F1 mice infected with polyomavirus or lymphocytic choriomeningitis virus display an accelerated and/or enhanced form of lupus (34). In addition, the anti-DNA response and lupus pathology are attenuated in SLE-prone mice that bear the *xid* mutation (47). This observation is of particular relevance to a model for induction of anti-DNA antibodies developed in our laboratory, as the *xid* mutation leads to diminished production of antibacterial, antipolysaccharide antibodies (2).

POSSIBLE ROLE OF BACTERIAL POLYSACCHARIDES IN TRIGGERING THE ANTI-dsDNA RESPONSE IN HUMANS; COMBINATIONAL LIBRARY APPROACH

To elucidate the potential role of bacterial antigens in triggering the anti-dsDNA response in SLE, we decided to sample the B-cell repertoire following antibacterial vaccination with pneumococcal polysaccharides.

We have generated combinatorial libraries from the spleens of lupus patients who had been given a pneumococcal vaccine 10 days before splenectomy. While the combinatorial library is formed, in principle, by the random association of heavy and light chains of pre-existing antibodies, studies that have compared antibodies derived from combinatorial libraries to antibodies produced by B-cell hybridomas or transformed B-cell lines demonstrate striking similarities of heavy- and light-chain pairing (8, 10, 37, 42). Thus, we believe that the combinatorial library is reasonably representative of the in vivo B-cell repertoire. In order to explore the relationship between protective and pathogenic antibodies, we selected antibodies that express the 3I idiotype present on both anti-DNA and antipneumococcal antibodies. Since the affinity of antipneumococcal antibodies for carbohydrate epitopes is generally low, we opted to select antibodies that express an idiotypic specificity, an approach more likely to yield the desired antibodies. After affinity panning for Fab fragments that express the 3I idiotype, the selected clones were tested for binding to the bacterial antigens: immunizing polysaccharides and phosphorylcholine (PC) coupled to a protein carrier. PC is a major hapten of the bacterial cell wall polysaccharide present on all strains of *Streptococcus pneumoniae* and of the capsular polysaccharide of some serotypes of *S. pneumoniae*. In addition, we tested these clones for binding to dsDNA. Of 30 clones that displayed reactivity to any of the antigens described above, 8 possessed unique sequences for the heavy and light chains and the remaining were clonal repeats. Most of the clones displayed significant binding to at least one of the bacterial antigens, PC or pneumococcal polysaccharide, and four of them showed strong cross-reactivity with dsDNA (Fig. 1) (23). Reciprocal inhibition assays with dsDNA and pneumococcal polysaccharide performed with three of the cross-reactive antibodies demonstrated that dsDNA and polysaccharide bind at the same or a proximal site on these antibodies (Fig. 2). The fact that Fab fragments bind to both dsDNA and pneumococcal polysaccharide and, more importantly, the fact that they bind to both antigens at the same (or proximal) binding site imply that the interaction with both antigens may occur through the same contact residues on the antibody.

Furthermore, the inhibition assays performed with cross-reactive Fab fragments over a wide range of inhibitor concentrations indicate that dsDNA and pneumococcal polysaccharide, both polyvalent antigens, appear to bind with a high avidity and in a similar fashion throughout the range of inhibitor concentrations. A major question concerning cross-reactivity is how different antigens bind with similar strengths to the same antibody. What are the structural features shared by dsDNA and pneumococcal polysaccharide that contribute to their recognition by antibodies? Studies of protein-DNA interactions including antibody-DNA interactions provide valuable information regarding the DNA-protein interface (20, 39, 56). DNA-protein binding usually requires a more polar interface and a far greater number of hydrogen bonds, both direct and water mediated, than protein-protein interactions. The polar nature of the interface complements the negative charge on the surface of the DNA molecule. Arginine (R), which is hydrophilic and positively charged, has the greatest disposition for the DNA-protein interface, followed by the polar residues threonine (T), asparagine (N), and serine (S). Other polar amino acids such as glycine (G) and glutamine (Q) and two basic amino acids, lysine (K) and histidine (H), also provide contact residues for DNA binding. Proteins that interact with DNA in a dimeric or multimeric form, such as immunoglobulin, will provide large interfaces with multiple hydrogen bonds. In addition, in contrast to protein-DNA binding in monomeric form, which usually displays tightly packed, single interaction sites, the DNA-protein interactions in multimeric forms display sites that are larger, less tightly packed, and highly segmented (20). We postulate that this type of interaction site is well suited to the polyvalent nature not only of DNA but of polysaccharides as well. Two studies regarding the binding properties of

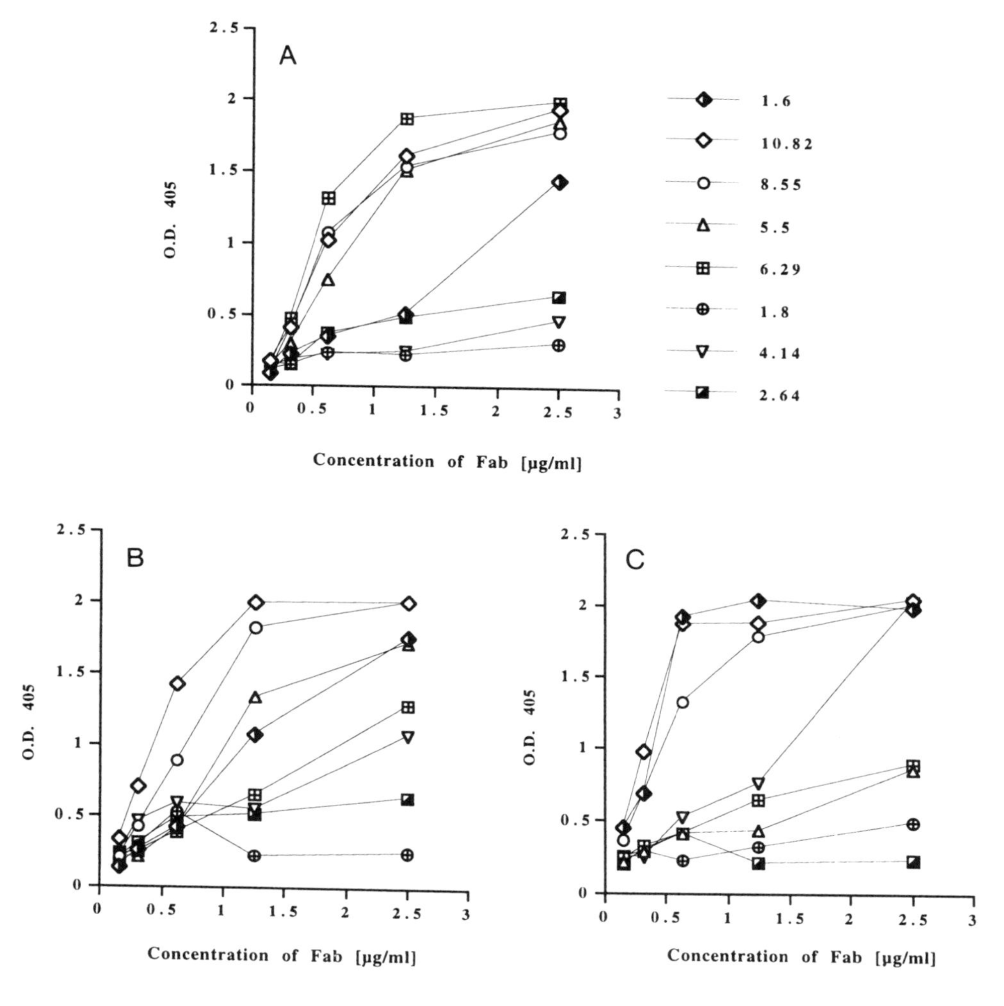

Figure 1. Binding of affinity-purified Fab fragments to bacterial and self-antigens. Panels show binding of Fab to pneumococcal polysaccharide (A), PC (B), and dsDNA (C). (Modified from reference 23).

antipolysaccharide antibodies support this conjecture. Surface plasmon resonance analysis of an antipolysaccharide antibody in both divalent and monovalent forms (immunoglobulin G [IgG] and scFv, respectively) demonstrated a 100-fold increase in functional affinity of the divalent molecule (29). Furthermore, vaccine studies with different polysaccharide conjugates revealed a relationship between the immunogenicity of the vaccine and the length of the polysaccharide conjugate (25, 36).

Finally, the contact residues that have been implicated in polysaccharide binding in the antibody binding site are strikingly close to residues known to facilitate DNA binding, including R, N, Q, and H (7). Crystal structure analysis of two antipolysaccharide antibodies

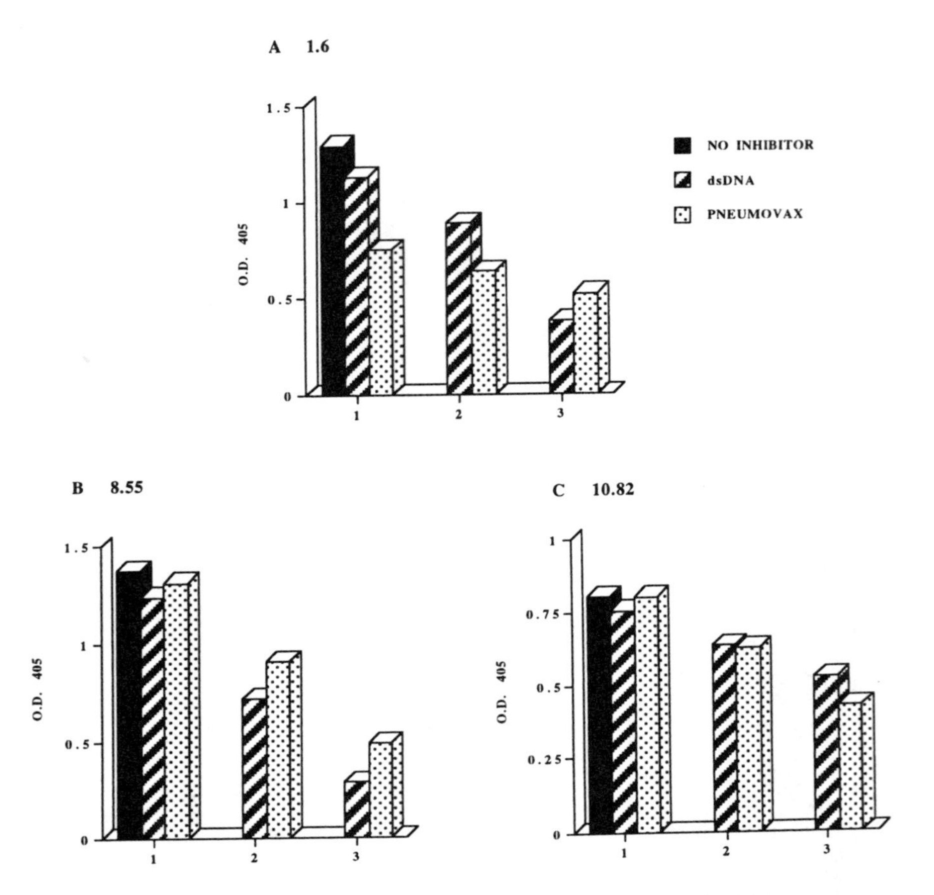

Figure 2. Competitive inhibition ELISA (modified from reference 23). The Fab fragment concentration was 1.25 μg/ml. The concentrations of dsDNA as soluble inhibitor were as follows: 0.00165 μg/ml (bar set 1), 0.0165 μg/ml (bar set 2), and 0.165 μg/ml (bar set 3). The concentrations of the pneumococcal polysaccharide were as follows: 0.002 μg/ml (bar set 1), 0.02 μg/ml (0.2 μg/ml for clone 8.55) (bar set 2), and 0.2 μg/ml (20 μg/ml for clone 8.55) (bar set 3). Values represent the mean optical densities (O.D.) for duplicate samples in the absence or presence of inhibitor.

disclosed an enrichment of aromatic residues, mainly tyrosines (Y), in the antibody binding site (9, 43). The DNA binding sites of anti-DNA antibodies are also enriched in aromatic residues.

It has been proposed, on the basis of crystallographic and modeling studies, that anti-dsDNA antibodies form a relatively flat binding surface (5). The characteristic deep pocket of antibodies that bind to single-stranded DNA, which is formed by variable heavy-chain (V_H) CDR2 and variable light-chain (V_L) CDR1 loops, cannot accommodate dsDNA. A shorter V_H CDR2 loop in combination with a longer V_HCDR3 loop in anti-dsDNA antibodies results in a flatter binding site (5). Since the Fab fragments examined in our laboratory bind to pneumococcal polysaccharide and dsDNA, it would be interesting to obtain their crystal structures to learn how cross-reactive antibodies accommodate dsDNA and pneumococcal polysaccharide.

The combinatorial library revealed preferential utilization of the V_H3 family, which may be a common trait of an antibacterial (antipolysaccharide) and anti-dsDNA response (23). The V_H3 heavy-chain family is overrepresented in these cross-reactive, dsDNA, and polysaccharide binding Fab fragments, as six of eight fragments are encoded by V_H3 family members. In addition, V_H3 genes encode five of six lupus-derived DNA-binding Fab fragments that have previously been reported (4). The human V_H3 heavy-chain family, consisting of 19 functional genes, represents almost half of the total number of 39 functional genes of the human heavy-chain gene locus (31) and is frequently expressed in the human immune response. However, in an analysis of 159 human antibodies derived from combinatorial libraries (86 antibodies) and from hybridomas (73 antibodies), only 43% expressed V_H3 heavy chains (33).

An interesting question that is raised by these observations is whether biased expression of V_H3 and cross-reactivity with dsDNA is a part of the antibacterial response in nonautoimmune individuals or whether it is lupus specific. There is only one hybridoma-derived, fully sequenced, antipneumococcal antibody, Dob1, that originated in a nonautoimmune individual (49). Interestingly, Dob1 uses the same V_H3-15 germ line gene as two of the cross-reactive Fab fragments from the combinatorial library. Furthermore, partial sequences of antibodies purified from the serum of nonautoimmune individuals given a pneumococcal vaccine demonstrated V_H3 gene usage in 10 of 12 preparations (49). Preferential use of V_H3 family genes in the antipneumococcal response in nonautoimmune individuals indicates, therefore, that the biased use of these genes in a lupus patient is a part of the normal, not aberrant, immune response. The same predominant V_H3 usage has been demonstrated in antibodies reactive with another bacterial capsular polysaccharide, that of *Haemophilus influenzae,* studied by other investigators (1, 46).

These data obtained from the combinatorial library demonstrate that cross-reactive antibacterial, anti-DNA antibodies can be found in a lupus patient following vaccination with pneumococcal polysaccharide. If potentially pathogenic cross-reactivity may develop with bacterial infection, does this offer some advantage to the host? Does cross-reactivity with DNA reflect a beneficial side to molecular mimicry, an immunological readiness for bacterial invasion? This raises the question of whether self-antigen is involved in shaping the preimmune repertoire. It has been suggested previously (3) that lupus-associated, mutated, high-affinity, dsDNA-binding antibodies may originate from natural autoantibodies present in preimmune serum of all individuals. Are these natural autoantibodies also precursors for protective antibacterial responses? Natural autoantibodies are usually of low affinity. They are unmutated, IgM antibodies and are expressed early in the fetal repertoire (3). Most of our cross-reactive Fab fragments (six of eight) are IgM. Furthermore, six of the eight heavy-chain genes used by these antibodies are abundantly expressed in the fetal repertoire. What distinguishes natural autoantibodies from these Fab fragments is somatic mutation. All but one heavy-chain genes display somatic mutations, suggesting selection in germinal centers. Whether germinal center maturation alters the specificity and affinity for self-antigen or foreign antigen of these antibodies is not known. Site-directed mutagenesis that reverts some of the genes to germ line sequences will provide useful information regarding the origin of the cross-reactivity.

This study discloses a structural and immunological relationship between dsDNA and bacterial polysaccharide. It remains to be shown that these antibodies were induced by immunization and that nonautoimmune individuals also generate these antibodies but have mechanisms to prevent their continued expression. We would speculate, however, that cross-reactive antigens like bacterial polysaccharide can excite production of anti-self-reactivity in all individuals and can lead to unregulated production of anti-self-reactivity in a susceptible host.

POSSIBLE ROLE OF SOMATIC MUTATION IN CREATING OR MODIFYING SELF-REACTIVITY DURING THE ANTIBACTERIAL RESPONSE

We have also studied the murine immune response to PC to look for a cross-reactive anti-self, antibacterial response. In this model, we have been able to show that somatic mutation can play a major role in generating cross-reactivity and anti-self-specificity. The anti-PC response in the BALB/c nonautoimmune mouse strain is predominantly encoded by a single heavy- and light-chain combination, S107, which has the V_H1-DFL16.1-J_H1 heavy chain and the V_κ22-J_κ5 light chain, and bears the T15 idiotype (16, 35). This antibody binds to PC in its germ line configuration, and there is no affinity maturation that leads to the appearance of antibodies with affinities for PC higher than that of the canonical unmutated antibody. The T15 antibody is produced in vitro by the S107 cell line. Several years ago, we identified a subclone of S107, U4, which contains a single amino acid mutation, an alanine (A)-to-aspartic acid (D) change in residue 35 of CDR1 of the S107 heavy-chain gene (12). The mutant line no longer binds to foreign antigen, PC, but has acquired binding to self-antigen, dsDNA. On the basis of this observation, we attempted in vivo studies to look for the generation of specificity for DNA during the course of a response to PC.

While we can detect a transient elevation of the titer of anti-DNA antibody in the serum of nonautoimmune BALB/c mice after PC-carrier immunization (44), our early attempts to recover DNA-reactive B cells from nonautoimmune mice immunized with PC were unsuccessful, because autoreactive B cells undergo programmed cell death or apoptosis. Only after interfering with the regulatory mechanisms of the immune system were we able to obtain anti-DNA, anti-PC cross-reactive hybridomas (27). Interestingly, five anti-PC, anti-dsDNA cross-reactive antibodies obtained during the immune response to PC were found to display structural and functional characteristics that differ from those of canonical anti-PC response. None of them is encoded by the S107 V_H1-DFL16.1-J_H1 heavy-chain segment (Table 1) that encodes the dominant T15 antibody. The typical V_κ22-J_κ5 light-chain combination is also absent. All five antibodies bind to both PC-keyhole limpet hemocyanin (KLH) and dsDNA; in addition, all stain cell nuclei in an antinuclear antibody assay. Moreover, all these cross-reactive antibodies have pathogenic potential as they cause proteinuria when injected into severe combined immunodeficient (SCID) mice and display typical SLE-type glomerular deposition in kidney sections. Three of these cross-reactive antibodies provide protection to mice challenged with a lethal dose of the WU2 strain of *S. pneumoniae*. Although this protection was equivalent to that conferred by the canonical IgM, T15-positive anti-PC antibody, these atypical antibodies comprise a very minor portion of the normal anti-PC response. Molecular analysis of these antibodies showed that the cross-reactivity could be germ line gene encoded and could be part of a preimmune repertoire (27). Cross-reactive antibodies that display both protective and

Table 1. V_H gene family usage in cross-reactive antibodies[a]

Clone	Heavy-chain family	Isotype	No. of mutations (no. of bases sequenced)
244.3g	S107	IgM	0 (301)
245F.6g	J558	IgM	1 (288)
246B.4g	J558	IgM	0 (288)
1410E.10e	J558	IgM	51 (294)
1411E.3b	Q52	IgM	10 (291)
b4H8	S107	IgM	3 (275)
b3B2	J558	IgM	ND[b]
b4D3	7183	IgM	8 (231)
b10-33	Q52	IgM	21 (269)
b5-32	J558	IgM	26 (200)
b6-14	J558	IgM	4 (200)
b8-8	Q52	IgM	16 (268)
b13-12	J558	IgM	9 (225)
b9-5	S107	IgM	2 (270)
b19-3	J558	IgM	4 (220)
b5-38	SM7	IgM	2 (212)

[a] Modified from references 27 and 41. The sequences were compared to that of the most homologous germ line gene in GenBank.
[b] ND, not determined.

potentially pathogenic specificities might be a transient part of an antibacterial reactivity in a normal host but are regulated to maintain self-tolerance. The restriction in V-gene usage that is so characteristic of the anti-PC serum response does not reflect an inability to form antibodies with other V_H and V_L gene segments that bind to PC and protect against a lethal infection but, rather, may help maintain self-tolerance.

The transient elevation of an anti-self-response (44) and the conversion of protective antibody into a self-reactive antibody in vitro by a single base pair change (12) led us to postulate that anti-DNA specificity may be routinely generated in vivo during an antibacterial response. The anti-DNA specificity is generated both by activation of cross-reactive antibodies and by the process of somatic mutation but is most likely downregulated by the normal host. Three major mechanisms have been postulated to account for downregulation of autoreactive B cells: deletion, anergy or functional inactivation, and receptor editing, which involves replacement of a self-reactive antibody by new immunoglobulin gene rearrangements. Autoreactive cells destined for deletion, anergy, or receptor editing could be a valuable source of information regarding the frequency and the nature of self-reactivity.

Apoptosis involves a strictly regulated sequence of events that lead to cell death. Studies of B-cell fate and B-cell apoptosis led to the identification of the Bcl-2 family of proteins (22). Expression of Bcl-2 has been shown to be essential in B-cell survival, and the sera of some mouse strains with constitutive expression of Bcl-2 in B cells demonstrate increased titers of autoantibodies (48). Furthermore, Bcl-2 helps mediate survival of autoreactive B cells in double transgenic mice which express both Bcl-2 and autoantibody (26).

We exploited these known properties of Bcl-2 to devise a methodology to sample potentially autoreactive B cells destined for apoptosis (40). NSO is a myeloma cell line routinely used to generate B-cell hybridomas. We transfected this line with cDNA that encodes the Bcl-2 protein under the control of the cytomegalovirus promoter and the Eμ immunoglobulin heavy-chain enhancer and demonstrated that Bcl-2 expression rescues NSO cells from staurosporin-, aminopterin-, and nutrient depletion-induced apoptosis.

Two studies were then performed to assess whether the Bcl-2-expressing fusion partner would aid in the analysis of the autoreactive B-cell population. In the first study we demonstrated that Bcl-2 could rescue autoreactive B cells from a transgenic mouse that expressed the heavy chain of an anti-DNA antibody (40). Nonautoimmune mouse strains that harbor the R4A-γ2b transgene, which encodes the heavy chain of a pathogenic, high-affinity dsDNA-reactive antibody, maintain tolerance and have low serum IgG2b anti-dsDNA titers (32). We have shown, however, the existence of a lipopolysaccharide-inducible, anergic population of IgG2b anti-DNA B cells in the spleens of these mice and have also demonstrated an apoptotic cell population (19, 32). Fusions were performed with splenocytes of naive R4A transgenic mice by using wild-type NSO and Bcl-2-transfected NSO cells. While none of 103 IgG2b-secreting hybridomas obtained from fusions with wild-type NSO cells produced antibodies which bound to dsDNA, 16 of 248 derived from Bcl-2-transfected NSO cells produced antibodies which bound to dsDNA with a high affinity (40).

In a second study, we analyzed the immune response in BALB/c mice following PC-KLH immunization (41). B-cell fusions were performed at different time points following immunization: day 2, day 5, day 10, and day 11. While all of the fusions generated PC-specific hybridomas, only fusions with Bcl-2-transfected NSO cells and only those performed on day 10 or 11 following immunization produced a large number of cross-reactive, PC- and dsDNA-binding hybridomas. We recovered only a single cross-reactive hybridoma from fusion with wild-type NSO cells, while approximately 40% of PC-binding clones from fusions with Bcl-2-transfected NSO cells displayed cross-reactivity to dsDNA. The heavy-chain gene utilization in these cells was quite diverse, and only 1 of 11 clones possessed the canonical anti-PC heavy- and light-chain gene rearrangements (Table 1). This clone, F101b9-5, is identical in heavy-chain and light-chain sequences to the canonical T15-positive antibody with the exception of a 2-bp difference at the D-J junction of the heavy chain and a single base pair difference in FR2 of the light chain. It differs from the T15-positive prototype by a single amino acid replacement, Y to H, in CDR3 of the heavy chain and a single amino acid change, E to D, at residue 47 in FR2 of the light chain. The antibody binds to PC with an apparent affinity similar to that of the germ line-encoded antibody but has acquired the ability to bind to dsDNA and thus represents the in vivo analog of the previously described U4 mutant. Three of five cross-reactive antibodies including F101b9-5 displayed glomerular deposition in SCID mice, thus demonstrating the potential pathogenicity of antibodies that arise routinely and at a high frequency in the nonautoimmune host following immunization with foreign antigen (41). As these antibodies display somatic mutations and can be found only 10 days or more following immunization, we have speculated that somatic mutation is, in fact, critical in generating the anti-self-specificity in many of these antibodies.

That study demonstrated the importance of somatic mutation in shifting the immune response in vivo from a protective, antibacterial response to a response that is cross-reactive with self-antigen and that has potential lupus-like pathogenicity. Furthermore, the study suggests that this potentially pathogenic response, which comprises up to 40% of the protective, antibacterial response, is routinely downregulated by apoptosis.

PROTEIN ANTIGENS AS A POTENTIAL SOURCE OF ANTI-dsDNA RESPONSE IN LUPUS: PEPTIDE MIMOTOPES

The question that we addressed next was the possibility that protein antigens also act as immunogens in the anti-DNA response. Peptide libraries have been particularly useful in identifying cross-reactive antigens with no obvious sequence homology to the initial antigen. To search for a protein antigen that can serve as a surrogate DNA antigen, we screened a phage display decapeptide library (52) using two closely related antibodies: R4A and 52b3. The first antibody, R4A, is, as described previously (44), a BALB/c-derived monoclonal IgG2b anti-dsDNA antibody that causes glomerular deposition in nonautoimmune mice. The second antibody, 52b3, is a mutant of R4A generated by site-directed mutagenesis and contains three amino acid substitutions, R to Q at residue 66, T to A at residue 82b, and R to S at residue 83 (21). These substitutions in FR3 of the heavy chain result in a rather unexpected 10-fold increase in the level of dsDNA binding and in a shift in tissue pathology, from glomerular to tubular deposition (21). Several rounds of affinity selection on the phage library were performed with R4A and 52B3 (15). Binding of R4A and 52b3 to selected phage was confirmed by enzyme-linked immunosorbent assay (ELISA). Peptide inserts of phage clones that displayed specific binding were sequenced, and two consensus motifs were identified, one for each antibody (Table 2). Although these motifs show some similarity (D/E × D/E), R4A and 52b3 antibodies display only minimal cross-reactivity with 52b3- and R4A-selected phages, respectively. Calf thymus DNA inhibits binding of both R4A and 52b3 to the appropriately selected phage. Peptides DWEYSVWLSM and RHEDGDWPRV inhibited dsDNA binding to R4A and 52b3, respectively, suggesting that DNA and peptide occupy the same or closely adjacent binding sites. R4A binding to dsDNA was also inhibited by DWEYS, a smaller consensus motif synthesized as both the L and the D isoforms. Peptide DWEYS in the D or L configuration was tested for potential blocking of the deposition of nephritogenic R4A antibody in vivo. Most serum proteases are specific for L amino acid enantiomers: D enantiomers are, therefore, more resistant to proteolytic degradation and might be expect-

Table 2. Sequences of phage clones selected by anti-dsDNA antibodies[a]

Antibody	Sequence														
R4A						D	W	E	Y	S	V	W	L	S	N
				L	Y	F	E	D	Y	R	C	E	L		
						D	W	D	Y	G	A	L	M	W	A
				Y	S	D	W	D	Y	S	E	G	L		
	S	T	E	H	S	E	A	D	L	W					
Consensus						D/E	W	D/E	Y	S/G					
52b3			R	H	E	D	G	D	W	P	R	V			
			L	L	D	D	G	F	W	P	R	V			
			C	G	V	D	G	R	W	P	R	V			
			S	L	I	S	D	E	W	P	R	W			
						D	G	E	W	P	R	E	G	W	S
			E	D	L	E	G	E	W	P	M	R			
			S	L	D	E	L	D	W	D	S	M			
Consensus						D/E	G	D/E	W	P	R				

[a]Modified from reference 15.

ed to be more effective therapeutic reagents (6). SCID mice that received either the L or the D form of DNA mimotope displayed similar titers of the injected IgG2b R4A. Only mice injected simultaneously with R4A antibody and DWEYS peptide in the D form, however, showed no antibody deposition in the glomeruli, suggesting that the resistance to proteolytic cleavage of the D isomer allowed blockage of renal deposition of R4A (15). These experiments clearly demonstrate that protein antigen can, in fact, mimic polynucleotide antigen, dsDNA, and that carefully designed DNA mimotopes might have therapeutic applications.

After demonstrating that peptide antigens can indeed serve as DNA mimics both in vitro and in vivo, we investigated whether peptide antigens can induce an SLE-like immune response. We immunized BALB/c mice with DWEYSVWLSM peptide in a multimeric form on a polylysine backbone (38). The results of these immunizations were striking. DWEYSVWLSM peptide induced not only a peptide-specific response but also an anti-dsDNA response. The response is specific, as anti-PC, anti-KLH, and antilysozyme titers are not increased after immunization. Anti-dsDNA antibodies can be detected 14 days after immunization and continue to increase through day 49. The anti-dsDNA titer in the serum of peptide-immunized BALB/c mice is similar to the anti-DNA response in 6-month-old (NZB × NZW)F1 lupus-prone mice. The dominant isotype in both the antipeptide and the anti-DNA responses is IgG1, with a presence of IgG2a and IgG2b, but at much lower levels; in contrast, the anti-DNA response of a typical (NZB × NZW)F1 lupus-prone mouse is dominated by IgG2a and IgG2b.

The serum of BALB/c mice immunized with peptide also displays other lupus-like specificities, i.e., anticardiolipin and antihistone specificities. More importantly, staining of the kidney sections of peptide-immunized mice killed at day 49 confirmed the potential pathogenicities of the elicited antibodies. The breakdown of tolerance demonstrated in BALB/c mice seems to be strain specific, as several other inbred strains of mice, C57BL/6 (H-2^b), A/J (H-2^a), AKR/J (H-2^k), CBA/Ca (H-2^k), or even DBA/2 (H-2^d), which has the same MHC haplotype as BALB/c mice, did not mount either an antipeptide response or an anti-dsDNA response comparable to that seen in BALB/c mice (38).

We demonstrated in these experiments that peptide antigens can mimic dsDNA in binding to anti-DNA antibodies and in blocking anti-DNA antibody deposition in the kidneys. More importantly, we showed that this mimicry can trigger an autoimmune response that leads to tissue injury in nonautoimmune hosts. We also showed that breaking of tolerance in this model with a DNA surrogate antigen is genetically determined and may involve both MHC and non-MHC loci.

These studies demonstrate that a peptide can mimic DNA as an antigen and can induce production of anti-DNA antibodies. It does not prove that there is a protein containing the DWEYS peptide that does or that can trigger lupus-like autoreactivity. Several bacterial enzymes contain the consensus sequence, although it is not clear that an antigenic conformation of the peptide is conserved in these proteins. Nor is it clear that the peptide is present at an exposed site where it would be accessible to antibody. The most interesting eukaryotic protein to contain the peptide sequence is the glutamate receptor, but again, it is not yet known if anti-DNA antibody can access and bind to the motif when it is present in the intact protein.

SUMMARY

The studies described in this chapter demonstrate that molecular mimicry between DNA and bacterial antigens can occur in mice and humans and that mimicry between DNA and a model peptide antigen can occur in mice. DNA-binding antibodies cross-reactive with PC or polysaccharide can be identified, and immunization with PC or polysaccharide can induce antibacterial, anti-DNA cross-reactive B cells. The cross-reactivity may be germ line gene encoded or may be generated by somatic mutation of the antibody. Nonautoimmune hosts can generate B cells with potentially pathogenic autospecificities, and the antigenic relatedness of DNA to carbohydrate antigens is sufficient to activate cross-reactive B cells. In humans, it remains to be determined if this cross-reactivity is potentially pathogenic; in mice, it clearly is. Immunization with a peptide mimotope of DNA can also induce lupus-like autoreactivity in mice. The immune system, in general, maintains self-tolerance, despite molecular mimicry through mechanisms that remain to a large degree unknown. Sometimes, as with peptide immunization in BALB/c mice, tolerance is breached and pathogenic autoreactivity ensues. Learning how to refocus the immune response away from cross-reactive epitopes to epitopes unique to foreign antigens may be critical in learning to intervene in autoimmune disease.

REFERENCES

1. **Adderson, E. E., P. G. Shackelford, A. Quinn, P. M. Wilson, M. W. Cunningham, R. A. Insel, and W. L. Carroll.** 1993. Restricted immunoglobulin VH usage and VDJ combinations in the human response to Haemophilus influenzae type b capsular polysaccharide. Nucleotide sequences of monospecific anti-Haemophilus antibodies and polyspecific antibodies cross-reacting with self-antigens. *J. Clin. Investig.* **91:**2734–2743.
2. **Amsbaugh, D. F., C. T. Hansen, B. Prescott, P. W. Stashak, R. Asofsky, and P. J. Baker.** 1974. Genetic control of the antibody response to type 3 pneumococcal polysaccharide in mice. II. Relationship between IgM immunoglobulin levels and the ability to give an IgM antibody response. *J. Exp. Med.* **139:**1499–1512.
3. **Avrameas, S., and T. Ternynck.** 1993. The natural autoantibodies system: between hypotheses and facts. *Mol. Immunol.* **30:**1133–1142.
4. **Barbas, S. M., H. J. Ditzel, E. M. Salonen, W. P. Yang, G. J. Silverman, and D. R. Burton.** 1995. Human autoantibody recognition of DNA. *Proc. Natl. Acad. Sci. USA* **92:**2529–2533.
5. **Barry, M. M., C. D. Mol, W. F. Anderson, and J. S. Lee.** 1994. Sequencing and modeling of anti-DNA immunoglobulin Fv domains. Comparison with crystal structures. *J. Biol. Chem.* **269:**3623–3632.
6. **Bessalle, R., A. Kapitkovsky, A. Gorea, I. Shalit, and M. Fridkin.** 1990. All-D-magainin: chirality, antimicrobial activity and proteolytic resistance. *FEBS Lett.* **274:**151–155.
7. **Bundle, D. R.** 1998. Ab combining sites and oligosaccharide determinants studied by competitive binding, sequencing and X-ray crystallography. *Pure Appl. Chem.* **61:**1171–1180.
8. **Caton, A. J., and H. Koprowski.** 1990. Influenza virus hemagglutinin-specific antibodies isolated from a combinatorial expression library are closely related to the immune response of the donor. *Proc. Natl. Acad. Sci. USA* **87:**6450–6454.
9. **Cygler, M., D. R. Rose, and D. R. Bundle.** 1991. Recognition of a cell-surface oligosaccharide of pathogenic Salmonella by an antibody Fab fragment. *Science* **253:**442–445.
10. **Czerwinski, M., D. Siemaszko, D. L. Siegel, and S. L. Spitalnik.** 1998. Only selected light chains combine with a given heavy chain to confer specificity for a model glycopeptide antigen. *J. Immunol.* **160:**4406–4417.
11. **Desai, D. D., M. R. Krishnan, J. T. Swindle, and T. N. Marion.** 1993. Antigen-specific induction of antibodies against native mammalian DNA in nonautoimmune mice. *J. Immunol.* **151:**1614–1626.
12. **Diamond, B., and M. D. Scharff.** 1984. Somatic mutation of the T15 heavy chain gives rise to an antibody with autoantibody specificity. *Proc. Natl. Acad. Sci. USA* **81:**5841–5844.

13. **el-Roiey, A., O. Sela, D. A. Isenberg, R. Feldman, B. C. Colaco, R. C. Kennedy, and Y. Shoenfeld.** 1987. The sera of patients with Klebsiella infections contain a common anti-DNA idiotype (16/6) Id and anti-polynucleotide activity. *Clin. Exp. Immunol.* **67:**507–515.
14. **Gauntt, C. J., H. M. Arizpe, A. L. Higdon, H. J. Wood, D. F. Bowers, M. M. Rozek, and R. Crawley.** 1995. Molecular mimicry, anti-coxsackievirus B3 neutralizing monoclonal antibodies, and myocarditis. *J. Immunol.* **154:**2983–2995.
15. **Gaynor, B., C. Putterman, P. Valadon, L. Spatz, M. D. Scharff, and B. Diamond.** 1997. Peptide inhibition of glomerular deposition of an anti-DNA antibody. *Proc. Natl. Acad. Sci. USA* **94:**1955–1960.
16. **Gearhart, P. J., N. D. Johnson, R. Douglas, and L. Hood.** 1981. IgG antibodies to phosphorylcholine exhibit more diversity than their IgM counterparts. *Nature* **291:**29–34.
17. **Gilkeson, G. S., A. M. Pippen, and D. S. Pisetsky.** 1995. Induction of cross-reactive anti-dsDNA antibodies in preautoimmune NZB/NZW mice by immunization with bacterial DNA. *J. Clin. Investig.* **95:**1398–1402.
18. **Grayzel, A., A. Solomon, C. Aranow, and B. Diamond.** 1991. Antibodies elicited by pneumococcal antigens bear an anti-DNA-associated idiotype. *J. Clin. Investig.* **87:**842–846.
19. **Iliev, A., L. Spatz, S. Ray, and B. Diamond.** 1994. Lack of allelic exclusion permits autoreactive B cells to escape deletion. *J. Immunol.* **153:**3551–3556.
20. **Jones, S., P. van Heyningen, H. M. Berman, and J. M. Thornton.** 1999. Protein-DNA interactions: a structural analysis. *J. Mol. Biol.* **287:**877–896.
21. **Katz, J. B., W. Limpanasithikul, and B. Diamond.** 1994. Mutational analysis of an autoantibody: differential binding and pathogenicity. *J. Exp. Med.* **180:**925–932.
22. **Korsmeyer, S. J.** 1999. BCL-2 gene family and the regulation of programmed cell death. *Cancer Res.* **59:**1693–1700.
23. **Kowal, C., A. Weinstein, and B. Diamond.** 1999. Molecular mimicry between bacterial and self antigen in a patient with systemic lupus erythematosus. *Eur. J. Immunol.* **29:**1901–1911.
24. **Lafer, E. M., J. Rauch, C. J. Andrzejewski, D. Mudd, B. Furie, R. S. Schwartz, and B. D. Stollar.** 1981. Polyspecific monoclonal lupus autoantibodies reactive with both polynucleotides and phospholipids. *J. Exp. Med.* **153:**897–909.
25. **Laferriere, C. A., R. K. Sood, J. M. de Muys, F. Michon, and H. J. Jennings.** 1998. Streptococcus pneumoniae type 14 polysaccharide-conjugate vaccines: length stabilization of opsonophagocytic conformational polysaccharide epitopes. *Infect. Immun.* **66:**2441–2446.
26. **Lang, J., B. Arnold, G. Hammerling, A. W. Harris, S. Korsmeyer, D. Russell, A. Strasser, and D. Nemazee.** 1997. Enforced Bcl-2 expression inhibits antigen-mediated clonal elimination of peripheral B cells in an antigen dose-dependent manner and promotes receptor editing in autoreactive, immature B cells. *J. Exp. Med.* **186:**1513–1522.
27. **Limpanasithikul, W., S. Ray, and B. Diamond.** 1995. Cross-reactive antibodies have both protective and pathogenic potential. *J. Immunol.* **155:**967–973.
28. **Livneh, A., E. Gazit, and B. Diamond.** 1994. The preferential expression of the anti-DNA associated 8.12 idiotype in lupus is not genetically controlled. *Autoimmunity* **18:**1–6.
29. **MacKenzie, C. R., T. Hirama, S. J. Deng, D. R. Bundle, S. A. Narang, and N. M. Young.** 1996. Analysis by surface plasmon resonance of the influence of valence on the ligand binding affinity and kinetics of an anti-carbohydrate antibody. *J. Biol. Chem.* **271:**1527–1533.
30. **Madaio, M. P., S. Hodder, R. S. Schwartz, and B. D. Stollar.** 1984. Responsiveness of autoimmune and normal mice to nucleic acid antigens. *J. Immunol.* **132:**872–876.
31. **Matsuda, F., K. Ishii, P. Bourvagnet, K. Kuma, H. Hayashida, T. Miyata, and T. Honjo.** 1998. The complete nucleotide sequence of the human immunoglobulin heavy chain variable region locus. *J. Exp. Med.* **188:**2151–2162.
32. **Offen, D., L. Spatz, H. Escowitz, S. Factor, and B. Diamond.** 1992. Induction of tolerance to an IgG autoantibody. *Proc. Natl. Acad. Sci. USA* **89:**8332–8336.
33. **Ohlin, M., and C. A. Borrebaeck.** 1996. Characteristics of human antibody repertoires following active immune responses in vivo. *Mol. Immunol.* **33:**583–592.
34. **Oldstone, M. B. A.** 1972. Virus induced autoimmune disease: viruses in the production and prevention of autoimmune disease, p. 469–475. *In* S. B. Day and R. A. Good (ed.), *Membranes and Viruses in Immunopathology.* Academic Press, Inc., New York, N.Y.

35. **Perlmutter, R. M., B. Berson, J. A. Griffin, and L. Hood.** 1985. Diversity in the germline antibody repertoire. Molecular evolution of the T15 VN gene family. *J. Exp. Med.* **162:**1998–2016.
36. **Pichichero, M. E., S. Porcelli, J. Treanor, and P. Anderson.** 1998. Serum antibody responses of weanling mice and two-year-old children to pneumococcal-type 6A-protein conjugate vaccines of differing saccharide chain lengths. *Vaccine* **16:**83–91.
37. **Portolano, S., G. D. Chazenbalk, J. S. Hutchison, S. M. McLachlan, and B. Rapoport.** 1993. Lack of promiscuity in autoantigen-specific H and L chain combinations as revealed by human H and L chain "roulette". *J. Immunol.* **150:**880–887.
38. **Putterman, C., and B. Diamond.** 1998. Immunization with a peptide surrogate for double-stranded DNA (dsDNA) induces autoantibody production and renal immunoglobulin deposition. *J. Exp. Med.* **188:**29–38.
39. **Radic, M. Z., and M. Weigert.** 1994. Genetic and structural evidence for antigen selection of anti-DNA antibodies. *Annu. Rev. Immunol.* **12:**487–520.
40. **Ray, S., and B. Diamond.** 1994. Generation of a fusion partner to sample the repertoire of splenic B cells destined for apoptosis. *Proc. Natl. Acad. Sci. USA* **91:**5548–5551.
41. **Ray, S. K., C. Putterman, and B. Diamond.** 1996. Pathogenic autoantibodies are routinely generated during the response to foreign antigen: a paradigm for autoimmune disease. *Proc. Natl. Acad. Sci. USA* **93:**2019–2024.
42. **Reason, D. C., T. C. Wagner, and A. H. Lucas.** 1997. Human Fab fragments specific for the *Haemophilus influenzae* b polysaccharide isolated from a bacteriophage combinatorial library use variable region gene combinations and express an idiotype that mirrors in vivo expression. *Infect. Immun.* **65:**261–266.
43. **Rose, D. R., M. Przybylska, R. J. To, C. S. Kayden, R. P. Oomen, E. Vorberg, N. M. Young, and D. R. Bundle.** 1993. Crystal structure to 2.45 Å resolution of a monoclonal Fab specific for the Brucella A cell wall polysaccharide antigen. *Protein Sci.* **2:**1106–1113.
44. **Shefner, R., G. Kleiner, A. Turken, L. Papazian, and B. Diamond.** 1991. A novel class of anti-DNA antibodies identified in BALB/c mice. *J. Exp. Med.* **173:**287–296.
45. **Shoenfeld, Y., Y. Vilner, A. R. Coates, J. Rauch, G. Lavie, D. Shaul, and J. Pinkhas.** 1986. Monoclonal anti-tuberculosis antibodies react with DNA, and monoclonal anti-DNA autoantibodies react with Mycobacterium tuberculosis. *Clin. Exp. Immunol.* **66:**255–261.
46. **Silverman, G. J., and A. H. Lucas.** 1991. Variable region diversity in human circulating antibodies specific for the capsular polysaccharide of Haemophilus influenzae type b. Preferential usage of two types of VH3 heavy chains. *J. Clin. Investig.* **88:**911–920.
47. **Steinberg, B. J., P. A. Smathers, K. Frederiksen, and A. D. Steinberg.** 1982. Ability of the xid gene to prevent autoimmunity in (NZB × NZW)F1 mice during the course of their natural history, after polyclonal stimulation, or following immunization with DNA. *J. Clin. Investig.* **70:**587–597.
48. **Strasser, A., S. Whittingham, D. L. Vaux, M. L. Bath, J. M. Adams, S. Cory, and A. W. Harris.** 1991. Enforced BCL2 expression in B-lymphoid cells prolongs antibody responses and elicits autoimmune disease. *Proc. Natl. Acad. Sci. USA* **88:**8661–8665.
49. **Sun, Y., M. K. Park, J. Kim, B. Diamond, A. Solomon, and M. H. Nahm.** 1999. Repertoire of human antibodies against the polysaccharide capsule of *Streptococcus pneumoniae* serotype 6B. *Infect. Immun.* **67:**1172–1179.
50. **Tsao, B. P., K. Ohnishi, H. Cheroutre, B. Mitchell, M. Teitell, P. Mixter, M. Kronenberg, and B. H. Hahn.** 1992. Failed self-tolerance and autoimmunity in IgG anti-DNA transgenic mice. *J. Immunol.* **149:**350–358.
51. **Unni, K. K., K. E. Holley, F. C. McDuffie, and J. L. Titus.** 1975. Comparative study of NZB mice under germfree and conventional conditions. *J. Rheumatol.* **2:**36–44.
52. **Valadon, P., G. Nussbaum, L. F. Boyd, D. H. Margulies, and M. D. Scharff.** 1996. Peptide libraries define the fine specificity of anti-polysaccharide antibodies to Cryptococcus neoformans. *J. Mol. Biol.* **261:**11–22.
53. **Westerink, M. A., P. C. Giardina, M. A. Apicella, and T. Kieber-Emmons.** 1995. Peptide mimicry of the meningococcal group C capsular polysaccharide. *Proc. Natl. Acad. Sci. USA* **92:**4021–4025.
54. **Wloch, M. K., A. L. Alexander, A. M. Pippen, D. S. Pisetsky, and G. S. Gilkeson.** 1997. Molecular properties of anti-DNA induced in preautoimmune NZB/W mice by immunization with bacterial DNA. *J. Immunol.* **158:**4500–4506.
55. **Zhao, Z. S., F. Granucci, L. Yeh, P. A. Schaffer, and H. Cantor.** 1998. Molecular mimicry by herpes simplex virus-type 1: autoimmune disease after viral infection. *Science* **279:**1344–1347.
56. **Zouali, M.** 1997. The structure of human lupus anti-DNA antibodies. *Methods* **11:**27–35.

Molecular Mimicry, Microbes, and Autoimmunity
Edited by M. W. Cunningham and R. S. Fujinami

Chapter 11

Peptide Mimicry of the Polysaccharide Capsule of *Cryptococcus neoformans*

David O. Beenhouwer, Philippe Valadon, Rena May, and Matthew D. Scharff

Organisms that possess a polysaccharide capsule, such as *Streptococcus pneumoniae, Haemophilus influenzae* type b, *Neisseria meningitidis,* group B streptococcus, and *Cryptococcus neoformans,* are extremely important clinically because they are responsible for a majority of serious infections and for most deaths from meningitis (65). The capsule is a major virulence factor, and acapsular strains are typically nonpathogenic. Serotype-specific anticapsular antibody confers protection against infection with these organisms. Purified capsular polysaccharide vaccines against several types of *S. pneumoniae* and *N. meningitidis* strains are effective in preventing infection with homologous serotypes in older children and adults (for a review, see reference 5). However, these vaccines are not effective in protecting those individuals most commonly infected and least able to resist the ravages of infection, namely, infants, people over age 65, and those with impaired immunity, primarily because polysaccharides are poorly immunogenic and typically induce a T-cell-independent response that lacks immunological memory (32). Given their medical significance and the fact that, except for cryptococcus, the rate of antibiotic resistance is rapidly increasing in this group of pathogens, vaccines that protect susceptible individuals against encapsulated organisms are desperately needed.

An effective vaccine can have a dramatic effect. Until recently, *H. influenzae* type b was the most common cause of bacterial meningitis in children, causing more than 20,000 cases annually in the United States. Effective glycoconjugate vaccines (capsular polysaccharide linked to an immunogenic carrier protein) were licensed in the United States in 1987 and have virtually eliminted *H. influenzae* type b as a cause of meningitis in this country (50), with only 54 cases reported in 1999 (12). A similar approach has been applied to *S. pneumoniae,* and in recent clinical trials, pneumococcal capsule-protein glycoconjugates appear to be much more effective in infants, who usually respond poorly to the 23-valent capsular polysaccharide vaccine in current use (3, 58). While glycoconjugate vaccines may be the answer, there is evidence that antibodies directed against certain epi-

David O. Beenhouwer • Departments of Cell Biology and Internal Medicine, Division of Infectious Diseases, Albert Einstein College of Medicine, Bronx, NY 10461. ***Philippe Valadon, Rena May, and Matthew D. Scharff*** • Department of Cell Biology, Albert Einstein College of Medicine, Bronx, NY 10461.

topes can block the efficacies of protective antibodies and that some antibodies can even enhance infection (8, 24, 28, 38, 48, 75). In addition, glycoconjugates may not be sufficiently immunogenic in elderly and immunosuppressed individuals (for a review, see reference 46). Faced with the need (i) to induce an effective, T-cell-dependent immune response even in immunocompromised individuals and (ii) to focus this response on epitopes that elicit protective immunity, we decided to investigate whether peptides that mimic carbohydrate capsular antigens could surmount these problems and serve as effective vaccines against encapsulated organisms.

The reasons for searching for a peptide mimic of a carbohydrate structure are manyfold. In comparison to peptide chemistry, which is well-defined and relatively easy to manipulate, carbohydrate chemistry is more complex. For example, while it is known that the major component of the cryptococcal capsule is glucuronoxylomannan (GXM) and that this molecule is estimated to have at least seven epitopes (14), the exact structure of GXM remains unknown and oligosaccharides for examination of the nature of these epitopes are not available. This limitation applies to the polysaccharide structures that surround many encapsulated organisms. Peptides are also attractive because phage peptide display provides a high-throughput technology that allows the selection of potential peptide mimetics from a very large diversity of peptide sequences (56). Most importantly, in contrast to the weak T-cell-independent response induced by carbohydrates, peptides are strongly immunogenic and induce a T-cell-dependent response, implying that peptide mimotopes of carbohydrate antigens could make very effective vaccines against an encapsulated organism. While carbohydrate vaccines are ineffective in young children and immunosuppressed individuals, particularly those infected with human immunodeficiency virus (HIV) (9), a peptide vaccine might induce protective immunity in these populations. At the same time, the relatively small molecular size of peptides provides the capacity to focus the immune response on a single "protective epitope" which has the ability to induce antibodies that prolong the life of the host, thus avoiding the production of blocking or enhancing antibodies. Peptide vaccines have other potential advantages. Their composition can be exactly defined, identical preparations are easily reproduced, they are chemically and environmentally stable, and they can be made resistant to proteolysis in vivo if D rather than L amino acids are used (69). In addition, the delayed-release mechanisms available for peptide molecules might allow a single immunization to suffice as both primary and booster vaccinations (19, 31, 57).

Despite their seeming advantages, only one peptide vaccine (against foot-and-mouth disease in cattle [6]) is in general use, and while some have been moderately effective in clinical trials, none have been approved for use in humans. There are several possibilities for this lack of success. As with most small molecules, peptides less than 20 amino acids in length are often poorly immunogenic. This can usually be overcome by coupling the peptide to a carrier molecule; however, the conjugation process may alter the peptide so that it appears antigenically different to the immune system. Carrier molecules may also be responsible for epitope-specific suppression (for a review, see reference 21). This phenomenon occurs when sequential immunization first with a carrier molecule and then with a hapten attached to the same carrier results in suppression of the hapten-specific antibody response. In addition, since peptides can achieve multiple conformations, there is the potential for induction of pathogenic autoantibodies. Recent progress, however, has been made in overcoming many of these problems, including improved methods of conjugation and better adjuvants (4). In one of the most promising advances, several peptides can readily be joined to a synthetic backbone to form a multivalent vaccine (61). These so-called multiple antigenic peptides (MAPs) can use identical peptides which, by binding to several receptors at the same time, might engage appropriate B cells more effectively. MAPs can be made from different peptides that could represent multiple B-cell epitopes, which would be valuable in the case of some of the encapsulated organisms that have multiple serotypes, such as *S. pneumoniae,* where immunity is not cross-protective (68). Finally, a MAP can be created from peptides that bind to both B- and T-cell epitopes, thus avoiding the need for a carrier molecule (62, 63).

The concept of molecular mimicry gained prominence in immunology with Jerne's (27) observation that idiotypic networks can bear the "image" of the antigen either directly or as a "mirror" inscribed in the antibody binding site. Strong evidence for molecular mimicry of ligands came from the induction of networks of molecular interactions with anti-idiotypic antibodies capable of simulating biological receptors (for a review, see reference 64). Some of these ligands were not proteins, demonstrating that anti-idiotypic antibodies could successfully imitate nonprotein antigens. From structural studies of different complexes between antilysozyme antibody, lysozyme, and anti-idiotypic antibodies (23), it has been inferred that mimics reproduce the binding interactions between the antigen and the antibody, thus validating Jerne's (27) original idea of a molecular image. It is not clear that such a model would be true for small molecules, especially when the contact surface is not as large and flat as that of lysozyme.

Convincing evidence that peptidic molecules could be immunological mimics of saccharide structures came from studies involving binding site-specific anti-idiotypic antibodies generated from antibodies directed against carbohydrate epitopes of group A streptococcus, the capsular polysaccharide of *Escherichia coli* K13 (59), GD2 and GD3 gangliosides (13, 52), the O-specific side chain of *Pseudomonas aeruginosa* lipopolysaccharide (53, 54), and the major surface glycoproteins of *Trypanosoma cruzi* (51), *Schistosoma mansoni* (42), and *H. influenzae* (47). In these studies, specific anticarbohydrate antibodies were induced following immunization with anti-idiotypic antibodies, and in some cases protection against infection with the organism that bears the parent antigen was observed (42, 47, 59). While anti-idiotypic antibodies have worked successfully in some murine models, these antibodies are difficult to obtain routinely; because they are often generated in animal systems, they carry a high risk of causing allergic reactions if they are administered to humans, and to date none has been approved for use as a vaccine. To overcome some of these problems, Westerink and colleagues (72) have used a peptide from complementarity-determining region 3 (CDR3) of an anti-idiotypic antibody complexed to proteasomes to elicit a protective immune response to meningococcal group C capsular polysaccharide in mice.

Recently, a number of laboratories have examined whether phage peptide libraries could be used to identify mimics for polysaccharides; and while several have been identified, including those for concanavalin A (41, 55), Lewis Y antigens (25, 30), HIV type 1 (2), GDlα (26), glycosphingolipids (60), *Shigella flexneri* O antigen (45), and GXM (77), little is known about the potential of these peptides to direct the immune response. These efforts have, moreover, identified certain amino acid sequences that may be important in conferring carbohydrate mimicry. A two-aromatic-amino-acid repeat is found in peptides that bind to concanavalin A (YPY) (41, 55), peptides that mimic the Lewis Y antigen

(WLY [25] and YRY, YPY, and WRY [30]), a peptide mimic of GDlα (WHW), and a peptide that binds to an antibody against the meningococcal group C capsular polysaccharide (YRY) (72), arguing that a particular peptide structure is important for polysaccharide mimicry (22). On the other hand, high prevalences of tryptophans and tyrosines have been shown to occur in peptides that bind to polystyrene (1), and heavy-chain CDR3s are also rich in aromatic amino acids, particularly tyrosines (43). A more precise understanding of the binding of peptides and saccharides at the molecular level is required in order to determine whether the occurrence of the motif ArXAr (where Ar stands for aromatic amino acid) in mimics of saccharide structures is due to molecular mimicry or simply reflects an advantage provided by aromatic rings for peptide-antibody interactions.

Given that there are only a few reports to date of peptide mimics that effectively induce anticarbohydrate responses and even fewer instances of peptide mimics that elicit high-titer antibody, it may be that a mimotope must be structurally similar to the original antigen and that peptides are simply unable to induce anticarbohydrate antibodies. Vargas-Madrazo and colleagues (70), after looking at the available antibody sequence data and three-dimensional structures, have found that most antibodies can be organized into 10 classes on the basis of the structures of their CDRs and that combinations of these structures tend to bind to different types of antigens (e.g., polysaccharides, proteins, haptens, and peptides). Of the three structural classes that bound to peptide antigens, one was found in a number of polysaccharide binding antibodies, one was associated with polysaccharide binding antibodies less frequently, and one class did not contain any antibodies that bound to polysaccharides. This work suggests that some antibodies which recognize peptides can bind to polysaccharides as well, while others may not.

In the following sections of this chapter, we discuss the identification of peptide mimics of the carbohydrate capsule of *C. neoformans* and their potential use as vaccines and as tools in understanding the nature of protective and nonprotective epitopes. It is our hope that our experience with *C. neoformans* can be successfully applied to other encapsulated organisms.

MODEL

C. neoformans is an environmentally ubiquitous encapsulated yeast that causes a life-threatening meningoencephalitis in 8% of AIDS patients in the United States and as many as 30% of HIV-infected people in Africa (16, 17, 78); it is the most common cause of central nervous system infections in New York City. While the disease is treatable with antifungal agents such as amphotericin B and fluconazole, it carries a high mortality rate and survivors require lifelong suppressive therapy with fluconazole. There are four serotypes of *C. neoformans* (serotypes A, B, C, and D); serotypes A and D cause the most cases of *C. neoformans* infection among humans in the United States (11). The abundant polysaccharide capsule that surrounds the organism is composed primarily of GXM and is a major virulence factor: it resists phagocytosis, inhibits leukocyte migration (20), can cause immune paralysis (37), and can enhance HIV infection (44). Experimental infection with *C. neoformans* can be established in mice by the intratracheal, intravenous, intraperitoneal, and intracerebral routes. With an inoculum of 10^4 to 10^6 organisms, the infection, while indolent in most strains of mice, is progressively fatal over several weeks. Murine infection models have allowed researchers to elucidate the nature of the immune response to *C. neo-*

formans. It is now clear that complement natural killer cells, macrophages, T cells, and a T_H1 response are involved in protection against infection (for a review, see reference 11).

Initially, the role of the humoral response in resistance to infection was unclear, as immune sera did not protect naive animals against infection (10). Studies conducted in our laboratory and initiated by Arturo Casadevall and Jean Mukherjee provided a library of mouse monoclonal antibodies (MAbs), all of which were derived from single B cells but which differed from each other in isotype and by a few amino acid substitutions in their variable regions that resulted in differences in fine specificity (33). These MAbs were obtained from BALB/c mice after infection with cryptococcus or immunization with a GXM-tetanus toxoid (GXM-TT) conjugate and were found to be highly restricted in variable gene usage to variable heavy-chain (V_H) V_H7183 and variable kappa-chain (V_κ) $V_\kappa5.1$ antibodies (33, 40). The restricted nature of this response is similar to that reported for other polysaccharide antigens (7). Studies with these antibodies have shown that most murine immunoglobulin M (IgM) and IgG MAbs directed against GXM enhanced the in vitro phagocytosis of *C. neoformans,* reduced serum polysaccharide levels, and protected mice against cryptococcal infection (34, 36). However, two IgM MAbs generated in these experiments, MAbs 13F1 and 21D2, did not protect mice against infection (34). Using immunofluorescence and immunoelectron microscopy, Gabriel Nussbaum and colleagues (38) determined that these two antibodies bound to the serotype D cryptococcal capsule in a punctate pattern, while other antibodies that were protective and that presumably recognized a different epitope bound in an annular pattern (Fig. 1). At the same time, it was found that IgG3 antibodies did not prolong the lives of or diminish the fungal burdens in lethally infected mice; that is, they were not protective and might even have enhanced the infection (39). This was an important observation given the goal of developing an effective vaccine against *C. neoformans,* in that the typical murine response to T-cell-independent antigens is primarily IgM and IgG3. If hybridomas that produce these IgG3 antibodies were switched in culture to the other IgG isotypes (IgG1, IgG2b, and IgG2a), these MAbs, which had the same variable region as the nonprotective IgG3, were protective. When nonprotective MAb 13F1 was mixed with protective MAb 12A1 or when nonprotective MAb IgG3 3E5 was administered to mice along with its protective switch variant, IgG1 MAb 3E5 (39), the efficacy of the protective antibody was lost (Fig. 2). These experiments provided two explanations for the initial finding that polyclonal sera were not protective: (i) that there were epitopes on the capsule of *C. neoformans* that could induce blocking antibodies and (ii) that the IgG3 isotype was somehow involved in promoting infection with cryptococcus. These studies also suggested that a vaccine which induced a humoral response focused on a protective epitope and on the most protective IgG isotypes would be more efficacious than GXM-TT.

RESULTS

To search for an effective peptide vaccine against *C. neoformans,* one of the most protective MAbs, 2H1 (35), was used to screen two peptide libraries displayed on the pIII coat protein of phage fd (Fig. 3): one with a six-amino-acid insert containing 2×10^8 peptides created by Scott and Smith (56) and another with a 10-amino-acid insert containing 6×10^8 peptides constructed by Philippe Valadon and colleagues in our laboratory (66). By using three or four successive rounds of screening with progressively lower concentrations

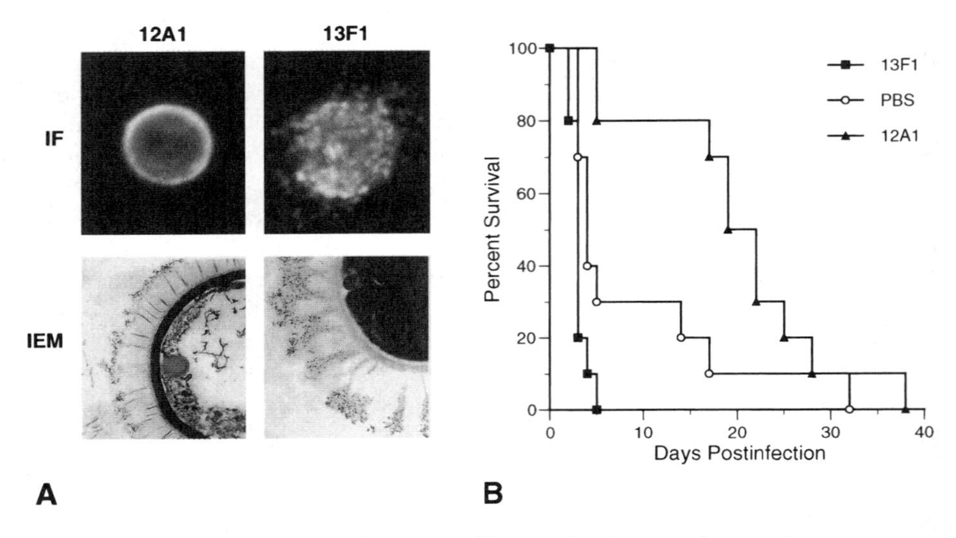

Figure 1. (A) Indirect immunofluorescence (IF; top row) and immunoelectron microscopy (IEM; bottom row) reveal differences in antibody localization within the capsule of *C. neoformans* serotype D. Protective MAb 12A1 (left column) binds to the outer rim of the capsule in an annular pattern, and nonprotective antibody 13F1 (right column) binds throughout the capsule in a punctate fashion. (B) Survival of A/J mice treated with MAbs or phosphate-buffered saline (PBS; controls) and infected with *C. neoformans* serotype D. MAb 12A1 has a statistically significant protective effect, whereas MAb 13F1 is no better than phosphate-buffered saline and may actually enhance infection (38).

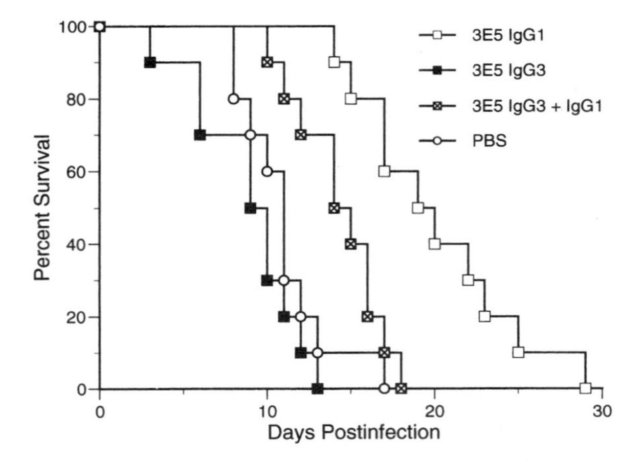

Figure 2. Survival of A/J mice treated with MAbs or phosphate-buffered saline (PBS; controls) and infected with *C. neoformans* serotype D. MAb IgG3 3E5 does not protect mice against infection, whereas its IgG1 switch variant is protective. When these two MAbs are administered simultaneously, the protective efficacy of IgG1 is lost (39). Similar results have been obtained when MAbs directed against protective and nonprotective epitopes are mixed (data not shown).

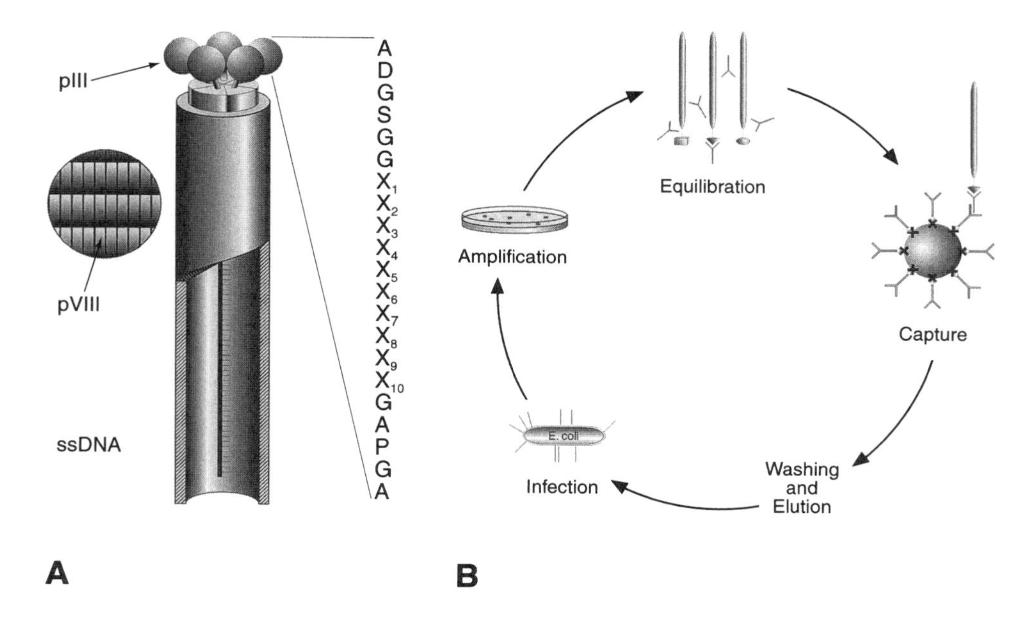

Figure 3. (A) Schematic diagram of filamentous bacteriophage fd depicting two sites for display of peptides: the major coat protein, pVIII, and the minor coat protein, pIII. Five molecules of pIII are expressed on one end of the phage, whereas approximately 2,700 molecules of pVIII are on the cylindrical surface of the viral capsid (29). The vast difference in valence allows detection of both high- and low-affinity target peptides, respectively. An example of a 10-mer random peptide inserted into the pIII minor coat protein used to obtain mimics of GXM is shown. ssDNA, single-stranded DNA. (B) To screen a random phage library for a peptide of desired specificity, phage are electroporated into susceptible bacteria (infection) and allowed to grow (amplification). Bacteria and impurities are removed and phages are incubated with a screening MAb (equilibration). Bound phage are then captured on streptavidin-coated magnetic beads bearing biotinylated antibodies directed against the isotype of the screening MAb (capture). Selected phage are then purified (washing and elution) and passed through successive rounds of panning with lower concentrations of screening antibody and shorter equilibration times to obtain peptides with specificity and high affinity for the screening antibody.

of MAb 2H1 and shorter capture times, 45 phage clones were recovered for analysis, and from these clones it was possible to identify by sequence homology three motifs that dominated the libraries (Table 1). The synthetic peptides that represented these motifs (indicated by capital letters) were examined in detail: PA1 (gLQYTPSWMLVg), P601E (dgaSYSWMYEa), and P514 (ADWADWLDYPg). The highest-affinity peptide (peptide PA1) bound to Fab′ of MAb 2H1 with a dissociation constant (K_d) of 295 nM, as determined by measurement of surface plasmon resonance (66). Competition experiments revealed that all three peptides inhibited binding of MAb 2H1 to phage representing these motifs, indicating that there was a shared binding site on MAb 2H1 that was able to accommodate at least three different short linear peptides (Fig. 4A). All three peptides blocked MAb 2H1 binding to GXM, with 50% inhibitory doses of 35 μM for PA1, 500 μM for P601E, and 1,325 μM for P514 (66). A control peptide, P315

Table 1. Three motifs derived from several peptides isolated from hexa- and decapeptide libraries screened by 2H1[a]

Peptide[b]	Sequence[c]											ELISA result (OD)[d]
			(E)	**T**	**P**	**X**	**W**	**M/L**	**M/L**			
PA1	L	Q	Y	T	P	S	W	M	L	V		>2.0
					W/Y	**X**	**W**	**M/L**	**Y**	**E**		
P601E				S	Y	S	W	M	Y	E		0.69
			D	**W**	**X**	**D**	**W**					
P514		A	D	W	A	D	W	L	D	Y	P	1.68

[a]Data are from reference 66.
[b]Synthetic peptides representing each motif.
[c]The motif sequence is indicated in bold-face; a representative peptide sequence is shown directly beneath the motif sequence.
[d]Optical density (OD) by ELISA of MAb 2H1 for phage bearing the indicated peptide.

(CKVMVHDPHSLA), had no effect on binding at concentrations as high as 2.5 mM (Fig. 4B), showing that the peptide binding site on MAb 2H1 at least partially overlapped that of GXM. Thus, these peptides were deemed GXM mimetics in that they inhibited the binding of the anti-GXM antibodies to GXM.

To examine the fine specificity of nonprotective antibody 13F1 versus that of protective antibody 12A1, we screened the two phage peptide display libraries with these antibodies (38), which differ by only seven amino acids in the CDRs and six amino acids in the framework regions. MAb 13F1 bound to phage bearing the motif WLW(E)W (Table 2). Peptides in this group were specific for MAb 13F1; they did not react by enzyme-linked immunosorbent assay (ELISA) with MAb 12A1 or MAb 21D2, another nonprotective antibody. MAb 12A1 selected the motif W/YXW/YXW/Y(E), which is similar to the motif for P601E (Table 1). In fact, one clone in this group bore an insert whose sequence was

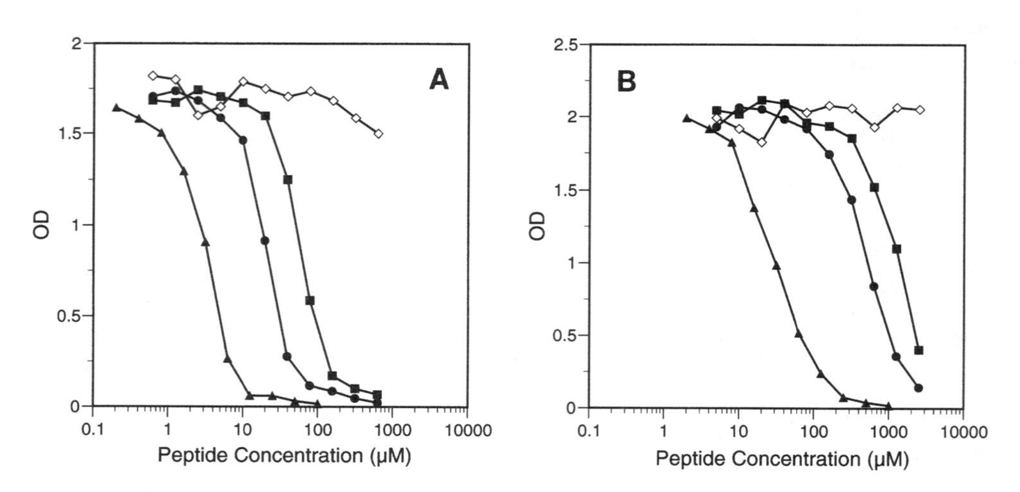

Figure 4. ELISA data showing free peptides in competition with Fab of MAb 2H1 for binding to phage 601E (A) and GXM (B) (66). P315 is an irrelevant control peptide (CKVMVHDPHSLA). Results similar to those presented in panel A were found for phage 514 and A1 (data not shown). OD, optical density. Symbols: ◇, P315; ■, P514, ●, P601E; ▲, PA1.

Table 2. Peptides selected by either protective MAb 12A1 or nonprotective MAb 13F1[a]

Sequence[b]													Selection[c]	ELISA result (OD)[d]		
														12A1	13F1	21D2
					W/Y	**X**	**W/Y**	**X**	**W/Y**	**(E)**						
					Y	D	W	L	M	F			12A1	0.7	—[e]	—
					Y	D	W	L	Y	E[f]			12A1	0.24	—	—
			Y	F	W	N	Y	S	Y	E	V	F	12A1	>2	—	—
					W	**L**	**W**	**(E)**	**W**							
			F	D	W	L	W	L	W	D	T		12A1	—	0.9	—
			L	Q	W	L	W	E	W	P	R	T	13F1	—	1.61	—
			L	D	W	L	W	E	W	A	E	Q	13F1	—	>2.0	—
		T	H	D	W	L	W	E	W	A	S		13F1	—	0.26	—
L	L	D	Y	G	W	L	W	M	W				13F1	—	1.47	—

[a]Data are from reference 38.
[b]Motifs are shown in boldface.
[c]MAb used to screen phage libraries.
[d]Optical density (OD) by ELISA of MAbs 12A1, 13F1, and 21D2 (another nonprotective antibody) for phage bearing the indicated peptide.
[e]—, less than or equal to background optical density (~0.100).
[f]This clone has a sequence identical to one isolated by MAb 2H1.

identical to that of a peptide selected by MAb 2H1. The peptides in this group did not cross-react with MAb 13F1 or MAb 21D2. However, one peptide selected by MAb 12A1 reacted only with MAb 13F1, suggesting that the binding sites of these antibodies share some similar features, which was further supported by the fact that other phage clones selected by either MAb 13F1 or MAb 12A1 fell into a third motif, V/AV/IWXWXW (not shown in Table 2), and were generally cross-reactive with all three antibodies.

We wished to determine if the mimetic peptides selected by MAb 2H1 were mimotopes that could stimulate the production of protective anti-GXM antibodies in immunized animals. There appeared to be at least two distinct strategies for selection of the peptide with the highest likelihood of being a useful mimotope: (i) choose the peptide that reacted with the largest number of anti-GXM MAbs, with the hope that this peptide would best complement the "generic" anti-GXM binding site and interact with the largest number of important areas in this site, or (ii) select the peptide that had the highest affinity for the most protective MAbs since it would have the best chance of engaging B cells that make antibodies against protective epitopes on GXM. Obviously, these strategies are not mutually exclusive and ignore major aspects such as the role of T cells and the restrictions of antigen presentation. Nevertheless, we evaluated the binding of these peptides to a variety of anti-GXM MAbs. P514 bound to the fewest MAbs with the lowest relative affinity, P601E bound to a wider array of antibodies than PA1, but PA1 bound with the highest relative affinity (66). Most importantly, none of these peptides bound to nonprotective MAb 13F1 or MAb 21D2. This illustrates how peptides might be used to screen polyclonal responses or large numbers of MAbs.

In an attempt to explore the two strategies described above, we immunized mice with the P601E and PA1 peptides conjugated to keyhole limpet hemocyanin (KLH) with glutaraldehyde. P601E did not induce significant anti-GXM titers (<1:50) (67), while PA1 induced low anti-GXM titers (~1:200) (Fig. 5). In both cases, the anticarbohydrate antibody response was only a small portion of the antipeptide antibody response generated by immunization (~1:12,800).

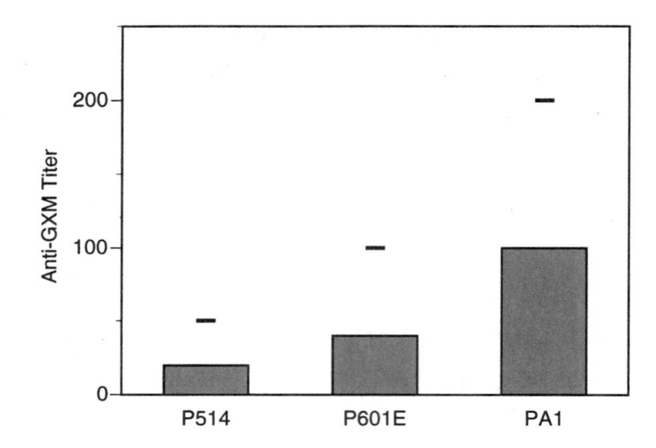

Figure 5. Anti-GXM titers in BALB/c mice 14 days after secondary immunization with peptide conjugated to KLH. Mean titers are represented by gray bars; peak titers are indicated by horizontal lines. In comparing these peptides, it appears that as their affinity for MAb 2H1 increases they induce higher titers of anti-GXM antibodies.

Peptide binding MAbs were obtained from mice immunized with P601E with the hope that their molecular structure would provide an explanation as to why they did not bind to GXM. One (MAb 4A11) of 12 antibodies sequenced was quite similar to MAb 2H1 in that it used the same $V_\kappa5.1$ $J_\kappa1$ light chain and 7183 V_H chain, differing only in its J_H region, J_H3 versus J_H2 (Table 3) (67). However, this resulted in the absence of an arginine and an aspartic acid at the DJ junction that is present in all of the anti-GXM antibodies that we have sequenced (33). While the other 11 MAbs used a V_L chain ($V_\kappa45.1$) with only a four-amino-acid difference from $V_\kappa5.1$, they used different V_H chains, primarily V36-60. This relatively restricted response was encouraging in that this peptide appeared to bind to only a few types of antibodies, suggesting that it presented only one or a few epitopes. While MAb 4A11 recognized the PA1 peptide in addition to the P601E peptide that was used to elicit them, the other 11 antibodies, which used a different V_H than MAb 2H1 did, recognized only the immunizing peptide P601E. We used MAb 4A11 and three other antipeptide antibodies representing each of the anti-P601E MAb V_H-V_κ combinations to screen the decapeptide library. After only one round of selection we found that the anti-P601E MAbs bound to a peptide motif (ArXWMY) that is almost identical to those selected by MAb 2H1 (67). This confirmed that even though these MAbs elicited by the peptide did not bind to GXM, their binding sites were similar to that of MAb 2H1 and encouraged us to continue to seek a mimotope for GXM.

The inability to use peptide mimetics of polysaccharide antigens to elicit the production of high-titer antipolysaccharide antibody is apparently not unusual, although it is difficult to estimate the relative number of failures and successes. Rather than blindly screening the many peptides we had obtained by immunizing mice with each of them, we sought an explanation for our failure by determining the three-dimensional structure of the binding site of the 2H1 antibody with and without the peptide mimics in the binding site. Attempts to cocrystallize P601E with the Fab of MAb 2H1 resulted in structures that did

Table 3. Characteristics of MAbs obtained following immunization with P601E[a]

MAb	Isotype	V_H	D	J_H	V_κ	J_κ	Id[b]	ELISA result[c]
2H1[d]	γ1, κ	7183	Unknown	J_H2	$V_\kappa5.1$	$J_\kappa1$	+	514, 601, A1
4A11	γ1, κ	7183	DFL16.2	J_H3	$V_\kappa5.1$	$J_\kappa1$	–	601, A1
18	γ1, κ	V11	DFL16.1/.2/DSP2.2	J_H3	$V_\kappa45.1$	$J_\kappa1$	–	601
3A10	γ1, κ	V36.60	DSP2.5/.6/.7	J_H4	$V_\kappa45.1$	$J_\kappa2$	–	601
124	γ2b, κ	V36.60	DSP2.6/.7/.8	J_H4	$V_\kappa45.1$	$J_\kappa1$	+	601
544	γ1, κ	V36.60	DSP2.6/.7/.8	J_H4	$V_\kappa45.1$	$J_\kappa1$	+	601
752	γ2b, ND[e]	V36.60	DSP2.6/.7/.8	J_H4	$V_\kappa45.1$	$J_\kappa1$	+	601
431	γ2b, κ	V36.60	DQ52	J_H2	$V_\kappa45.1$	$J_\kappa1$	+	601
147	γ2a, κ	V36.60	DQ52	J_H4	$V_\kappa45.1$	$J_\kappa5$	+	601
409	γ1, κ	V36.60	DQ52	J_H4	$V_\kappa45.1$	$J_\kappa5$	+	601
546	γ1, κ	V36.60	DQ52	J_H4	$V_\kappa45.1$	$J_\kappa5$	+	601
234	γ2b, ND	V36.60	DSP2.x/DFL16.1	J_H4	$V_\kappa45.1$	$J_\kappa1$	+	ND
638	γ1, κ	V36.60	DSP2.x/DFL16.1	J_H4	$V_\kappa45.1$	$J_\kappa1$	–	601
316	γ1, κ	V36.60	DSP2.x/DFL16.1	J_H4	$V_\kappa45.1$	$J_\kappa1$	+	601
717	γ1, k	V36.60	DSP2.x/DFL16.1	J_H4	$V_\kappa45.1$	$J_\kappa1$	+	601

[a]Data are from reference 67.
[b]ELISA for rabbit anti-MAb 2H1 idiotype (Id) antibody.
[c]ELISA for phage bearing the indicated peptides.
[d]Data for MAb 2H1 are shown at the top for comparison.
[e]ND, not determined.

not contain the peptide (76). Higher-affinity peptide PA1 did cocrystallize with the Fab of MAb 2H1, and the complex was solved at 2.4 Å resolution (Fig. 6). Compared to the crystal structure of MAb 2H1 alone, it was noted that binding of peptide does not result in significant changes in the conformation of the antibody binding site (76). The antibody binding site of MAb 2H1 is a hydrophobic pocket delimited by CDR2 and CDR3 of the heavy chain and CDR1 and CDR3 of the light chain (Fig. 6A). PA1 can be seen nestled in the groove of the antibody binding site, with the carboxy-terminal end of the peptide associating with the heavy-chain CDRs and the amino-terminal part of the peptide interacting with the light-chain CDRs (76). When complexed to MAb 2H1, the peptide adopts a tightly coiled conformation composed of an inverse γ turn and a type II β turn. Other linear peptides bound to antibody molecules often possess some secondary structure, usually a type I or II β turn (71, 73, 74). It has been suggested that a β turn maximizes the total buried surface area of a peptide bound to an antibody, improving the fit with the antibody-binding pocket (49). PA1 has threefold more van der Waals interactions with the light chain of MAb 2H1 than the heavy chain (76). In addition, two hydrophobic cavities between the peptide and the surface of MAb 2H1 can be appreciated in the region of the M9 residue of the peptide and parts of the heavy-chain CDR (Fig. 6B). The molecular structure of the anti-P601E antibodies and the crystal structure of MAb 2H1 with PA1 provided an explanation for our inability to stimulate the production of anti-GXM antibodies: the peptide was binding to B cells with a light chain similar to that of MAb 2H1 but did not effectively engage B cells with the appropriate heavy chain.

Since we do not have the structure of GXM in complex with MAb 2H1, we cannot

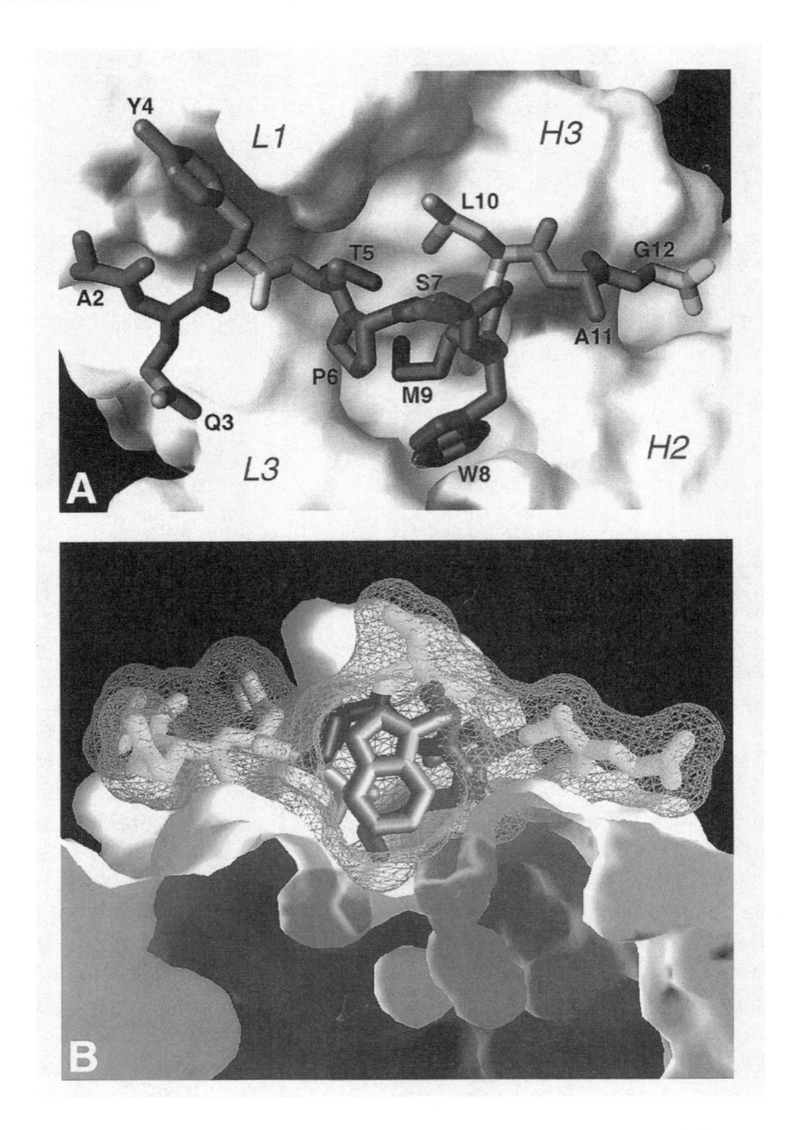

Figure 6. Crystal structure of PA1 bound to MAb 2H1. (For color version of figure, see Color Plates, p. 276.) (A) Looking down on PA1 resting in the binding site of 2H1, which is shown in white, with positively charged regions in blue and negatively charged areas in red. Antibody heavy-chain CDR2 and CDR3 and light-chain CDR1 and CDR3 are denoted by H2, H3, L1, and L3, respectively. (B) Side view showing the cutaway surface of MAb 2H1 in white and the molecular surface of PA1 as a lilac mesh. The orientation of the peptide is similar to that in panel A, and the residues corresponding to the PA1 motif are colored as follows: T5, blue-green; P6, purple; W8, pink; M9, orange; L10, green. The remainder of the peptide is shown in yellow. The surfaces of two cavities between the antibody and the bound peptide are colored yellow, green, and orange according to their proximity to the antibody light chain, antibody heavy chain, and peptide, respectively.

compare the binding interactions of the original antigen and the peptide mimetic with MAb 2H1. On the basis of structural studies of the O antigen from *Salmonella* serogroup B complexed with Fab, polysaccharide antigens make only a few contacts with the antibody binding sites compared to the number of interactions between protein antigens and antibodies (18). Because it seems unlikely that we can find a peptide which will exactly duplicate the interactions of the polysaccharide and the antibody CDRs, our observations suggest that, at least in the case of a pocket-like antibody, a peptide mimic will have to fit tightly in the area of the antibody surface and have as many contact sites as possible to be an effective mimotope that will elicit high titers of antibody to the original antigen.

While these studies did not completely resolve the higher affinity versus broader cross-reactivity issue, at a practical level it seemed logical to seek higher-affinity peptides with the thought that they should improve the completeness of the fit between antibody and peptide and thus work as better mimotopes. Since, in this case, the antibody binding site is rigid, such extensive molecular imprinting might be sufficient to generate molecular mimetism, at least from an immunological point of view. This tight fit may be even more important for nonpeptide antigens (15, 70). The relative lack of interaction of the PA1 peptide with the heavy-chain CDRs in the crystal structure suggested a strategy for the development of an effective mimotope by making sublibraries by using the PA1 motif as a fixed sequence at the amino-terminal end of the peptide and adding amino acids at the C-terminal end with the goal of increasing the interactions with the heavy-chain CDRs. While in principle this could be addressed by molecular modeling and rational drug design, we chose instead the more generally accessible approach of using the phage display library by fixing the PA1 motif and randomizing extensions of that motif at both the C-terminal and N-terminal ends.

To do this, we created an evolutionary sublibrary of PA1 using a degenerate six-amino-acid core peptide of the PA1 motif [TPXW(M/L)(M/L)] with the addition of six random amino acids on either end. The choice to randomize the amino-terminal end came from the fact that we knew very little about the influence of amino acids in this region on the overall structure of the peptide or its interaction with antibody. Seven rounds of screening with progressively decreasing MAb 2H1 concentrations were performed, and representative peptides were selected for further analysis. One of these, P206.1, inhibits binding of the Fab of MAb 2H1 at nearly 100-fold lower concentrations than PA1 (Table 4). Studies are under way to (i) determine whether this peptide induces high titers of anti-GXM antibodies and to (ii) define the molecular characteristics of the antibodies that are induced as well as determine their protective potentials.

Table 4. Evolution of a peptide mimotope for GXM

Peptide	Sequence	Affinity[a]	Titer[b]
P601E	S Y S W M Y E	3,000	<1:50
PA1	L Q Y T P S W M L V	300	1:200
P206.1	F G G E T F T P D W M M E V A I D N E	3	ND[c]

[a]For MAb 2H1, as estimated by ELISA with Fab of MAb 2H1.
[b]Maximum anti-GXM titers in BALB/c mice 14 days after secondary immunization with peptide conjugated to KLH.
[c]ND, not determined.

SUMMARY AND CONCLUSIONS

We have described our attempts to use MAbs and phage display libraries to identify peptides that will elicit antipolysaccharide antibodies to protective epitopes on the capsular polysaccharide of the fungus *C. neoformans.* The initial reason for identifying peptide mimetics was to use them to test the fine specificity of a large library of MAbs since oligosaccharides were not available. We fulfilled this goal and were thereby able to distinguish those antibodies that prolonged the life of lethally infected animals from those that did not. This result led us to the more long-term goal of creating a peptide vaccine that could be used to immunize and protect immunosuppressed individuals from infection with this organism. This required that we identify a peptide that would stimulate B cells that express an antigen binding site that would react with the protective epitope on GXM. Since the peptide mimetics that were useful in determining fine specificity did not stimulate high titers of anti-GXM antibodies, we either had to search for other peptides, perhaps in libraries with longer or constrained inserts, or to try to understand why our peptides were not working. The crystal structure of one of the peptides bound to MAb 2H1, combined with the molecular structure of the antipeptide MAbs elicited by another one of the peptides, provided a clue as to why our peptides were not good mimotopes. A comparison of the cocrystal of PA1 and the Fab of MAb 2H1 to other published structures with peptides in the antibody binding site revealed that PA1 had fewer interactions with the heavy-chain CDRs than other peptides (76). Since the antipeptide antibodies used the same variable light chain (V_L) but a different V_H than MAb 2H1 did, it seemed logical to use the power of phage libraries to screen for peptides that might bind better to the heavy-chain CDRs. This led us to the sublibrary approach, which has provided us with much higher affinity peptides. In addition, this strategy could be applied to any situation in which a mimic did not work as a mimotope and would not require either a crystal structure or the molecular structure of non-antigen binding MAbs as guides.

These studies forced us to confront a number of issues that are relevant to immunological mimicry in general and for peptide vaccines for nonprotein antigens in particular. One such question is whether a mimic must make the same contacts with the antibody binding site as the original antigen (23). Furthermore, must the chemical natures of these contacts be the same? While such a situation would be ideal, even with the enormous number of structural options provided by combinatorial technologies such as phage display, this may be hard to achieve if the mimic is protein based and the antigen is a polysaccharide or nucleic acid. In addition, if we are trying to achieve a higher-affinity interaction than usually occurs with polysaccharide antigens, it will be necessary to create a mimic that has additional or perhaps stronger interactions with the antibody binding site than the original antigen. In fact, if we can figure out a way to engage and stimulate the replication of B cells that express the exact same binding site as MAb 2H1 and then cause those B cells to undergo somatic mutation, there is the possibility that we not only can use the mimic to immunize against the polysaccharide antigen but also can achieve affinity maturation and higher-affinity antibodies than are usually produced against polysaccharide antigens. However, since this would require multiple rounds of selection in vivo, the mimotope would have to accurately reflect the conformation of the original antigen or the mutations would lead to a loss of binding of the polysaccharide. Although all of this seems difficult to achieve, it was this goal that led us to use sublibraries with the hope that peptides with higher affinity for the antibody would also represent a better mirror image of the binding site of the 2H1 antibody.

Certain properties of this particular antigen-antibody system may in fact favor this approach. The 2H1 antibody does not change its conformation when it interacts with the peptide mimic, making it unlikely that changes in the site occur when it interacts with GXM. This suggests that the peptide is more likely to adjust its conformation to fit into the site, and the conformation of the peptide in the MAb 2H1 binding site suggests that it does indeed fold into that site. In addition, the antibody response to GXM is highly restricted, suggesting that B cells with the correct heavy- and light-chain V region combination and CDR3s are usually present, making it possible to engage those B cells. Even with these advantages, it is unclear whether it will be possible to identify a peptide that will selectively stimulate those B cells that have the DJ junctional arginine-aspartate in the CDR3 of the heavy chain that is found in all of the anti-GXM antibodies that we have sequenced so far (33). Should this be possible, the use of evolutionary peptide libraries could make it feasible to make useful peptide vaccines even when the crucial epitope is not a protein.

Acknowledgments. D.O.B. is supported by NIH grant AI-01434 from the National Institutes of Health (NIH), P.V. was supported in part by the Philippe Foundation, R.M. is supported by NIH grant 5T32AG00194, and M.D.S. is supported by grants CA-39838, AR-44192, AI-42297, and AI-43937 and the Harry Eagle Chair provided by the National Women's Division of the Albert Einstein College of Medicine. We thank Jin Oh and Jieru Zhang for technical support for these studies as well as Arturo Casadevall and Liise-anne Pirofski for helpful comments in preparing this chapter.

REFERENCES

1. **Adey, N. B., A. H. Mataragnon, J. E. Rider, J. M. Carter, and B. K. Kay.** 1995. Characterization of phage that bind plastic from phage-displayed random peptide libraries. *Gene* **156:**27–31.
2. **Agadjanyan, M., P. Luo, M. A. Westerink, L. A. Carey, W. Hutchins, Z. Steplewski, D. B. Weiner, and T. Kieber-Emmons.** 1997. Peptide mimicry of carbohydrate epitopes on human immunodeficiency virus. *Nat. Biotechnol.* **15:**547–551.
3. **Ahman, H., H. Kayhty, H. Lehtonen, O. Leroy, J. Froeschle, and J. Eskola.** 1998. *Streptococcus pneumoniae* capsular polysaccharide-diphtheria toxoid conjugate vaccine is immunogenic in early infancy and able to induce immunologic memory. *Pediatr. Infect. Dis. J.* **17:**211–216.
4. **Arnon, R., and M. H. Van Regenmortel.** 1992. Structural basis of antigenic specificity and design of new vaccines. *FASEB J.***6:**3265–3274.
5. **Austrian, R.** 1985. Polysaccharide vaccines. *Ann. Inst. Pasteur Microbiol.* **136B:**295–307.
6. **Bittle, J. L., R. A. Houghten, H. Alexander, T. M. Shinnick, J. G. Sutcliffe, R. A. Lerner, D. J. Rowlands, and F. Brown.** 1982. Protection against foot-and-mouth disease by immunization with a chemically synthesized peptide predicted from the viral nucleotide sequence. *Nature* **298:**30–33.
7. **Bona, C.** 1993. Molecular characteristics of anti-polysaccharide antibodies. *Springer Semin. Immunopathol.* **15:**103–118.
8. **Butterworth, A. E., R. Bensted-Smith, A. Capron, M. Capron, P. R. Dalton, D. W. Dunne, J. M. Grzych, H. C. Kariuki, J. Khalife, and D. Koech.** 1987. Immunity in human *Schistosomiasis mansoni:* prevention by blocking antibodies of the expression of immunity in young children. *Parasitology* **94:**(Pt. 2):281–300.
9. **Carson, P. J., R. L. Schut, M. L. Simpson, J. O'Brien, and E. N. Janoff.** 1995. Antibody class and subclass responses to pneumococcal polysaccharides following immunization of human immunodeficiency virus-infected patients. *J. Infect. Dis.* **172:**340–345.
10. **Casadevall, A.** 1995. Antibody immunity and invasive fungal infections. *Infect. Immun.* **63:**4211–4218.
11. **Casadevall, A., and J. R. Perfect.** 1998. *Cryptococcus neoformans.* ASM Press, Washington, D.C.

12. **Centers for Disease Control and Prevention.** 1999. Impact of vaccines universally recommended for children: United States, 1990–1998. *Morbid. Mortal. Weekly Rep.* **48:**243–248.
13. **Chapman, P. B., and A. N. Houghton.** 1991. Induction of IgG antibodies against GD3 ganglioside in rabbits by an anti-idiotypic monoclonal antibody. *J. Clin. Invest.* **88:**186–192.
14. **Cherniak, R., and J. B. Sundstrom.** 1994. Polysaccharide antigens of the capsule of *Cryptococcus neoformans. Infect. Immun.* **62:**1507–1512.
15. **Chothia, C., A. M. Lesk, E. Gherardi, I. M. Tomlinson, G. Walter, J. D. Marks, M. B. Llewelyn, and G. Winter.** 1992. Structural repertoire of the human VH segments. *J. Mol. Biol.* **227:**799–817.
16. **Clumeck, N., J. Sonnet, H. Taelman, F. Mascart-Lemone, M. De Bruyere, P. Vandeperre, J. Dasnoy, L. Marcelis, M. Lamy, and C. Jonas.** 1984. Acquired immunodeficiency syndrome in African patients. *N. Engl. J. Med.* **310:**492–497.
17. **Currie, B. P., and A. Casadevall.** 1994. Estimation of the prevalence of cryptococcal infection among patients infected with the human immunodeficiency virus in New York City. *Clin. Infect. Dis.* **19:**1029–1033.
18. **Cygler, M., D. R. Rose, and D. R. Bundle.** 1991. Recognition of a cell-surface oligosaccharide of pathogenic Salmonella by an antibody Fab fragment. *Science* **253:**442–445.
19. **Diwan, M., A. Misra, R. K. Khar, and G. P. Talwar.** 1997. Long-term high immune response to diphtheria toxoid in rodents with diphtheria toxoid conjugated to dextran as a single contact point delivery system. *Vaccine* **15:**1867–1871.
20. **Dong, Z. M., and J. W. Murphy.** 1995. Effects of the two varieties of *Cryptococcus neoformans* cells and culture filtrate antigens on neutrophil locomotion. *Infect. Immun.* **63:**2632–2644.
21. **Elliott, J. I.** 1990. Epitope-specific regulation and the literature. *Scand. J. Immunol.* **32:**193–195.
22. **Evans, S. V., D. R. Rose, R. To, N. M. Young, and D. R. Bundle.** 1994. Exploring the mimicry of polysaccharide antigens by anti-idiotypic antibodies. The crystallization, molecular replacement, and refinement to 2.8 Å resolution of an idiotope-anti-idiotope Fab complex and of the unliganded anti-idiotope Fab. *J. Mol. Biol.* **241:**691–705.
23. **Fields, B. A., F. A. Goldbaum, X. Ysern, R. J. Poljak, and R. A. Mariuzza.** 1995. Molecular basis of antigen mimicry by an anti-idiotope. *Nature:* **374:**739–742.
24. **Grzych, J. M., M. Capron, C. Dissous, and A. Capron.** 1984. Blocking activity of rat monoclonal antibodies in experimental schistosomiasis. *J. Immunol.* **133:**998–1004.
25. **Hoess, R., U. Brinkmann, T. Handel, and I. Pastan.** 1993. Identification of a peptide which binds to the carbohydrate-specific monoclonal antibody B3. *Gene* **128:**43–49.
26. **Ishikawa, D., H. Kikkawa, K. Ogino, Y. Hirabayashi, N. Oku, and T. Taki.** 1998. GD1alpha-replica peptides functionally mimic GD1alpha, an adhesion molecule of metastatic tumor cells, and suppress the tumor metastasis. *FEBS Lett.* **441:**20–24.
27. **Jerne, N. K.** 1974. Towards a network theory of the immune system. *Ann. Immunol. (Paris)* **125C:**373–389.
28. **Joiner, K. A., R. Scales, K. A. Warren, M. M. Frank, and P. A. Rice.** 1985. Mechanism of action of blocking immunoglobulin G for *Neisseria gonorrhoeae. J. Clin. Invest.* **76:**1765–1772.
29. **Kay, B. K., J. Winter, and J. McCafferty.** 1996. *Phage Display of Peptides and Proteins.* Academic Press, Inc., San Diego, Calif.
30. **Luo, P., M. Agadjanyan, J. Qiu, M. A. Westerink, Z. Steplewski, and T. Kieber-Emmons.** 1998. Antigenic and immunological mimicry of peptide mimotopes of Lewis carbohydrate antigens. *Mol. Immunol.* **35:**865–879.
31. **Men, Y., B. Gander, H. P. Merkle, and G. Corradin.** 1996. Induction of sustained and elevated immune responses to weakly immunogenic synthetic malarial peptides by encapsulation in biodegradable polymer microspheres. *Vaccine* **14:**1442–1450.
32. **Mond, J. J., A. Lees, and C. M. Snapper.** 1995. T cell-independent antigens type 2. *Annu. Rev. Immunol.* **13:**655–692.
33. **Mukherjee, J., A. Casadevall, and M. D. Scharff.** 1993. Molecular characterization of the humoral responses to *Cryptococcus neoformans* infection and glucuronoxylomannan-tetanus toxoid conjugate immunization. *J. Exp. Med.* **177:**1105–1116.
34. **Mukherjee, J., G. Nussbaum, M. D. Scharff, and A. Casadevall.** 1995. Protective and nonprotective monoclonal antibodies to *Cryptococcus neoformans* originating from one B cell. *J. Exp. Med.* **181:**405–409.
35. **Mukherjee, J., M. D. Scharff, and A. Casadevall.** 1992. Protective murine monoclonal antibodies to *Cryptococcus neoformans. Infect. Immun.* **60:**4534–4541.

36. **Mukherjee, S., S. C. Lee, and A. Casadevall.** 1995. Antibodies to *Cryptococcus neoformans* glucuronoxylomannan enhance antifungal activity of murine macrophages. *Infect. Immun.* **63:**573–579.
37. **Murphy, J. W., and G. C. Cozad.** 1972. Immunological unresponsiveness induced by cryptococcal capsular polysaccharide assayed by the hemolytic plaque technique. *Infect. Immun.* **5:**896–901.
38. **Nussbaum, G., W. Cleare, A. Casadevall, M. D. Scharff, and P. Valadon.** 1997. Epitope location in the *Cryptococcus neoformans* capsule is a determinant of antibody efficacy. *J. Exp. Med.* **185:**685–694.
39. **Nussbaum, G., R. Yuan, A. Casadevall, and M. D. Scharff.** 1996. Immunoglobulin G3 blocking antibodies to the fungal pathogen *Cryptococcus neoformans. J. Exp. Med.* **183:**1905–1909.
40. **Nussbaum, G., S. Anandasabapathy, J. Mukherjee, M. Fan, A. Casadevall, and M. D. Scharff.** 1999. Molecular and idiotypic analyses of the antibody response to *Cryptococcus neoformans* glucuronoxylomannan-protein conjugate vaccine in autoimmune and nonautoimmune mice. *Infect. Immun.* **67:**4469–4476.
41. **Oldenburg, K. R., D. Loganathan, I. J. Goldstein, P. G. Schultz, and M. A. Gallop.** 1992. Peptide ligands for a sugar-binding protein isolated from a random peptide library. *Proc. Natl. Acad. Sci. USA* **89:**5393–5397.
42. **Olds, G. R., and T. F. Kresina.** 1987. Protective anti-idiotype vaccine against *Schistosoma mansoni* infection. *Trans. Assoc. Am. Physicians* **100:**215–221.
43. **Padlan, E. A.** 1990. On the nature of antibody combining sites: unusual structural features that may confer on these sites an enhanced capacity for binding ligands. *Proteins* **7:**112–124.
44. **Pettoello-Mantovani, M., A. Casadevall, T. R. Kollmann, A. Rubinstein, and H. Goldstein.** 1992. Enhancement of HIV-1 infection by the capsular polysaccharide of *Cryptococcus neoformans. Lancet* **339:**21–23.
45. **Phalipon, A., A. Folgori, J. Arondel, G. Sgaramella, P. Fortugno, R. Cortese, P. J. Sansonetti, and F. Felici.** 1997. Induction of anti-carbohydrate antibodies by phage library-selected peptide mimics. *Eur. J. Immunol.* **27:**2620–2625.
46. **Pirofski, L. A., and A. Casadevall.** 1998. Use of licensed vaccines for active immunization of the immunocompromised host. *Clin. Microbiol. Rev.* **11:**1–26.
47. **Reason, D. C., M. Y. Kitamura, and A. H. Lucas.** 1994. Induction of a protective human polysaccharide-specific antibody response in hu-PBL SCID mice by idiotypic vaccination. *J. Immunol.* **152:**5009–5013.
48. **Rice, P. A., and D. L. Kasper.** 1982. Characterization of serum resistance of *Neisseria gonorrhoeae* that disseminate. Roles of blocking antibody and gonococcal outer membrane proteins. *J. Clin. Investig.* **70:**157–167.
49. **Rini, J. M., U. Schulze-Gahmen, and I. A. Wilson.** 1992. Structural evidence for induced fit as a mechanism for antibody-antigen recognition. *Science* **255:**959–965.
50. **Robbins, J. B., R. Schneerson, P. Anderson, and D. H. Smith.** 1996. The 1996 Albert Lasker Medical Research Awards. Prevention of systemic infections, especially meningitis, caused by *Haemophilus influenzae* type b. Impact on public health and implications for other polysaccharide-based vaccines. *JAMA* **276:**1181–1185.
51. **Sacks, D. L., L. V. Kirchhoff, S. Hieny, and A. Sher.** 1985. Molecular mimicry of a carbohydrate epitope on a major surface glycoprotein of *Trypanosoma cruzi* by using anti-idiotypic antibodies. *J. Immunol.* **135:**4155–4159.
52. **Saleh, M. N., J. D. Stapleton, M. B. Khazaeli, and A. F. LoBuglio.** 1993. Generation of a human anti-idiotypic antibody that mimics the GD2 antigen. *J. Immunol.* **151:**3390–3398.
53. **Schreiber, J. R., K. L. Nixon, M. F. Tosi, G. B. Pier, and M. B. Patawaran.** 1991. Anti-idiotype-induced, lipopolysaccharide-specific antibody response to *Pseudomonas aeruginosa.* II. Isotype and functional activity of the anti-idiotype-induced antibodies. *J. Immunol.* **146:**188–193.
54. **Schreiber, J. R., M. Patawaran, M. Tosi, J. Lennon, and G. B. Pier.** 1990. Anti-idiotype-induced, lipopolysaccharide-specific antibody response to *Pseudomonas aeruginosa. J. Immunol.* **144:**1023–1029.
55. **Scott, J. K., D. Loganathan, R. B. Easley, X. Gong, and I. J. Goldstein.** 1992. A family of concanavalin A-binding peptides from a hexapeptide epitope library. *Proc. Natl. Acad. Sci. USA* **89:**5398–5402.
56. **Scott, J. K., and G. P. Smith.** 1990. Searching for peptide ligands with an epitope library. *Science* **249:**386–390.
57. **Singh, M., X. M. Li, H. Wang, J. P. McGee, T. Zamb, W. Koff, C. Y. Wang, and D. T. O'Hagan.** 1997. Immunogenecity and protection in small-animal models with controlled-release tetanus toxoid microparticles as a single-dose vaccine. *Infect. Immun.* **65:**1716–1721.

58. **Sorensen, R. U., L. E. Leiva, P. A. Giangrosso, B. Butler, F. C. Javier III, D. M. Sacerdote, N. Bradford, and C. Moore.** 1998. Response to a heptavalent conjugate *Streptococcus pneumoniae* vaccine in children with recurrent infections who are unresponsive to the polysaccharide vaccine. *Pediatr. Infect. Dis. J.* **17:**685–691.
59. **Stein, K. E., and T. Soderstrom.** 1984. Neonatal administration of idiotype or antiidiotype primes for protection against *Escherichia coli* K13 infection in mice. *J. Exp. Med.* **160:**1001–1011.
60. **Taki, T., D. Ishikawa, H. Hamasaki, and S. Handa.** 1997. Preparation of peptides which mimic glycosphingolipids by using phage peptide library and their modulation on beta-galactosidase activity. *FEBS Lett.* **418:**219–223.
61. **Tam, J. P.** 1988. Synthetic peptide vaccine design: synthesis and properties of a high-density multiple antigenic peptide system. *Proc. Natl. Acad. Sci. USA* **85:**5409–5413.
62. **Tam, J. P., P. Clavijo, Y. A. Lu, V. Nussenzweig, R. Nussenzweig, and F. Zavala.** 1990. Incorporation of T and B epitopes of the circumsporozoite protein in a chemically defined synthetic vaccine against malaria. *J. Exp. Med.* **171:**299–306.
63. **Tam, J. P., and Y. A. Lu.** 1989. Vaccine engineering: enhancement of immunogenicity of synthetic peptide vaccines related to hepatitis in chemically defined models consisting of T- and B-cell epitopes. *Proc. Natl. Acad. Sci. USA* **86:**9084–9088.
64. **Taub, R., and M. I. Greene.** 1992. Functional validation of ligand mimicry by anti-receptor antibodies: structural and therapeutic implications. *Biochemistry* **31:**7431–7435.
65. **Tunkel, A. R., and W. M. Scheld.** 1995. Acute meningitis, p. 831–865. *In* G. L. Mandell, J. E. Bennett, and R. Dolin (ed.), *Principles and Practice of Infectious Diseases,* 4th ed., vol. 1. Churchill Livingstone, New York, N.Y.
66. **Valadon, P., G. Nussbaum, L. F. Boyd, D. H. Margulies, and M. D. Scharff.** 1996. Peptide libraries define the fine specificity of anti-polysaccharide antibodies to *Cryptococcus neoformans. J. Mol. Biol.* **261:**11–22.
67. **Valadon, P., G. Nussbaum, J. Oh, and M. D. Scharff.** 1998. Aspects of antigen mimicry revealed by immunization with a peptide mimetic of *Cryptococcus neoformans* polysaccharide. *J. Immunol.* **161:**1829–1836.
68. **van Dam, J. E., A. Fleer, and H. Snippe.** 1990. Immunogenicity and immunochemistry of *Streptococcus pneumoniae* capsular polysaccharides. *Antonie Leeuwenhoek* **58:**1–47.
69. **Van Regenmortel, M. H., G. Guichard, N. Benkirane, J. P. Briand, S. Muller, and F. Brown.** 1998. The potential of retro-inverso peptides as synthetic vaccines. *Dev. Biol. Stand.* **92:**139–143.
70. **Vargas-Madrazo, E., F. Lara-Ochoa, and J. C. Almagro.** 1995. Canonical structure repertoire of the antigen-binding site of immunoglobulins suggests strong geometrical restrictions associated to the mechanism of immune recognition. *J. Mol. Biol.* **254:**497–504.
71. **Verdaguer, N., M. G. Mateu, J. Bravo, E. Domingo, and I. Fita.** 1996. Induced pocket to accommodate the cell attachment Arg-Gly-Asp motif in a neutralizing antibody against foot-and-mouth-disease virus. *J. Mol. Biol.* **256:**364–376.
72. **Westerink, M. A., P. C. Giardina, M. A. Apicella, and T. Kieber-Emmons.** 1995. Peptide mimicry of the meningococcal group C capsular polysaccharide. *Proc. Natl. Acad. Sci. USA* **92:**4021–4025.
73. **Wien, M. W., D. J. Filman, E. A. Stura, S. Guillot, F. Delpeyroux, R. Crainic, and J. M. Hogle.** 1995. Structure of the complex between the Fab fragment of a neutralizing antibody for type 1 poliovirus and its viral epitope. *Nat. Struct. Biol.* **2:**232–243.
74. **Wilson, I. A., and R. L. Stanfield.** 1994. Antibody-antigen interactions: new structures and new conformational changes. *Curr. Opin. Struct. Biol.* **4:**857–867.
75. **Yi, X. Y., A. J. Simpson, R. de Rossi, and S. R. Smithers.** 1986. The presence of antibody in mice chronically infected with *Schistosoma mansoni* which blocks in vitro killing of schistosomula. *J. Immunol.* **137:**3955–3958.
76. **Young, A. C., P. Valadon, A. Casadevall, M. D. Scharff, and J. C. Sacchettini.** 1997. The three-dimensional structures of a polysaccharide binding antibody to *Cryptococcus neoformans* and its complex with a peptide from a phage display library: implications for the identification of peptide mimotopes. *J. Mol. Biol.* **274:**622–634.
77. **Zhang, H., Z. Zhong, and L. A. Pirofski.** 1997. Peptide epitopes recognized by a human anti-cryptococcal glucuronoxylomannan antibody. *Infect. Immun.* **65:**1158–1164.
78. **Zuger, A., E. Louie, R. S. Holzman, M. S. Simberkoff, and J. J. Rahal.** 1986. Cryptococcal disease in patients with the acquired immunodeficiency syndrome. Diagnostic features and outcome of treatment. *Ann. Intern. Med.* **104:**234–240.

Molecular Mimicry, Microbes, and Autoimmunity
Edited by M. W. Cunningham and R. S. Fujinami

Chapter 12

Autoimmunity in Lyme Arthritis: Molecular Mimicry between OspA and LFA-1

Dawn M. Gross and Brigitte T. Huber

Lyme disease is a multifaceted illness, initiated upon infection with the spirochete *Borrelia burgdorferi.* While the majority of patients who develop Lyme arthritis can be treated successfully with appropriate antibiotic therapy (67–69) about 10% go on to develop what we have termed "treatment-resistant Lyme arthritis." These patients continue to have arthritis, typically affecting one or both knees, for months to even several years after multiple courses of oral and/or intravenous antibiotics (67–69). Moreover, such patients are usually negative by PCR for spirochetal DNA in joint fluid after antibiotic therapy, suggesting that the spirochete has been eliminated by this treatment (11, 35, 51, 67). Thus, the following question arises: Why does chronic arthritis persist in these patients when the causative agent has been eradicated?

The first indication that autoimmunity might play a role in treatment-resistant Lyme arthritis was a study of HLA-allele association with Lyme arthritis (35). Specifically, an increased frequency of HLA-DR4 was seen in conjunction with a lack of response to antibiotic therapy and the development of chronic disease (35). More recently, 22 patients with treatment-resistant Lyme arthritis have been HLA typed and have been shown to have an increased frequency of the HLA-DRB1*0401 allele ($P = 0.001$) (A. C. Steere and L. A. Baxter-Lowe, unpublished data). Furthermore, hypervariable region 3 of DRB1 chains homologous to DRB1*0401 at residues 67 to 74 are believed to confer susceptibility to rheumatoid arthritis (RA) (25, 49, 50). This sequence may be found in at least 15 different DRB1 alleles (Table 1). In an ongoing study, analysis of the DRB1 alleles present in patients with Lyme arthritis has suggested that most patients with prolonged treatment-resistant disease possess one of these homologous alleles (A. C. Steere and L. A. Baxter-Lowe, unpublished data). The question which naturally follows is: What antigen are these class II molecules presenting?

In people, a complex immune reaction develops in response to infection with *B. burgdorferi,* resulting in a gradual expansion of immunoreactivity to an increasing array of spirochetal proteins over a period of months to years (64). Over time, it has been shown that in approximately 70% of patients with Lyme arthritis, the final point of antibody response expansion is the development of immunoglobulin G (IgG) antibody to outer sur-

Dawn M. Gross and Brigitte T. Huber • Department of Pathology, Program in Immunology, Sackler School of Biomedical Sciences, Tufts University School of Medicine, Boston, MA 02111.

Table 1. DR molecules with peptide-binding cleft residues homologous to DRB1*0401

DRB1 molecule	Residues of HVR3 contained within the "shared epitope" domain of DRB1 (amino acids 67 to 74)[a]
*0401	L L E Q K R A A
*0101	L L E Q R R A A
*0102	L L E Q R R A A
*0104	L L E Q R R A A
*0404	L L E Q R R A A
*0405	L L E Q R R A A
*0408	L L E Q R R A A
*0409	L L E Q K R A A
*0413	L L E Q K R A A
*11011	F L E D R R A A
*1402	L L E Q R R A A
*1406	L L E Q R R A A
*1409	L L E Q R R A A
*1413	L L E Q R R A A
*1417	L L E Q R R A A
Murine H-2 I-E^s	F L E Q R R A A
*0402	**I** L E **D E** R A A

[a] The "shared epitope" hypothesis suggests that DR molecules which are conserved at residues 67 to 74 with DRB1*0401 bind to identical or highly similar peptides (25). Letters in boldface highlight nonconserved amino acid discrepancies between the non-RA-associated allele DRB1*0402 and the prototypic RA-associated allele DRB1*0401.

face protein A (OspA) and OspB of the spirochete (35, 36). The onset of this response appears late in disease and typically occurs near the beginning of prolonged episodes of arthritis (35). Furthermore, the risk for development of treatment-resistant Lyme arthritis has been shown to double when patients have the HLA-DR4 specificity and generate an anti-OspA antibody response (36). Thus, reactivity to OspA is a significant correlate of prolonged Lyme arthritis.

Despite intensive investigations by many research groups, we still do not understand the cellular and molecular mechanisms which lead to antibiotic treatment-resistant Lyme arthritis in certain individuals. No murine model for treatment-resistant Lyme arthritis has been established to date. The observations that development of treatment-resistant Lyme arthritis is correlated with the presence of DR4 alleles and persistent OspA reactivity late in disease have led us to hypothesize that patients with treatment-resistant Lyme arthritis have progressed into an autoimmune state by developing a cross-reactive response between OspA and a self-antigen. In order to substantiate our hypothesis, we set out to identify an autoantigen involved in the development of this localized, chronic immune reaction.

ANIMAL MODELS OF LYME ARTHRITIS

Numerous animal models of Lyme arthritis have been developed, including models with mice, rats, hamsters, guinea pigs, rabbits, dogs, and monkeys. Unfortunately, most of these animals develop only incremental aspects of the disease in humans. For instance, rabbits are susceptible to infection with *B. burgdorferi,* yet the only manifestation of the disease which resembles infection in humans is the development of erythema migrans (39, 45). Guinea pigs infected with a *Borrelia* isolate obtained from a dog develop myocarditis

(12), whereas dogs become lame and may develop lesions within the skin, joints, and lymph nodes, particularly if they were infected when they are young (1). Rhesus monkeys infected with 10^8 *B. burgdorferi* organisms via syringe inoculation are less likely to develop erythema migrans than monkeys infected via tick bite (53). Neither route of infection results in arthritic disease (53).

The development of arthritis, in particular, has most commonly been studied in laboratory mice and hamsters. Age and genotype appear to influence the susceptibility of mice to infection with *B. burgdorferi.* When 3-day-old C3H/He, SWR, C57BL/6, SJL, or BALB/c mice are infected via a peritoneal inoculation of *B. burgdorferi,* each of these mouse strains develops arthritis within 30 days. This is likely due to the immature status of the immune system in these animals, thereby giving them phenotypes similar to those of infected severe combined immunodeficient (SCID) mice (see below). Yet, if these same mice are infected at 3 weeks of age, only C3H and SWR mice develop arthritis. Finally, if these mice are infected even later in life (12 weeks of age), the arthritis which develops in C3H mice is more mild than that which develops in 3-week-old mice (2).

Several studies have suggested that susceptibility to disease severity is linked to the major histocompatibility complex (MHC) (33, 57, 59). Mice with the $H\text{-}2^k$ haplotype on a variety of backgrounds develop severe arthritis, while mice with the $H\text{-}2^{b,r,s}$ haplotype develop various degrees of arthritis of different durations. Finally, mice with the $H\text{-}2^d$ haplotype show no signs of arthritis at any time point (33, 57, 59). On the other hand, mice of all MHC haplotypes produce similar patterns of protective antibodies after infection, thereby implicating a critical role for T cells in the ensuing inflammatory reaction (59, 66). Interestingly, *Borrelia* species of different genotypes are not equally infectious. *B. garinii* and *B. afzelii* are able to infect nearly 100% of challenged C3H mice. This is in contrast to *B. japonica,* which is able to infect an average of only 20% of challenged mice (33).

Although susceptibility to severe arthritis is a dominant, genetically determined trait, the genetics responsible for this trait are not clear and may not be solely MHC linked, in contrast to what has been proposed above. Yang et al. (72) have shown that F^1 offspring from a C3H (susceptible) and BALB/c (resistant) cross develop severe arthritis. Yet, congenic studies with C3H and C57BL/6 mice did not demonstrate an association with arthritis susceptibility and the *H-2* locus (72). It is thought that, in comparison with the resistant BALB/c mice, C3H mice either are more permissive for spirochetal growth or are less efficient at clearing the infection (72). These seemingly discrepant findings might be due to a number of differences in the experimental designs used by Yang et al. (72) versus those used by Schaible et al. (59). As mentioned above, the age of the mice when they are tested can significantly influence the animal's susceptibility to infection with *B. burgdorferi* (6). The ages of the mice used in each of these studies differed considerably (5 weeks versus 6 to 10 weeks, respectively). Furthermore, each group used distinct strains of *B. burgdorferi* (N40 versus ZS7, respectively), which can dramatically affect immunoreactivities in both MHC-dependent and -independent patterns (see below) (24). Finally, the number of spirochetes injected and the route of inoculation differed substantially (2×10^5 injected intradermally into the shaven back versus 1×10^8 injected subcutaneously into the base of the tail, respectively). As discussed previously (5, 44), this can influence the expression pattern of spirochetal proteins and the resulting immune response. All of these parameters can modify the type of immune response mounted in an animal and, therefore, alter the course of *B. burgdorferi* infection, likely explaining these opposing findings.

It is important to point out, however, that all immunocompetent mouse strains clear the initial arthritic reaction and, although they remain carriers of a low level of *B. burgdorferi* spirochetes for life, do not suffer from chronic inflammatory disease (3, 7, 72). In addition, the *B. burgdorferi*-induced transient Lyme arthritis in rodents cannot be transferred with immune T cells in the absence of an infection. It is likely, therefore, that the mouse is unable to elicit an autoimmune reaction as a consequence of the inflammatory response to the spirochete. One possibility is that this species lacks the self-determinant which evokes the cross-reactive response. It may be this feature that renders the wild mouse a natural carrier of *B. burgdorferi.*

Several mice transgenic (tg) for or deficient in various proteins thought to influence the course of Lyme disease have been studied. Mice transgenic for OspA or OspB (23) for HLA.DRB1*0401 (22) have been described as developing Lyme arthritis equivalent to that in their normal littermates. Importantly, the DR4-tg studies used mice which express murine I-A^f class II, in addition to the human-mouse chimeric transgene. Therefore, the result may not be an accurate representation of the influence that DR4 has on the course of Lyme disease. Complement C5-deficient mice developed infection and regression of disease indistinguishable from that in their littermates, suggesting that C5-mediated complement activation is not required for the antibody-mediated protection from Lyme disease (7). Finally, immunodeficient SCID mice infected with *B. burgdorferi* develop severe arthritis, carditis, and hepatitis (55, 58), which are preventable if anti-OspA antibodies are administered prior to infection (56) and which are treatable if antibodies are administered after induction of arthritis (4). These results suggest that acquired immunity is not required for development of acute arthritis.

Demonstration of T-cell involvement in the pathogenesis of severe destructive Lyme arthritis has been described in a hamster model of Lyme disease. Lymph node T cells from hamsters vaccinated with 10^8 formalin-inactivated spirochetes in adjuvant are able to confer susceptibility to severe destructive arthritis when transferred into naive hamsters challenged with 10^6 viable *B. burgdorferi* organisms (41). The same is not true when naive recipients are either infused with normal T cells and challenged with viable spirochetes or infused with primed T cells and challenged with dead *B. burgdorferi* cells (41). These results imply that a gene product which is actively expressed by the spirochete when it is inside a mammalian host is involved in the T-cell-mediated propagation of destructive arthritis. Hence, T cells seem to be responsible for the development of destructive arthritis, as opposed to conferring protection from disease (18).

In summary, animal models of *B. burgdorferi* infection have yielded different results with few unambiguous answers. Yet, some trends are apparent. Innate immunity appears to play a protective role in acute disease, as demonstrated in the SCID mouse model. Antibodies clearly provide protection from disease if administered prior to infection, and T cells seem to have either a beneficial or a detrimental role, depending upon the animal infected and the type of response generated. This will be discussed in more detail below.

IMMUNOREACTIVITY IN LYME DISEASE

The subset of $CD4^+$ T-helper (Th) cells activated during an infection determines the efficiency with which the host is able to mount a protective immune response. This is achieved by the production of characteristic cytokine profiles. Th1 cells, which are capable of secreting interleukin 2 (IL-2), lymphotoxin, and gamma interferon (IFN-γ), elicit an inflammatory response, thereby regulating antiviral responses and immunity to intracellular pathogens. Alternatively, Th2 cells, which produce IL-4, IL-5, IL-10, and IL-13, mediate humoral immunity but inhibit cell-mediated inflammatory responses. Therefore, the type of Th cells induced in response to invasion by a particular pathogen can have a significant effect on the host's ability to successfully combat the infection (10, 13, 32, 47, 48, 62, 65).

Analysis of the T-cell subsets involved in the development of transient Lyme arthritis in mice reveals a picture reminiscent of murine *Leishmania* or human leprosy, in which the Th response determines the severity of the disease (30, 71). BALB/c (H-2^d) mice, which show only a mild arthritis when they are infected with *B. burgdorferi,* develop a predominant Th2 response. In contrast, C3H/HeJ (H-2^k) mice, which mount severe arthritis when they are infected with *B. burgdorferi,* have a dominant Th1 response (38, 43). When these T-cell responses are reversed by administration of anticytokine blocking antibodies, the severity of the arthritis is exacerbated or ameliorated in the respective mouse strains (43).

Our work has explored the population of active cells involved in the propagation of treatment-resistant Lyme arthritis. We have previously shown that Th1 cells dominate the immune response in the synovial fluid of patients with Lyme arthritis, as well as those with RA or other types of chronic inflammatory arthritis (26). In addition, we identified a population of Th1 cells specific for OspA of *B. burgdorferi* which is restricted to the synovial fluid of patients with Lyme arthritis. These OspA-reactive Th1 cells persist in patients with prolonged antibiotic treatment-resistant disease, suggesting the possibility of an autoimmune process.

T-cell lines from patients with treatment-resistant Lyme arthritis have previously been shown to preferentially recognize OspA in comparison with T-cell lines from patients with treatment-responsive disease (37). Attempts to map dominant T-helper-cell epitopes of OspA in humans with a variety of HLA-DR alleles (37), as well as in H-2^k mice (8, 76), have been reported. Although several T-cell epitopes of OspA have been identified (37), the number of patients tested to date is too small to draw conclusions regarding dominant epitopes for particular class II MHC alleles. In the H-2^k mouse a single, C-terminal peptide is able to prime the animal for a protective anti-OspA antibody response (8, 76).

An immunomodulatory role for synovial fluid $\gamma\delta$ T cells reactive against *B. burgdorferi* sonicate has been described in patients with Lyme arthritis (70). Vincent et al. (70) have shown that Fas-ligand^{high} $\gamma\delta$ T cells derived from the inflamed joint induce apoptosis of CD4$^+$ synovial T cells that express high levels of Fas antigen (Apo-1, CD95). Although the nature of the antigen reactivity demonstrated by the cells is unclear, their role in regulation of joint inflammation via Fas-mediated apoptosis is intriguing, as novel therapies for treatment-resistant Lyme arthritis are clearly needed.

The association of treatment-resistant Lyme arthritis with immune reactivity to OspA, as well as the presence of particular HLA-DR alleles, led us to hypothesize that patients with treatment-resistant Lyme arthritis have progressed into an autoimmune state by developing a cross-reactive response between OspA and a self-antigen. In order to substantiate our hypothesis, we set out to identify an autoantigen involved in the development of this localized immune reaction.

Using transgenic mice for HLA-DRB1*0401, we defined the immunodominant OspA Th-cell epitope for this prototypic class II allele which is linked to treatment-resistant Lyme arthritis. We made use of class II-deficient mice transgenic for a chimeric

DRB1*0401 molecule (DR4-tg), previously shown to functionally interact with murine $CD4^+$ T cells (34). Any $CD4^+$ T-cell response generated in these mice can be attributed directly to the presence of the DR4 molecule. The ElisaSpot assay was used to measure antigen-specific T-cell reactivity, a sensitive and efficient technique that allows the detection at the single-cell level of cytokine production which may occur in the absence of proliferation. We defined the immunodominant T-cell epitope of OspA ($OspA_{165-173}$) in the context of the prototypic disease-associated HLA-DR allele, DRB1*0401. A peptide homology search revealed that human lymphocyte function-associated antigen 1 (hLFA-1; CD11a/CD18, integrin $\alpha_L\beta_2$) possesses a highly related T-cell epitope (hLFA-1$\alpha_{L332-340}$) that could serve as a potential autoantigen. This was verified by our observation that T cells from patients with chronic Lyme arthritis respond in vitro to both whole OspA and hLFA-1, as well as to the DRB1*0401-defined immunodominant epitopes of $OspA_{165-173}$ and hLFA-1$\alpha_{L332-340}$. We suggest that OspA, in the context of DR4-homologous alleles, triggers development of an autoimmune response to hLFA-1, which leads to treatment-resistant Lyme arthritis (27). Thus, treatment-resistant Lyme arthritis is the first model for a human chronic inflammatory arthritide in which an initiating antigen and a cross-reactive autoantigen are defined.

INFECTION AND AUTOIMMUNITY

It is well established that a population of autoreactive T and B cells escapes thymic selection and deletion, as these cells can be detected in the circulation of healthy individuals (14, 15). Why these potentially pathogenic lymphocytes do not induce disease in all people at all times has remained elusive. The factors which control the immunoregulatory balance between a protective, pathogen-eliminating immune reaction and a damaging, self-destructive autoreactive response are not well-characterized, although it has been widely speculated that infectious agents might be capable of distorting this equilibrium (52, 61). It is plausible that an infectious process can influence the immune system, such that the immunoregulatory network is no longer able to keep the autoreactive cells in check, thereby resulting in a break in self-tolerance. Nevertheless, no direct evidence currently exists for an infectious etiology of autoimmune diseases such as RA, systemic lupus erythromatosus, or type I diabetes mellitus. In the case of Lyme disease, however, the infectious agent is known. Despite this information, the nature of the ensuing chronic immune response is not well-defined. Identification of a specific autoantigen would allow treatment-resistant Lyme arthritis to be classified as an autoimmune disease.

Much attention has been focused on the elucidation of the specific amino acid residues involved in class II peptide binding and T-cell receptor (TCR) recognition. Most notably, the crystal structure of DRB1*0401 with a collagen II peptide in the binding pocket has recently been resolved, providing for the first time a visual confirmation of the in vitro and in vivo models designed to depict these interactions (19). Peptide residues residing in pockets p1, p4, and p6 of the MHC binding groove have been suggested to be of significance for determination of peptide specificity, as well as DR allelic discrimination (28, 29). $OspA_{165-173}$ and hLFA-1$\alpha_{L332-340}$ peptides fulfill all the suggested requirements for binding specificity to RA-associated alleles (28, 29). Consistent with these findings, we have demonstrated the ability of 15-mer peptides, $OspA_{164-178}$ and hLFA-1$\alpha_{L331-345}$ (each containing the core 9-mer epitope sequences), to bind to DRB1*0401 in vitro. Furthermore, we have shown that 20-mers of these peptides act as immunoreactive epitopes in vivo (27). Conversely, neither of these peptides is predicted to bind to the non-RA-associated DRB1*0402 allele, as they both achieve negative binding scores according to the *0402 binding algorithm (Table 2). This is mainly due to the negatively charged Glu residue at position p4, which is not tolerated in the *0402 binding pocket (29). Hence, $OspA_{165-173}$ and hLFA-1$\alpha_{L332-340}$ are predicted to be presented specifically by RA-associated alleles, notably, the same alleles associated with the development of treatment-resistant Lyme arthritis (L. A. Steere and L. A. Baxter-Lowe, unpublished data).

While the genetic predisposition for the development of treatment-resistant Lyme arthritis has been correlated to DR4, we cannot rule out other genetic, environmental, and infectious factors which might play a role in the development of this prolonged form of arthritis. We found that some patients with treatment-resistant disease neither possess RA-associated alleles nor generate T-cell responses to hLFA-1, possibly implicating additional autoantigens involved in disease development (27).

Finally, mLFA-1$\alpha_{L332-340}$ fulfills all of the DRB1*0401-binding requirements, achieves a predicted DR4-binding score of 5.4, and, not surprisingly, was shown to bind to DRB1*0401 in vitro (50% inhibitory concentration, 0.9005 mM). Yet, DR4-tg mice do not appear to be susceptible to the development of treatment-resistant chronic Lyme arthritis (22). At least two possibilities might explain this finding. First, the DR4-tg mice used by Feng et al. (22) posses the H-2^f-class II molecule I-A, in addition to the chimeric I-E^d DR4 molecule. Hence, it is possible that these mice are able to efficiently tolerize their immune response to mLFA-1. Furthermore, these mice were infected via subcutaneous injection of 10^4 cultured spirochetes and were never found to produce anti-OspA or anti-OspB antibodies throughout the 180-day study. Since anti-OspA reactivity is associated with development of treatment-resistant Lyme arthritis, it is likely that the route and/or type of infection used did not accurately mimic the natural infection seen in humans. Finally, subtle sequence differences do exist between human and murine LFA-1$\alpha_{L332-340}$, which might explain the inability of mice to develop treatment-resistant disease. Specifically, the p2 residue present in hLFA-1$\alpha_{L332-340}$ and $OspA_{165-173}$ is altered in mLFA-1$\alpha_{L332-340}$ (Val→Ala). Since p2 is predicted to influence the TCR contact (19) rather than the MHC-peptide interaction, we propose that cross-reactivity cannot develop between OspA and mLFA, not as a result of insufficient DR4 binding but, rather, as a result of distinct MHC-peptide-TCR interactions. This suggests that the ability of mLFA-1 to activate T cells and/or influence thymic selection and the TCR repertoire in the mouse is distinct from what appears to be the case for hLFA-1 in people.

Table 2. $OspA_{165-173}$ and hLFA-1$\alpha_{L332-340}$ are not predicted to bind to DRB1*0402[a]

Parameter	p1			p4		p6				Total DR4*0402-binding score
$OspA_{165-173}$ sequence	Y	V	L	E	G	T	L	T	A	−1.0
Residue score			0.2	−2.3	−1.4	1.5	1.0			
hLFA-1$\alpha_{L332-340}$ sequence	Y	V	I	E	G	T	S	K	Q	−2.1
Residue score			0.5	−2.3	−1.4	1.5	−0.4			

[a] The individual residue score and total binding scores were determined by the DR4 algorithm (28).

DEVELOPMENT OF AUTOIMMUNITY IN PATIENTS WITH TREATMENT-RESISTANT LYME ARTHRITIS: THE MODEL

hLFA-1 is an attractive candidate autoantigen for several reasons: hLFA-1 is an adhesion molecule found on lymphocytes, with its highest level of expression being on activated T cells. It functions as a mediator of cell-cell interactions during an inflammatory response (17, 20, 40). It is most abundant on T and B cells, with its highest level of expression occurring on activated T lymphoblasts (17, 20, 40, 73). In addition, LFA-1 is present on macrophages after stimulation with lipopolysaccharide or IFN-γ (17, 20, 40). The ligand for hLFA-1, intercellular adhesion molecule 1 (ICAM-1), is upregulated on fibroblasts, endothelial cells, and synoviocytes when it is activated with IL-1, IFN-γ, tumor necrosis factor alpha, and, of particular importance, *B. burgdorferi* (9, 20, 21, 31, 42, 60, 63). Hence, the modulation of ICAM-1 expression via cytokine production in an inflammatory response would not only facilitate margination and extravasation of T cells to the site of inflammation but would also promote adherence of the localized T cells to connective tissue cells and thereby increase the efficiency of T-cell functions (17, 20, 40). This could ensue into a "vicious cycle" of inflammatory reaction, recruitment, and propagation of a local autoinflammatory response via cross-reaction of OspA-reactive T cells with LFA-1.

On the basis of the DR4-restricted OspA T-cell epitope mapping data presented here, as well as on the basis of previous work which described immune reactivity and cytokine production in response to infection with *B. burgdorferi,* we propose a model for immune reaction to *B. burgdorferi* that results in the development of an autoimmune response against hLFA-1 (Fig. 1): *B. burgdorferi* enters the host and disseminates to multiple tissues, possibly by mechanisms involving OspA–ICAM-1 interactions (discussed below). Months later, a highly inflammatory immune response develops in the joint, and, as we and others have shown previously, this response is dominated by IFN-γ-producing Th1 cells which may consist of OspA-reactive cells (26, 74). We propose that the high local concentration of IFN-γ upregulates expression of LFA-1 and ICAM-1 (20, 31) on synoviocytes and synovial fibroblasts, as well as MHC class II molecules on the local professional and nonprofessional antigen-presenting cells (APCs) (B. Nepom, unpublished data). Others have shown that enhanced antigen presentation, including class II presentation of endogenous self-peptides, can occur during a prolonged immune response (46, 54). In our model we would speculate that the combination of elevated LFA-1 expression on T cells and macrophages plus MHC class II upregulation on APCs results in increased levels of LFA-1 peptide presentation by macrophages and synoviocytes which have processed either endogenous or phagocytosed LFA-1 (46, 54). Hence, propagation of disease is initiated, such that even after elimination of the spirochetes by antibiotic therapy, the OspA-primed T cells remain activated by stimulation with LFA-1. The release of inflammatory cytokines by these activated T cells and macrophages would then result in tissue damage and joint destruction (16).

Although it is premature to state that this autoimmune response is the sole cause of the immune dysregulation seen in treatment-resistant Lyme arthritis, it provides one explanation of how chronic inflammation might persist in the absence of spirochetes and why antibiotic therapy is an ineffective treatment for such patients. Recently, a definitive study by Zhao et al. (75) demonstrated the role of molecular mimicry in the development of the herpes simplex virus type 1 KOS strain-induced murine model of autoimmune herpes stro-

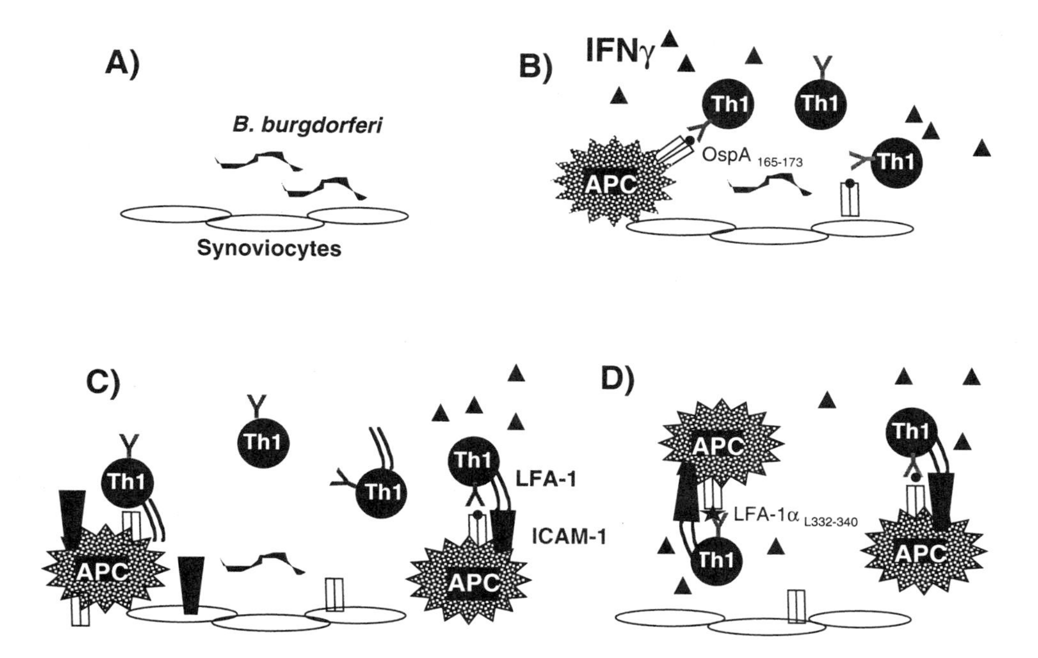

Figure 1. Model of how autoimmunity develops in patients with treatment-resistant Lyme arthritis. *B. burgdorferi* migrates to the joint and induces a localized Th1-dominant response with OspA specificity. The unbalanced production of IFN-γ induces upregulation of MHC class II on both classical and atypical APCs. Furthermore, LFA-1 expression becomes enhanced on T cells, as well as macrophages, as does ICAM-1 expression due to the presence of IFN-γ and *B. burgdorferi* itself. This results in processing and presentation of LFA-1, which leads to activation of OspA-primed T cells, even in the absence of *B. burgdorferi*.

mal keratitis (75). The identification of hLFA-1 as an autoantigen in treatment-resistant Lyme arthritis represents the first model of an arthritic autoimmune disease in which the initiating bacterial antigen and the persistent autoantigen are known.

CONCLUDING REMARKS

It is important to mention that treatment-resistant Lyme arthritis appears to be preventable when patients are treated early in the course of disease with appropriate antibiotic therapy. That is, if the spirochetal infection is prevented from disseminating into the joint or is eradicated before a strong proinflammatory response can develop in the localized space of the synovial fluid, the vicious cycle of chronic inflammation is prevented before molecular mimicry and autoimmunity can develop. Similarly, it seems likely that one could halt disease progression by downmodulating the IFN-γ proinflammatory response with anti-inflammatory, pro-IL-4 agents.

REFERENCES

1. **Appel, M. J., S. Allan, R. H. Jacobson, T. L. Lauderdale, Y. F. Chang, S. J. Shin, J. W. Thomford, R. J. Todhunter, and B. A. Summers.** 1993. Experimental Lyme disease in dogs produces arthritis and persistent infection. *J. Infect. Dis.* **167:**651–664.
2. **Barthold, S., D. S. Beck, G. M. Hansen, G. A. Terwilliger, and K. D. Moody.** 1990. Lyme borreliosis in selected strains and ages of laboratory mice. *J. Infect. Dis.* **162:**133–138.
3. **Barthold, S. W., M. S. de Souza, J. L. Janotka, A. L. Smith, and D. H. Persing.** 1993. Chronic Lyme borreliosis in the laboratory mouse. *Am. J. Pathol.* **143:**959–971.
4. **Barthold, S. W., M. deSouza, and S. Feng.** 1996. Serum-mediated resolution of Lyme arthritis in mice. *Lab. Investig.* **74:**57–67.
5. **Barthold, S. W., E. Fikrig, L. K. Bockenstedt, and D. H. Persing.** 1995. Circumvention of outer surface protein A immunity by host-adapted *Borrelia burgdorferi. Infect. Immun.* **63:**2255–2261.
6. **Barthold, S. W., K. D. Moody, and D. S. Beck.** 1990. Susceptibility of laboratory rats to isolates of *Borrelia burgdorferi* from different geographic areas. *Am. J. Trop. Med. Hyg.* **42:**596–600.
7. **Bockenstedt, L. K., S. Barthold, K. Deponte, N. Marcantonio, and F. Kantor.** 1993. *Borrelia burgdorferi* infection and immunity in mice deficient in the fifth component of complement. *Infect. Immun.* **61:**2104–2107.
8. **Bockenstedt, L. K., E. Fikrig, S. W. Barthold, R. A. Flavell, and F. S. Kantor.** 1996. Identification of a *Borrelia burgdorferi* OspA T cell epitope that promotes anti-OspA IgG in mice. *J. Immunol.* **157:**5496–5502.
9. **Boggemeyer, E., T. Stehle, U. E. Schaible, M. Hahne, D. Vestweber, and M. M. Simon.** 1994. *Borrelia burgdorferi* upregulates the adhesion molecules E-selectin, P-selectin, ICAM-1 and VCAM-1 on mouse endothelioma cells in vitro. *Cell Adhesion Commun.* **2:**145–157.
10. **Bottomly, K.** 1988. A functional dichotomy in $CD4^+$ T lymphocytes. *Immunol. Today* **9:**268–274.
11. **Bradley, J. F., R. C. Johnson, and J. L. Goodman.** 1994. The persistence of spirochetal nucleic acids in active Lyme arthritis. *Ann. Intern. Med.* **120:**487–489.
12. **Breitschwerdt, E. B., F. J. Geoly, D. J. Meuten, J. F. Levine, P. Howard, B. C. Hegarty, and L. C. Stafford.** 1996. Myocarditis in mice and guinea pigs experimentally infected with a canine-origin Borrelia isolate from Florida. *Am. J. Vet. Res.* **57:**505–511.
13. **Cherwinski, H. M., J. H. Schumacher, K. D. Brown, and T. R. Mosmann.** 1987. Two types of mouse helper T cell clone. III. Further differences in lymphokine synthesis between Th1 and Th2 clones revealed by RNA hybridization, functionally monospecific bioassays, and monoclonal antibodies. *J. Exp. Med.* **166:**1229–1244.
14. **Cohen, I. R.** 1992. The cognitive paradigm and the immunological homunculus. *Immunol. Today* **13:**490–494.
15. **Cohen, I. R.** 1992. The cognitive principle challenges clonal selection. *Immunol. Today* **13:**441–444.
16. **Corcoran, M. L., W. G. Stetler-Stevenson, P. D. Brown, and L. M. Wahl.** 1992. Interleukin 4 inhibition of prostaglandin E2 synthesis blocks interstitial collagenase and 92-kDa type IV collagenase/gelatinase production by human monocytes. *J. Biol. Chem.* **267:**515–519.
17. **Davignon, D., E. Martz, T. Reynolds, K. Kurainger, and T. A. Springer.** 1981. Monoclonal antibody to a novel lymphocyte function associated antigen (LFA-1): mechanism of blockade of T lymphocyte-mediated killing and effects on the T and B lymphocyte functions. *J. Immunol.* **127:**590–595.
18. **de Souza, M. S., A. L. Smith, D. S. Beck, G. A. Terwilliger, E. Fikrig, and S. W. Barthold.** 1993. Long-term study of cell-mediated responses to *Borrelia burgdorferi* in the laboratory mouse. *Infect. Immun.* **61:**1814–1822.
19. **Dessen, A., C. M. Lawrence, S. Cupo, D. M. Zaller, and D. C. Wiley.** 1997. X-ray crystal structure of HLA-DR4 (DRA*0101, DRB1*0401) complexed with a peptide from human collagen II. *Immunity* **7:**473–481.
20. **Dustin, M. L., R. Rothlein, A. K. Bhan, C. A. Dinarello, and T. A. Springer.** 1986. Induction by IL 1 and interferon-gamma: tissue distribution, biochemistry, and function of a natural adherence molecule (ICAM-1). *J. Immunol.* **137:**245–254.
21. **Ebnet, K., K. D. Brown, U. K. Siebenlist, M. M. Simon, and S. Shaw.** 1997. *Borrelia burgdorferi* activates nuclear factor-kappa B and is a potent inducer of chemokine and adhesion molecule gene expression in endothelial cells and fibroblasts. *J. Immunol.* **158:**3285–3292.
22. **Feng, S., S. W. Barthold, L. K. Bockenstedt, D. M. Zaller, and E. Fikrig.** 1995. Lyme disease in human DR4Dw4-transgenic mice. *J. Infect. Dis.* **172:**286–289.

23. **Fikrig, E., H. Tao, M. Chen, S. W. Barthold, and R. A. Flavell.** 1995. Lyme borreliosis in transgenic mice tolerant to *B. burgdorferi* OspA or B. *J. Clin. Investig.* **96:**1706–1714.
24. **Golde, W. T., T. R. Burkot, S. Sviat, M. G. Keen, L. W. Mayer, B. J. Johnson, and J. Piesman.** 1993. The major histocompatibility complex-restricted response of recombinant inbred strains of mice to natural tick transmission of *Borrelia burgdorferi. J. Exp. Med.* **177:**9–17.
25. **Gregersen, P. K., J. Silver, and R. J. Winchester.** 1987. The shared epitope hypothesis. An approach to understanding the molecular genetics of susceptibility to rheumatoid arthritis. *Arthritis Rheum.* **30:**1205–1213.
26. **Gross, D. M., A. C. Steere, and B. T. Huber.** 1998. T helper 1 response is dominant and localized to the synovial fluid in patients with Lyme arthritis. *J. Immunol.* **160:**1022–1028.
27. **Gross, D. M., T. Forsthuber, M. Tary-Lehmann, C. Etling, K. Ito, Z. Najy, J. Field, A. C. Steere, and B. T. Huber.** 1998. Identification of LFA-1 as a candidate autoantigen in treatment-resistant Lyme arthritis. *Science* **281:**703–706.
28. **Hammer, J., F. Gallazzi, E. Bono, R. W. Karr, J. Guenot, P. Valsasnini, Z. A. Nagy, and F. Sinigaglia.** 1995. Peptide binding specificity of HLA-DR4 molecules: correlation with rheumatoid arthritis association. *J. Exp. Med.* **181:**1847–1855.
29. **Hammer, J., P. Valsasnini, K. Tolba, D. Bolin, J. Higelin, B. Takacs, and F. Sinigaglia.** 1993. Promiscuous and allele-specific anchors in HLA-DR-binding peptides. *Cell* **74:**197–203.
30. **Heinzel, F. P., M. D. Sadick, S. S. Mutha, and R. M. Locksley.** 1991. Production of interferon gamma, interleukin 2, interleukin 4, and interleukin 10 by $CD4^+$ lymphocytes in vivo during healing and progressive murine leishmaniasis. *Proc. Natl. Acad. Sci. USA* **88:**7011–7015.
31. **Henninger, D. D., J. Panes, M. Eppihimer, J. Russell, M. Gerritsen, D. C. Anderson, and D. N. Granger.** 1997. Cytokine-induced VCAM-1 and ICAM-1 expression in different organs of the mouse. *J. Immunol.* **158:**1825–1832.
32. **Hsieh, C. S., S. E. Macatonia, A. O'Garra, and K. M. Murphy.** 1995. T cell genetic background determines default T helper phenotype development in vitro. *J. Exp. Med.* **181:**713–721.
33. **Ishii, N., E. Isogai, H. Isogai, K. Kimura, T. Nishikawa, N. Fijii, and H. Nakajima.** 1995. T cell response to *Borrelia garinii, Borrelia afzelii,* and *Borrelia japonica* in various congenic mouse strains. *Microbiol. Immunol.* **39:**929–935.
34. **Ito, K., H. J. Bian, M. Molina, J. Han, J. Magram, E. Saar, C. Belunis, D. R. Bolin, R. Arceo, R. Campbell, F. Falcioni, D. Vidovic, J. Hammer, and Z. A. Nagy.** 1996. HLA-DR4-IE chimeric class II transgenic, murine class II-deficient mice are susceptible to experimental allergic encephalomyelitis. *J. Exp. Med.* **183:**2635–2644.
35. **Kalish, R. A., J. M. Leong, and A. C. Steere.** 1993. Association of treatment-resistant chronic Lyme arthritis with HLA-DR4 and antibody reactivity to OspA and OspB of *Borrelia burgdorferi. Infect. Immun.* **61:**2774–2779.
36. **Kalish, R. A., J. M. Leong, and A. C. Steere.** 1995. Early and late antibody responses to full-length and truncated constructs of outer surface protein A of *Borrelia burgdorferi* in Lyme disease. *Infect. Immun.* **63:**2228–2235.
37. **Kamradt, T., B. Lengl-Janssen, A. F. Strauss, G. Bansal, and A. C. Steere.** 1996. Dominant recognition of a *Borrelia burgdorferi* outer surface protein A peptide by T helper cells in patients with treatment-resistant Lyme arthritis. *Infect. Immun.* **64:**1284–1289.
38. **Keane-Myers, A., and S. P. Nickell.** 1995. Role of IL-4 and IFN-gamma in modulation of immunity to *Borrelia burgdorferi* in mice. *J. Immunol.* **155:**2020–2028.
39. **Kornblatt, A. N., A. C. Steere, and D. G. Brownstein.** 1984. Experimental Lyme disease in rabbits: spirochetes found in erythema migrans and blood. *Infect. Immun.* **46:**220–223.
40. **Larson, R. S., and T. A. Springer.** 1990. Structure and function of leukocyte integrins. *Immunol. Rev.* **114:**181–217.
41. **Lim, L. C., D. M. England, B. K. DuChateau, N. J. Glowacki, and R. F. Schell.** 1995. *Borrelia burgdorferi*-specific T lymphocytes induce severe destructive Lyme arthritis. *Infect. Immun.* **63:**1400–1408.
42. **Marlin, S. D., and T. A. Springer.** 1987. Purified intercellular adhesion molecule-1 (ICAM-1) is a ligand for lymphocyte function-associated antigen 1 (LFA-1). *Cell* **51:**813–819.
43. **Matyniak, J. E., and S. L. Reiner.** 1995. T helper phenotype and genetic susceptibility in experimental Lyme disease. *J. Exp. Med.* **181:**1251–1254.
44. **Montgomery, R. R., S. E. Malawista, K. J. Feen, and L. K. Bockenstedt.** 1996. Direct demonstration of antigenic substitution of *Borrelia burgdorferi* ex vivo: exploration of the paradox of the early immune response to outer surface proteins A and C in Lyme disease. *J. Exp. Med.* **183:**261–269.

45. **Moody, K. D., S. W. Barthold, and G. A. Terwilliger.** 1990. Lyme borreliosis in laboratory animals: effect of host species and in vitro passage of *Borrelia burgdorferi*. *Am. J. Trop. Med. Hyg.* **43:**87–92.
46. **Moreno, J., D. A. Vignali, F. Nadimi, S. Fuchs, L. Adorini, and G. J. Hammerling.** 1991. Processing of an endogenous protein can generate MHC class II-restricted T cell determinants distinct from those derived from exogenous antigen. *J. Immunol.* **147:**3306–3313.
47. **Mosmann, T. R., H. Cherwinski, M. W. Bond, M. A. Giedlin, and R. L. Coffman.** 1986. Two types of murine helper T cell clone. I. Definition according to profiles of lymphokine activities and secreted proteins. *J. Immunol.* **136:**2348–2357.
48. **Mosmann, T. R., and R. L. Coffman.** 1989. TH1 and TH2 cells: different patterns of lymphokine secretion lead to different functional properties. *Annu. Rev. Immunol.* **7:**145–173.
49. **Nepom, G. T.** 1989. Determinants of genetic susceptibility in HLA-associated autoimmune disease. *Clin. Immunol. Immunopathol.* **53:**S53–S62.
50. **Nepom, G. T., and H. Erlich.** 1991. MHC class-II molecules and autoimmunity. *Annu. Rev. Immunol.* **9:**493–525.
51. **Nocton, J. J., F. Dressler, B. J. Rutledge, P. N. Rys, D. H. Persing, and A. C. Steere.** 1994. Detection of *Borrelia burgdorferi* DNA by polymerase chain reaction in synovial fluid from patients with Lyme arthritis. *N. Engl. J. Med.* **330:**229–234.
52. **Nygard, N. R., D. M. McCarthy, J. Schiffenbauer, and B. D. Schwartz.** 1993. Mixed haplotypes and autoimmunity. *Immunol. Today* **14:**53–56.
53. **Philipp, M. T., M. K. Aydintug, R. P. Bohm, Jr., F. B. Cogswell, V. A. Dennis, H. N. Lanners, R. C. Lowrie, Jr., E. D. Roberts, M. D. Conway, M. Karaçorlu, G. A. Peyman, D. J. Gubler, B. J. B. Johnson, J. Piesman, and Y. Gu.** 1993. Early and early disseminated phases of Lyme disease in the rhesus monkey: a model for infection in humans. *Infect. Immun.* **61:**3047–3059.
54. **Salemi, S., A. P. Caporossi, L. Boffa, M. G. Longobardi, and V. Barnaba.** 1995. HIVgp120 activates autoreactive CD4-specific T cell responses by unveiling of hidden CD4 peptides during processing. *J. Exp. Med.* **181:**2253–2257.
55. **Schaible, U. E., S. Gay, C. Museteanu, M. D. Kramer, G. Zimmer, K. Eichmann, U. Museteanu, and M. M. Simon.** 1990. Lyme borreliosis in the severe combined immunodeficiency (scid) mouse manifests predominantly in the joints, heart, and liver. *Am. J. Pathol.* **137:**811–820.
56. **Schaible, U. E., M. D. Kramer, K. Eichmann, M. Modolell, C. Museteanu, and M. M. Simon.** 1990. Monoclonal antibodies specific for the outer surface protein A (OspA) of *Borrelia burgdorferi* prevent Lyme borreliosis in severe combined immunodeficiency (scid) mice. *Proc. Natl. Acad. Sci. USA* **87:**3768–3772.
57. **Schaible, U. E., M. D. Kramer, C. W. Justus, C. Museteanu, and M. M. Simon.** 1989. Demonstration of antigen-specific T cells and histopathological alterations in mice experimentally inoculated with *Borrelia burgdorferi*. *Infect. Immun.* **57:**41–47.
58. **Schaible, U. E., M. D. Kramer, C. Museteanu, G. Zimmer, H. Mossmann, and M. M. Simon.** 1989. The severe combined immunodeficiency (scid) mouse. A laboratory model for the analysis of Lyme arthritis and carditis. *J. Exp. Med.* **170:**1427–1432.
59. **Schaible, U. E., M. D. Kramer, R. Wallich, T. Tran, and M. M. Simon.** 1991. Experimental *Borrelia burgdorferi* infection in inbred mouse strains: antibody response and association of H-2 genes with resistance and susceptibility to development of arthritis. *Eur. J. Immunol.* **21:**2397–2405.
60. **Schaible, U. E., D. Vestweber, E. G. Butcher, T. Stehle, and M. M. Simon.** 1994. Expression of endothelial cell adhesion molecules in joints and heart during *Borrelia burgdorferi* infection of mice. *Cell Adhesion Commun.* **2:**465–479. (Erratum, **3:**178, 1995.)
61. **Schwartz, B. D.** 1990. Infectious agents, immunity, and rheumatic diseases. *Arthritis Rheum.* **33:**457–465.
62. **Scott, P., and S. H. Kaufmann.** 1991. The role of T-cell subsets and cytokines in the regulation of infection. *Immunol. Today* **12:**346–348.
63. **Sellati, T. J., M. J. Burns, M. A. Ficazzola, and M. B. Furie.** 1995. *Borrelia burgdorferi* upregulates expression of adhesion molecules on endothelial cells and promotes transendothelial migration of neutrophils in vitro. *Infect. Immun.* **63:**4439–4447.
64. **Shanafelt, M. C., J. Anzola, C. Soderberg, H. Yssel, C. W. Turck, and G. Peltz.** 1992. Epitopes on the outer surface protein A of *Borrelia burgdorferi* recognized by antibodies and T cells of patients with Lyme disease. *J. Immunol.* **148:**218–224.
65. **Sher, A., and R. L. Coffman.** 1992. Regulation of immunity to parasites by T cells and T cell-derived cytokines. *Annu. Rev. Immunol.* **10:**385–409.

66. **Simon, M., U. E. Schaible, R. Wallich, and M. D. Kramer.** 1991. A mouse model for *Borrelia burgdorferi* infection: approach to a vaccine against Lyme disease. *Immunol. Today* **12:**11–16.
67. **Steere, A. C., E. Dwyer, and R. Winchester.** 1990. Association of chronic Lyme arthritis with HLA-DR4 and HLA-DR2 alleles. *N. Engl. J. Med.* **323:**219–223. (Erratum, **324:**129, 1991.)
68. **Steere, A. C., A. Gibofsky, M. E. Patarroyo, R. J. Winchester, J. A. Hardin, and S. E. Malawista.** 1979. Chronic Lyme arthritis. Clinical and immunogenetic differentiation from rheumatoid arthritis. *Ann. Intern. Med.* **90:**896–901.
69. **Steere, A. C., R. T. Schoen, and E. Taylor.** 1987. The clinical evolution of Lyme arthritis. *Ann. Intern. Med.* **107:**725–731.
70. **Vincent, M. S., K. Roessner, D. Lynch, D. Wilson, S. M. Cooper, J. Tschopp, L. H. Sigal, and R. C. Budd.** 1996. Apoptosis of Fashigh $CD4^+$ synovial T cells by borrelia-reactive Fas-ligand(high) gamma delta T cells in Lyme arthritis. *J. Exp. Med.* **184:**2109–2117.
71. **Yamamura, M., K. Uyemura, R. J. Deans, K. Weinberg, T. H. Rea, B. R. Bloom, and R. L. Modlin.** 1991. Defining protective responses to pathogens: cytokine profiles in leprosy lesions. *Science* **254:**277–279. (Erratum, **255:**12, 1992.)
72. **Yang, L., J. H. Weis, E. Eichwald, C. P. Kolbert, D. H. Persing, and J. J. Weis.** 1994. Heritable susceptibility to severe *Borrelia burgdorferi*-induced arthritis is dominant and is associated with persistence of large numbers of spirochetes in tissues. *Infect. Immun.* **62:**492–500.
73. **Yokota, A., N. Murata, O. Saiki, M. Shimizu, T. A. Springer, and T. Kishimoto.** 1995. High avidity state of leukocyte function-associated antigen-1 on rheumatoid synovial fluid T lymphocytes. *J. Immunol.* **155:**4118–4124.
74. **Yssel, H., M. C. Shanafelt, C. Soderberg, P. V. Schneider, J. Anzola, and G. Peltz.** 1991. *Borrelia burgdorferi* activates a T helper type 1-like T cell subset in Lyme arthritis. *J. Exp. Med.* **174:**593–601.
75. **Zhao, Z. S., F. Granucci, L. Yeh, P. A. Schaffer, and H. Cantor.** 1998. Molecular mimicry by herpes simplex virus-type 1: autoimmune disease after viral infection. *Science* **279:**1344–1347.
76. **Zhong, W., K. H. Wiesmuller, M. D. Kramer, R. Wallich, and M. M. Simon.** 1996. Plasmid DNA and protein vaccination of mice to the outer surface protein A of *Borrelia burgdorferi* leads to induction of T helper cells with specificity for a major epitope and augmentation of protective IgG antibodies in vivo. *Eur. J. Immunol.* **26:**2749–2757.

Molecular Mimicry, Microbes, and Autoimmunity
Edited by M. W. Cunningham and R. S. Fujinami

Chapter 13

Exploiting Molecular Mimicry in Targeting Carbohydrate Antigens

Gina Cunto-Amesty and Thomas Kieber-Emmons

Carbohydrate antigens are important, distinguishing landmarks on many infectious agents and neoplasms and, thus, potential targets for preventive vaccination. However, most carbohydrate antigens belong to the category of T-cell-independent (TI) antigens, reflecting their inability to stimulate major histocompatibility complex (MHC) class II-dependent T-cell help (65). Consequently, carbohydrates alone are not capable of inducing a sufficient anamnestic or secondary immune response and require extensive adjuvanticity to convert them into T-cell-dependent (TD) antigens. Here, we review aspects of an alternative approach to induction of responses that target carbohydrate antigens by using peptide or polypeptide surrogates that mimic carbohydrate structures. Such surrogate immunogens, being intrinsically TD antigens, may be manipulated to recruit cellular immune mechanisms that are not obtainable, by definition, by TI antigens. Surrogate immunogens that are TD can provide immunological memory related to vaccine composition, form, and delivery because peptides are intrinsically able to associate with MHC class II molecules.

The molecular nature of the mimicry between carbohydrates and surrogate antigens is not well understood. The idea of using surrogate antigens as immunogens requires that antigenic mimicry, accomplished by using amino acids in place of sugars, induces an immune-specific reactivity pattern against the nominal antigen (22, 35, 73, 107). The concept of mimicry holds enormous promise. A molecule that mimics a given antigen by eliciting a similar immune response is potentially useful as a vaccine and is defined as a mimotope. Carbohydrate-mimicking peptides could revolutionize vaccines against infectious pathogens (99). There are, however, issues related to their use in designing efficacious vaccines. Some are related to fundamental questions as to how mimicry occurs at the molecular level, and others are application oriented or are directed at important immunological mechanisms that are required to be elucidated if peptide mimotopes are to fulfill their promise as vaccines.

Fundamental model systems have been described for many types of carbohydrate antigens found on a variety of bacteria, fungi, some viruses, and tumor cells. In this chapter, emphasis is on studies that elucidate basic mechanisms or immune parameters associated with vaccination with surrogates of glycosphingolipid constituents that may be applied to a variety of antigenic systems. We primarily make use of results derived from

Gina Cunto-Amesty and Thomas Kieber-Emmons • Department of Pathology and Laboratory Medicine, Room 205, John Morgan Building, 36th and Hamilton Walk, Philadelphia, PA 19104-6082.

our own studies. We have shown that peptide mimotopes can induce cross-reactive carbohydrate responses that are able to protect mice from lethal bacterial infection (112) or can induce humoral responses with components that are capable of binding to envelope glycoproteins of the human immunodeficiency virus, neutralizing cell-free virus (3). More recently, we have shown that mimotopes can induce protection from graded-tumor challenges (45). These observations establish the feasibility of exploiting the molecular mimicry of carbohydrate antigens by peptides in the design of vaccine development strategies oriented to target carbohydrates.

CARBOHYDRATE-BASED VACCINES

The production of antipathogen vaccines capable of generating long-lasting immunity (represented in Table 1) is still a high priority worldwide. A major goal in effective vaccine development is the identification of antigens that can stimulate protective immune responses by eliciting both arms of the immune system. Many common bacteria cause disease because their polysaccharide (PS) capsules prevent phagocytosis. The idea that immunity directed against bacterial capsules must confer protection is based on the observation that antibodies (Abs) against capsular PS induce complement-mediated bactericidal activity, eliminating specific pathogens.

Currently, prevention of some bacterial infections is negated by the lack of effective immunization by the use of the available PS-derived vaccines, which have been shown to display poor immunogenicities. Conjugation of multivalent PS to potent immunological carriers, as represented by pneumococcal conjugate vaccines under development (Table 2) (43), is perceived to recruit T-cell help, evoking a more powerful immune response. However, the conjugation of PS to a carrier protein to elicit carrier-specific T- and B-cell responses does not necessarily enhance PS immunogenicity (59).

The development of a pneumococcal conjugate vaccine, for example, is complicated by the existence of at least 90 capsular serotypes. Consequently, separate vaccine formulations will be required to offer adequate protection against the most common serotypes among pediatric populations worldwide. From a functional perspective, conjugate vaccines have not proved to be more efficacious than the conventional PS vaccine in older adults (>50 years). Likewise, conjugate vaccine preparations have failed to induce a supe-

Table 1. Carbohydrate-based vaccines licensed or under development for bacteria and cancer[a]

Details	CE	LP	Example
Plain polysaccharide		X	*Haemophilus influenzae* type b
		X	Meningococcal (*Neisseria meningitidis*)
		X	Pneumococcal (*Streptococcus pneumoniae*)
Conjugate		X	*H. influenzae* type b
	X		Pneumococcal
	X		Group B streptococcal (*Streptococcus agalactiae*)
	X		GM2, melanoma
	X		GD2, melanoma
	X		LeY, ovarian
	X		sTn, breast

[a] CE, clinical evaluation; LP, licensed product.

Table 2. Pneumococcal conjugate vaccines in phase II and phase III trials

Vaccine	Serotype	Carrier	Manufacturer
PncCRM	4, 6B, 9V, 14, 18C, 19F, 23	CRM197	Wyeth-Lederle Vaccines Pediatrics
PncD	3, 4, 6B, 9V, 14, 18C, 19f, 23	Diphtheria toxoid	Connaught Laboratories
PncT	3, 4, 6B, 9V, 14, 18C, 19F, 23	Tetanus toxoid	Pasteur Mérieux Serums & Vaccines
PncOMPC	4, 6B, 9V, 14, 18C, 19F, 23F	Meningococcal outer membrane protein C	Merck Research Laboratories

rior Ab response in bone marrow transplant recipients and other immunocompromised adults. Finally, the primary Ab response induced by PS-protein conjugate vaccines continues to manifest TI-2 characteristics, such as dominance of the immunoglobulin G2 (IgG2) response over the IgG1 response in humans and a deficiency in IgG2a production in mice.

As with bacterium-derived carbohydrates, tumor-associated carbohydrate (TAC) antigens are notorious for eliciting diminished immune responses, of predominately IgM Abs, that display relatively low affinities for tumor-expressed carbohydrates (82). Different strategies for augmentation of the immunogenicity of TAC have been discussed on the basis of concepts developed for bacterial PS-protein conjugates (Table 1) (52). Representative difficulties for the design of TAC-based vaccines in general is that some TACs overexpressed on human cancers are poorly immunogenic in humans and mice, even when they are attached to immunological carrier proteins. Titers following booster immunizations with ganglioside GD3 conjugates, for example, were considered not to be as high as they are in response to classical TD antigens (32). In humans, attempts to induce an Ab response against GD3 by immunizing patients with the GD3 congeners 9-*O*-acetyl GD3 and GD3 conjugated to keyhole limpet hemocyanin (KLH) have been unsuccessful (84).

Another feature that may diminish responsiveness to TAC antigens is the native configuration of the immunizing antigen. For example, sialyl Tn (sTn)-KLH conjugates have proved to be effective in generating antitumor immune responses in mice and in breast cancer patients (53, 54, 57). Although such conjugates are usually immunogenic and are capable of producing anticarbohydrate Abs, some investigators have suggested that neoglycoproteins that contain sTn, in which the carbohydrate structures are clustered together, would make better immunogens (2, 119). Despite promising results, synthetic routes for the development of clustered carbohydrate epitopes are not straightforward. Mechanisms for the elicitation of more strict and specific immune responses by clustered glycoconjugates in vivo are not known.

MOLECULAR MIMICRY APPROACH TO AUGMENTATION OF ANTICARBOHYDRATE RESPONSES

There are numerous examples in which molecular mimicry of carbohydrate antigens by surrogates has been demonstrated. Anti-idiotypic (anti-Id) Abs have been developed to trigger pathogen-associated, antigen-specific immune responses to the PS capsules of *Escherichia coli* K13 (101) and *Escherichia coli* O111 (47), *Neisseria meningitidis* group C (111, 112), *Streptococcus pneumoniae* (16, 60), *Neisseria gonorrhoeae* lipooligosaccharide (LOS) (27), chlamydial exoglycolipid antigen (4), the lipopolysaccharide (LPS) of *Pseudomonas aeruginosa* (90, 91), and *P. aeruginosa* mucoid exopolysaccharide (92). Anti-Id Abs that mimic TAC antigens have been developed for fucose α(1-3)galactose [Fucα(1-3)Gal] (22), sialyl SSEA-1 (stage-specific embryonic antigen-1) (sialyl-Lewis X [s-LeX]) (107), sialyl-Lea (sLea) (25), carbohydrates associated with TAG-72 (88), Mbrl saccharide (110), GD2 (17, 18, 28, 85–87, 96), GD3 (15, 62), and GM3 (41), to mention a few.

The structural basis for mimicry by such anti-Id Abs is generally not known. In one instance it was found that both the variable heavy-chain and variable light-chain regions contribute to the antigenicity of an anti-Id Ab that mimicked the melanoma-associated ganglioside GM3 (41). We found that complementarity determining region (CDR) 3 (CDR3) of the heavy chain of an anti-Id Ab that mimicked the major C polysaccharide of *Neisseria meningitidis* contained the residue tract YYRYD that contributed to the mimicry of the nominal antigen (112). This peptide proved to elicit a protective immune response against bacterial challenge (112).

Crystallographic analysis of an Ab surrogate for a carbohydrate antigen was found to be unable to carry an internal image of the antigen and to induce PS-specific anti-anti-Ids, because the PS binding cleft on the isolating Ab was too narrow and deep to allow comprehensive contact with the binding site of the anti-Id (23). Insight as to how a peptide sequence may mimic a carbohydrate was forthcoming by analysis of a camel single-domain Ab that inhibits an enzyme by mimicking a carbohydrate substrate (106). It was shown that a single variable-domain fragment derived from a lysozyme-specific camel Ab that naturally lacks light chains mimics the oligosaccharide substrate functionally (inhibition constant for lysozyme, 50 nM) and structurally (lysozyme buried surface areas, hydrogen bond partners, and hydrophobic contacts are similar to those seen in sugar-complexed structures) (106). Most striking was the mimicry by the Ab's CDR3 loop (which is longer than murine or human Abs), especially Ala 104, which mimics the subsite C sugar 2-acetamido group. This group was previously identified as a key feature in the binding of lysozyme. Of particular note was the observation of the potential for the backbone of the CDR3 loop to parallel the three-dimensional position of the linkages of the carbohydrate substrate within the binding site.

PEPTIDE LIBRARY APPROACH TO IDENTIFYING CARBOHYDRATE MIMICS

The development of anti-Id Abs is not necessarily a simpler alternative to the development of PS conjugates, since retrieval of such Abs is relatively time-consuming. Also, elucidation of which ones among many possible anti-Id are in reality able to mimic TACs is not a straightforward task (83). Whereas Abs have demonstrated the ability to mimic various compounds, classic heavy- and light-chain Abs may be limited in their applications (117).

It is now evident that one Ab does not necessarily bind to a single antigen, but it may recognize antigens of different chemical structures (11, 19, 21, 29, 34, 49, 72, 77, 78, 93, 95, 100, 108, 109, 118). These findings draw attention to the molecular basis for cross-reactivity between molecules that are chemically different (16, 46, 98). Selection of particular peptide libraries is dependent on a number of factors. These include the sequence of the insert, the secondary structure, conformational restrictions, spatial relationships, and differential binding affinities.

Analysis of a larger repertoire of ligands reactive with an Ab-combining site might establish structure-function relationships not evident from analyses of binding to monoreactive Abs (11, 93, 95, 100). Since antigen selection operates on functional, not structural conformations, the known difficulties encountered with Ab mimics suggest that heterologous binding by unrelated molecular surfaces may be a common phenomenon in antigen-Ab interactions (51).

The binding site(s) of anticarbohydrate Abs mainly identifies the carbohydrate-mimicking peptides, as evidenced by results obtained by screening libraries or derived from studies of carbohydrate-mimicking anti-Id Abs. This observation translates into the hypothesis that peptide mimotopes select for particular anticarbohydrate Ab subsets (29). Immunization with peptides that bind to these subsets might nevertheless be extremely useful in enhancing the production of Abs against a particular carbohydrate epitope. Thus, an important criterion for selection of Abs for screening purposes is that the isolating Ab must display biological activity.

Selection of peptides may be less biased by affinity considerations, since Abs with moderate affinities are adequate for most studies. Previous work with anti-Id Abs indicates that factors like competition with a target carbohydrate antigen and high-affinity binding are often not sufficient for immunological mimicry and, at times, may not be necessary (64). The same may be true for peptide mimics. The fact that a peptide displays high-affinity binding to an isolating Ab or receptor does not mean that it is an adequate immunological mimic.

STRUCTURAL PROPERTIES OF PEPTIDE MIMOTOPES OF CARBOHYDRATES

Aromatic-aromatic interactions appear to play a role as mimics for a variety of carbohydrate-subunit interactions (34, 72, 77, 94, 97, 98, 108, 112). Some carbohydrate-binding proteins or Abs display a fine specificity for peptides that contain central (W/Y)XY residues (Table 3). This generalized motif feature is associated with peptides that are defined as mimics of glycosphingolipid constituents (103) and with epitopes from *Cryptococcus neoformans* (108). An all-D-amino acid peptide with the sequence Ac-YRYYGL-NH_2 (where Ac is acetyl) was recently identifed as a mimic for β-D-*N*-acetylglucosamine (GlcNAc) (77). This peptide was specifically recognized by the monoclonal antibody (MAb) HGAC 39.G3, with a relative affinity of 300 nM, as measured in a competitive binding assay. The contributions to overall specificity of the residues of the all-D-amino acid peptide in binding to MAb HGAC 39.G3 were examined with a series of truncation, L- for D-amino acid-substituted and retro analogs. Dimeric forms of the all-D-amino acid peptide were recognized with 10- to 100-fold greater affinities relative to that for the monomer. The all-D-amino acid peptide was found to inhibit MAb binding to an anti-Id Ab with approximately 1,000-fold greater affinity than GlcNAc did (77).

To evaluate the potential antigenic and immunological mimicry of the (W/Y)XY motif, we have designed peptides in which (W/Y)XY motifs are presented as tandem repeats (Table 4) and synthesized as multiple-antigen peptides (MAPs). This multivalent

Table 3. Peptide motifs that mimic carbohydrate structures

Motif	Carbohydrate	Structure[a]	Reference(s)
YYPY	Mannose	Methyl-α-D-mannopyranoside	72, 94
WRY	Glucose	α(1-4)Glucose	63, 70
PWLY	Lewis Y	Fucα1→2Galβ1→4(Fucα1→3)GlcNAc	34
YYRYD	Group C polysaccharide	α(2-9)Sialic acid	112

[a] Fuc, L-fucose.

Table 4. Peptides described in our previous studies

Designation	Sequence motif	Structure	Method of identification	MAb BR55-2 reactivity	Ab FH-6 reactivity	LeY inhibition
K61105	YPY	GGIYYPYDIYYPYDIYYPYD	Synthetic design	+	+	−
K61106	WRY	GGIYWRYDIYWRYDIYWRYD	Synthetic design	+++	+++	++
K61107	YRY	GGIYYRYDIYYRYDIYYRYD	Synthetic design	++	++	++
K61108	WLY	GGGAPWLYGGAPWLYAPWLY	Synthetic design	−	−	−
K61223	WLY	GGGAPWLYGAPWLYGAPWLY	Synthetic design	−	−	−
K61109	WRY	GGARVSFWRYSSFAPTY	Phage	−	−	−
K61110	WPY	GGGWPYLRFSPWVSPLG	Phage	−	−	−
K61111	WVF	GGAGRWVFSAPGVRSIL	Phage	−	−	−
K61104	FSLLW	IMILLIFSLLWFGGA	Phage	+	−	++

design is believed to impart peptide conformational properties reflective of the repetition of natively expressed carbohydrate moieties, presenting the peptides as multivalent immunogens and emulating potential epitope clustering that might represent a potential native configuration of naturally expressed carbohydrate antigens (119).

The peptides in Table 4 were tested to delineate reactivity patterns with the anti-Lewis Y (anti-LeY) MAb BR55-2 and anti-sLeX Ab FH-6 (56). BR55-2 and FH-6 have specific reactivities for their respective carbohydrate antigen targets expressed on human carcinoma cells. The neolactoseries structures LeY, LeX, sLeX, Lea, sLea, and Lewis B (Leb) all share a common epitope topography (12, 13, 36). Table 5 shows constituents that are structurally similar between these carbohydrates, which are mimicked by the (W/Y)XY motif (3). Consequently, MAbs raised against these carbohydrate antigens may cross-react with peptides that contain the (W/Y)XY motif.

MAb BR55-2 can bind to some peptides that contain the sequence tract (W/Y)RY, with binding mediated by the Arg residue that mimics the spatial position of Fucα(1-3), by contacting the same atoms within MAb BR55-2 as LeY does. The mimicry for Fucα(1-3) by the guanidinium group of Arg might be a basis for partial mimicry of LeY by (W/Y)RY-containing peptides. We have shown that the Arg-containing peptides K61106 and K61107 (Table 4) also bind to the anti-sLeX Ab FH-6 (56). sLeX shares the Fucα(1-3) moiety with the LeY homologue. However, the K61104 peptide did not display Ab FH-6 binding, implying that the FSLLW-containing peptide interaction with MAb BR55-2 is specific. The single substitution of Pro for Arg in the (W/Y)RY tract reduces MAb BR55-2 binding, further demonstrating that the specificity of binding can be determined by the identity of the peptide side chain constituents of the motif. These data provide strong evidence that peptides and carbohydrates can bind to the same Ab-binding site, while changes in peptide presentation lend to fine specificities of anticarbohydrate Abs. Secondary structure profiles based upon a neutral net prediction routine (Table 6) indicate that the structures of K61107 and K61106 are similar for all three tertiary structure classes, while the sequences of the MAb BR55-2-nonreactive peptides, like K61223 and K61105, are vastly different from the sequences of the MAb BR55-2- and FH-6-reactive peptides.

The interaction of MAb BR55-2 with some of the multivalent peptide mimetics was also studied in solution by competition (inhibition) assays by use of a biosensor approach (Table 7). Peptides K61104 and K61107 were used to inhibit the BR55-LeY interaction in the same way that LeY was used, but in the presence of 2 to 7% dimethyl sulfoxide, to achieve a complete solubilization of the peptides. It is of interest that in both cases, the LeY competition and the peptide competition for LeY, the off rate increases dramatically. It is

Table 5. Similarities among neolactoseries constituents

Chain type	Antigen	Structure[a]
Type 1	H-1	[Fucα(1-2)]Galβ(1-3)GlcNAc
	Lea	Galβ(1-3)GlcNAc[Fucα(1-4)]
	Leb	[Fucα(1-2)]Galβ(1-3)GlcNAc[Fucα(1-4)]
Type 2	H-2	[Fucα(1-2)]Galβ(1-4)GlcNAc
	LeX	Galβ(1-4)GlcNAc[Fucα(1-3)]
	LeY	[Fucα(1-2)]Galβ(1-4)GlcNAc[Fucα(1-3)]

[a] Structures of H, Lea, Leb, LeX and LeY blood group determinants. Fuc, L-fucose; Gal, D-galactose. These oligosaccharides are found at the nonreducing termini of sugar chains in glycolipids, glycoproteins, and mucins.

Table 6. Secondary structure profiles of various MAb BR55-2-reactive and nonreactive peptides[a]

Peptide	Sequence	Predicted structures within tertiary structure classes: None	All alpha	Alpha-beta	MAb BR55-2 reactivity
K61105	GGIYYPYDIYYPYDIYYPYD	- - - EEE - - - - - - - - - - - - - -	- - - - - - - - - - - - - - - - - - - -	- - - - - - - - - - - - - - - - - - - -	+
K61107	GGIYYRYDIYYRYDIYYRYD	- - - EEEEEEEEEEEEEE- - -	- - - - - H- - H- - H- - - - - - -	- - - E-EE - EE - - H - EE- - - -	+++
K61106	GGIYWRYDIYWRYDIYWRYD	- - - EEEEE - EEEEEEE- - - -	- - - - - H- - H- - - - HH- - - - -	- - - - - EH - EE -EH - EE- - - -	+++
K61223	GGGAPWLYGAPWLYGAPWLY	- - - - - E E - - - - - - - - - - - - -	- - - - - - - H- - - - - - - - - - - -	- - - - - - HH- - - HHH - - - - - -	–
K61109	ARVSFWRYSSFAPTY	- - - EEEE- - - - - - - -	- - - - HHH- - - - - - - -	- - - - - - - - - - - - - - -	–
K61104	IMILLIFSLLWFGGA	-EEEEEHHHHH- - - -	- HHHHHHHHH- - - - -	- EEEEEEEEEE- - - -	+++
PA1GXM	GLQYTPSWMLVG	- - - - - - - - E E - -	- - - - HHH- - - - -	- - - - - - - - - EE- -	–

[a] Secondary structure profiles were calculated by nnPREDICT at the server site at the University of California, San Francisco.

Table 7. Kinetic analysis of ligand binding to MAb BR55-2

Surface biotin LeY	Kinetic parameters: k_{a_1} (M^{-1} s^{-1})[a]	k_{a_2} (M^{-1} s^{-1})[a]	k_{d_1} (s^{-1})[a]	k_{d_2} (s^{-1})[a]	appK_{D_1} (M)	appK_{D_2} (M)
BR55-2	$1.09 \times 10^4 \pm 147$	26.4 ± 0.5	$9.4 \times 10^{-2} \pm 5.8 \times 10^{-4}$	$9.34 \times 10^{-6} \pm 0.2$	8.0×10^{-6}[b]	3.5×10^{-7}[b]
BR55-2 + soluble LeY			$1.5 \times 10^{-1} \pm 4 \times 10^{-4}$			3.7×10^{-7}[c]
BR55-2 + 2% dimethyl sulfoxide	$1.01 \times 10^4 \pm 260$	34.2 ± 0.7	$4.24 \times 10^{-2} \pm 1 \times 10^{-3}$	$4.61 \times 10^{-4} \pm 3.2 \times 10^{-6}$	4.2×10^{-6}[b]	1.345×10^{-9}[b]
With solution LeY						$\sim 2.0 \times 10^{-9}$[c]
With solution 104	$1.78 \times 10^4 \pm 1.0 \times 10^4$[d]		0.0319 ± 0.02[d]			2.1×10^{-9}[c]
With solution 107						1.2×10^{-9}[c]

[a] Values are means ± standard errors for the fit. Chi-square 2-5. k_d, rate constant; K_D, equilibrium constant.

[b] Calculated from k_d/k_a.

[c] The 50% effective concentration from the solution inhibition curve was fitted to a four-parameter equation, and the Cheng-Prusoff correction was used to estimate the apparent K_D (app K_D) from the 50% effective concentration.

[d] Kinetics of MAb BR55-2 binding in equilibrium with solution 104.

possible that the addition of a competitor disrupts the apparent multivalent Ab network, in which case the sites act independently and also seem to be affected by solution conditions.

The effect of changing the structural presentation of a motif in order to enhance its Ab recognition is exemplified by K61109 (Table 4). This peptide was isolated by MAb BR55-2, yet its MAP form does not react with BR55-2. Secondary profile analysis (Table 6) indicates that the sequence of K61109 shares some structural similarity to the sequence observed for the K61104 peptide. Not surprisingly, these data indicate that the conformations of the presented motifs can affect the antigenic properties of putative motifs. A complementary fit is required for Ab-antigen binding, and it has been suggested that a β-turn maximizes the total buried surface area of a peptide bound to an Ab, improving the fit within the Ab-binding pocket (26, 116). This general strategy has been argued for some time to have an influence on the functionality of anti-Id Abs (44, 46, 66, 67, 79, 113–115). The same idea has recently been suggested for peptides isolated from peptide display libraries (20). Anti-Id Abs can be viewed as nature's own peptide display library (108).

An important factor responsible for selection of phage during biopanning is the structure and conformation of the peptide sequence, which in turn is defined by the constituent amino acids. In the case of an unconstrained linear insert on the order of 15 residues, which can be the size of a CDR domain, the peptide is capable of assuming a number of molecular shapes. This multiplicity of shapes may lead to a differential presentation by the same peptide sequence and may considerably weaken the binding of the peptide to a given receptor. This would drastically alter the selection of phage. Amino acid analysis of long (20-mer) insert peptides, as identified by low-stringency phage screening, suggests that helical structures can also be correlated with Ab isolation and binding (102). While Ab loops can mimic helical structures (67), they do so in a conformationally restricted fashion. Conversely, the recognition process mediated by a β turn can be mimicked by an α helix (61).

It may be that the library screening process favors peptides that can assume conformations conducive to binding to the anticarbohydrate Ab when expressed on the phage protein but that do not necessarily achieve the same conformation in solution or when they are coupled to a carrier for immunization. The overall distribution of hydrophobic and hydrophilic elements, introduced by charged and aromatic amino acids in a given peptide sequence, would play a key role in affinity selection. Thus, inserts with unrelated sequences but similar surface presentations could be selected during affinity selection because amino acids with similar charges and structures can confer structural similarity. Antigen presentation on a phage does not fully equate with immunological mimicry. Immunization of mice with 19 phage display peptides that antigenically mimic the O-antigen (O-Ag) LPS from the human pathogen *Shigella flexneri* serotype 5a resulted in only two peptides that induced anti-O-Ag Abs (74). The immune responses for these two peptides were specific for the O-Ag of *S. flexneri* serotype 5a. The amino acid sequences of the immunogenic peptide mimics were YKPLGALTH (flanked by two Cys residues) and KVPPWARTA. However, it was not determined whether antisera to these peptides are bactericidal.

Other peptides derived from peptide display libraries and tested as immunogens include a mimic for LeY (55), a mimic for group B streptococcal type 3 capsular PS (76), a mimic for PS of *C. neoformans* (108), and a mimic for DNA (80). For the bacterial PS mimics, very few bactericidal responses, if any, were noted (74, 76, 108). Antitumor responses were observed for the LeY mimic (55), much like we observed previously (56).

Interestingly, systemic lupus erythematosus-like anti-double-stranded DNA reactivity could be generated in nonautoimmune mice by immunization with the peptide antigen mimic of DNA (80). Peptide-induced autoimmunity may prove to be useful for understanding the spreading of antigenic specificities targeted in systemic lupus erythematosus. However, most importantly, the demonstration that a peptide antigen can initiate a systemic lupus erythematosus-like immune response opens a new chapter on the potential antigenic stimuli that might trigger autoimmune diseases.

FINE Ab SPECIFICITIES FOR GSL-ASSOCIATED PEPTIDE MIMICS

Carbohydrates that influence the pathology and biology of pathogens are ubiquitous in nature. Immunochemical studies of the LOSs of the gram-negative bacteria *N. gonorrhoeae* and *N. meningitidis* indicate that they are antigenically and/or chemically identical to lacto-neoseries glycosphingolipids (GSLs) (5, 58, 68). The bacteria with LOSs which mimic the structures of GSLs that are also tumor associated are shown in Table 8. The terminal trisaccharide lactotriaose (Galβ1-4GlcNAcβ1-3Gal), a precursor of lacto-*N*-neotetraose, is common among LOSs from pathogenic *Neisseria* spp., since both serological and structural studies have shown that most meningococcal strains of serogroups B and C and most gonococcal strains possess this trisaccharide. In contrast, the O chains of a number of *Helicobacter pylori* strains exhibit mimicry for LeX and LeY blood group antigens (10). An additional type of mimicry from *N. gonorrhoeae* involves a pentasaccharide with an *N*-acetylgalactosamine (GalNAc) residue β1-3 linked to a terminal galactose (Gal) of lacto-*N*-neotetraose. *Haemophilus influenzae* also expresses similar sialylated LOSs, suggesting

Table 8. Glycosphingolipid constituents shared among bacteria and tumor cells

GSL series type	Structure	Bacterial species[a]
Lacto	Galβ1→4GlcNAcβ1→3Galβ1→4Glcβ1→Cer	*Neisseria gonorrhoeae* *Neisseria meningitidis* *Neisseria lactamica* *Neisseria cinera* *Haemophilus influenzae* type b (S) *Haemophilus influenzae* NT (S) *Haemophilus influenzae* biotype *aegyptius* *Haemophilus ducreyi* (S)
Globo	Galα1→4Galβ1→4Glcβ1→Cer	*Neisseria gonorrhoeae* *Neisseria meningitidis* *Haemophilus influenzae* type b *Haemophilus influenzae* NT *Branhamella catarrhalis*
Ganglio	GalNAcβ1→4Galβ1→4Glcβ1→Cer Galβ1→3GalNAcβ1→4Galβ1→4Glcβ1 GalNAcβ1→3Galβ1→4GlcNAcβ1→ 3Galβ1→4Glcβ1→Cer	*Neisseria gonorrhoeae*

[a] NT, nontypeable; S, smooth strain.

that this is a common pathogenic mechanism among bacteria. Antilactoneoseries MAbs bind to *Haemophilus ducreyi* and to type b and non-typeable *H. influenzae.*

Furthermore, *N. gonorrhoeae* 1291 has a LOS that contains a paragloboside-terminal oligosaccharide. Globotriaose (Galα1-4Galβ1-4Glc) has been identified as the terminal trisaccharide LOS of the pyocin-resistant mutant. This fragment is a homologue of Globo H (Fucα1→2Galβ1→3GalNAcβ1→3Galα1→4Galβ1→4Glc), a carbohydrate structure that shows enhanced expression on many human carcinomas (48).

Mapping of peptide epitopes with anticarbohydrate Abs can lend to the definition of fine specificities that can go undetected by screening against carbohydrate antigens alone. We have observed, for example, that the antiganglioside MAb ME36.1, which recognizes the GD2 and GD3 gangliosides on melanoma cells (104), partially cross-reacts with peptides that we have defined as mimics of the neolactoseries antigens (56). The core structure for GD2 is GalNAcβ1→4Gal[3←2αNeuNAc8←2aNeuNAc]β1→4Glcβ1→1Cer and GD3 has the structure NeuNAcα2→8NeuNAcα2→3Galβ1→4Glcβ1→1Cer. The Fab′ fragment of MAb ME36.1 has recently been solved (75), and a model of the GD2-MAb ME36.1 complex was built by a docking procedure analogous to that described in our own studies (75). This model indicates that MAb ME36.1 can interact with the GalNAc residue of GD2, lending to its preferred specificity for GD2 over GD3.

Using MAb ME36.1 in a phage display screening of a 15-mer library, we observed two peptides of particular interest (81). The first one, GVVWRYTAPVHLGDG, contains the WRY sequence tract, which may emulate portions of the lactoseries antigens, as suggested by our previous results with peptides that contain WRY. This peptide also contains an AP sequence tract that we have shown mimics GlcNAc in the anti-Le Y-binding site of B3. The second peptide, RNVPPIFNDVYWIAF, has also been isolated from a phage display screening as a mimic of lactotetraosylceramide (nLc4Cer) (103). A peptide with this exact sequence was shown to bind to an Ab that binds to nLc4Cer, binding also to the lectin of *Ricinus communis,* which recognizes preferentially the terminal β-Gal in the structure Galβ1-4GlcNAc of glycoconjugates (103). We have observed the VPP sequence tract in a peptide that was isolated with the anti-LeY Ab 15-6A but that is present in the C terminus of the peptide (81). Interestingly, a peptide with the sequence KVPPWARTA was isolated as a mimic of the O-Ag of *S. flexneri* (74). The role of the VPP sequence tract may be structural. The proline side chain may serve as an important structural feature to orient the aromatic side chains in a manner similar to the branched nature of the carbohydrates that it mimics.

Other sequence tracts that appear to mimic Galβ1-4GlcNAc in binding to the *R. communis* lectin include VPPYFTLMY, VPPCFTLMY, and VPPTFTLMY (103), which can be compared with the sequence tract VPPIFNDVY. However, it appears that VPPIFNDVY by itself does not bind to the lectin (103), while the RNVPPIFNDVYWIAF peptide does (103). The latter observation suggests that flanking residues play a role in the proper presentation of certain putative reactive sequences, a result that is implied in studies with YPY mimics, in which the flanking AGS sequence enhances binding of peptides to concanavalin A (42). These peptides are related by the FXXX(Y/F) motif, as observed in the peptide SFGSGFGGGY, which is also known to mimic GlcNAc (98). This peptide also binds to MAb BR55-2, indicating that this motif potentially mimics both GlcNAc and GalNAc-containing moieties. In addition, an aromatic variant, FXXXW, is observed in our LeY-mimicking peptide isolated by MAb BR55-2 (peptide K61104 in Table 4).

These results indicate that anticarbohydrate Abs can identify peptides that mimic particular carbohydrate configurations on what are otherwise dissimilar carbohydrates. The identification of multiple motifs reactive with an Ab may indicate that (i) peptides based on two different motifs isolated with the same MAb can mimic different structural topographies of the cognate carbohydrate, (ii) subsets of peptides may very likely represent nonoverlapping surfaces of the cognate antigen, and (iii) peptide binding to the MAb occurs at separate sites. These observations further emphasize that peptides can provide information about the fine specificities of anticarbohydrate Abs and about structural relationships that exist between GSL moieties that are currently underappreciated, as a consequence drawing attention to the power of peptides to map the fine specificities of Abs.

PEPTIDE MIMOTOPES CAN EMULATE CARBOHYDRATE BINDING

The failure to induce bactericidal activity by peptide mimics may be due to the lack of an adequate titer of Abs or by the fact that peptides are only partial mimics. The three-dimensional structures of Abs in complex with peptides should facilitate the structure-based design of peptides for their use as vaccines (26, 71, 116). Most studies designed to examine antigen-Ab binding have involved protein antigens or their idiotypes and anti-Id Abs. A limited number of studies on the binding of PS antigens to their Abs have been performed. Peptide mimics of PSs can be used to study protein-carbohydrate interactions, but the exact nature of most PS antigen-Ab interactions remains to be elucidated. The greater our understanding of PS or peptide binding at the molecular level, the more predictable peptide mimic design will become.

Specificity in protein-carbohydrate recognition is achieved by a combination of multivalency and geometry. General structural features associated with molecular recognition and interaction are the stacking of aromatic side chains against sugar rings, the presence of hydrogen bond networks in which sugar-OH groups act as both acceptors and donors, and the coordination of multiple hydrogen bonds with water molecules. Aromatic residues are postulated to play a role in binding to Abs by orienting peptides, by coming into direct contact with Ab functional groups, or by lending to the binding in a nonspecific fashion. Aromatic residues can mimic carbohydrate substituents in size and shape (3).

On the basis of three-dimensional analysis, it can be argued that the construction of an effective mimotope requires an improved fit between the bound peptide and the Ab, oriented to maximize particular interactions within the Ab heavy and light chains. To improve upon the ability of peptides to induce a more specific immune response, we have defined how LeY binds to anti-LeY Abs as a representative example. The crystal structure of anti-LeY Ab BR96, determined with BR96 in complex with a nonate-methyl ester LeY tetrasaccharide (39), provides a molecular basis for specific LeY antigenic recognition of some anti-LeY Abs (105) and an explanation for the ways that mimicry comes about. Our modeling study showed that the anti-LeY Abs B3 and BR55-2 share recognition features for the difucosylated type 2 lactoseries LeY structure similar to those displayed in a crystal structure of the anti-LeY Ab BR96. We observed that a major source of specificity for the LeY structure emanates from the interaction with the GlcNAc residue and from the nature of the structures extended at the reducing site of the fucosylated lactosamine (105).

The B3 Ab binds to the peptide sequence APWLYGPA, in which the putative pentapeptide sequence APWLY is critical for binding to the Ab (34). To establish how this putative sequence mimics LeY binding to Ab B3, we fit the pentapeptide sequence into the B3-combining site using the program LIGAND-DESIGN (LUDI Biosym Technologies) (71). This program searches a molecular library for fragments representative of the amino acids in the target peptide sequence. The program then positions the fragments within the combining site devoid of steric conflicts. The APWLY sequence was modeled in such a manner that the Trp, Tyr, Leu, and Ala residues occupied the same relative positions as the fragments identified by use of the program from LUDI. The judicious positioning of the peptide relied upon intermolecular interaction calculations, in which several potential binding modes of the peptide were ranked according to the stability of the complex. This approach parallels those used in studies associated with elucidation of low energy structures of carbohydrates for Abs and lectins (37, 38).

In the most stable conformation, we observed that the alanine, proline (A/P) residues occupied a position similar to that occupied by the LeY GlcNAc residue. This positioning indicates that the proline residue mimics the spatial position of the glucose unit of GlcNAc, while the Ala methyl group is positioned similarly to the terminal methyl group of GlcNAc's *N*-acetyl group. The Trp residue occupied a volume associated with the Fucα1-3 moiety, and the Leu residue occupied the volume of βGal. The Tyr residue occupied a position not associated with LeY binding to B3. The overall stable conformation of the APWLY structure within the Ab B3-combining site was of a β-turn character. This conformation itself lends the Tyr residue of the peptide for a potential interaction with several residues in the CDR2 of the heavy chain of B3 that include Asp H53, Ser H52, Ser H55, and Ser H56. These residues are different in MAb BR55-2, which does not bind to the monovalent APWLYGPA peptide, as assessed by enzyme-linked immunosorbent assay (Table 4). This analysis therefore indicated that while the isolating Ab contacted a peptide in a fashion similar to that in which it contacted LeY, sequence differences between Abs dictate the extent to which the peptide cross-reacts with other members of a panel, a conclusion drawn by Scott and coworkers (29). While not reactive with all anti-LeY reactive MAbs, this peptide does induce an anti-LeY immune response (55), drawing attention to the concept that peptide mimics can induce carbohydrate-reactive Ab subsets.

INDUCTION OF T-CELL RESPONSES BY PEPTIDE MIMICS

There is a considerable potential for T-cell cross-reactivity, and it is useful for studies on the role of molecular mimicry in the etiology of T-cell-mediated disease (33). A commonly proposed model for degenerate T-cell recognition (the molecular mimicry model) that is used to explain T-cell receptor (TCR) cross-reactivity suggests that the molecular surfaces of the recognized complexes are similar in shape or charge, or both, despite differences in their primary sequences. Rather than simple molecular mimicry, unpredictable arrays of common and differential contacts on the two class I complexes can be used for their recognition by the same TCR (120).

Is it possible that T cells can see peptides and carbohydrate antigens? In recent studies by McKenzie and coworkers (6, 7), a peptide mimic of the Galα(1,3)Gal sugar (DAHWESWL) was found in a phage display library that was also mimicked by MUC-1 peptides represented by the sequence APDTRPAP (6, 7). It was found that cytotoxic T cells (CTLs) that recognize the MUC-1 peptide can react with the Gal mimic. It was of interest that while mice make CTLs and little Ab to MUC-1, humans make the reverse immune response. It was found that the cross-reaction of the natural Galα(1,3)Gal Abs in humans to MUC-1 was likely responsible for the diversion. The molecular mechanisms for this cross-reactivity are interesting. It was suggested that Abs could recognize the ES residues in the Gal peptide mimic and the DT residues in the MUC-1 peptide. Conformational studies indicate that these respective homologous residues participate in the formation of turn structures in the respective peptide sequences (6). Consequently, anti-MUC-1 Abs can cross-react with the Gal mimic and vice versa. Since MUC-1 CTLs can recognize both peptides, this further indicates that molecular mimicry can operate at the T-cell level. However, the mechanistic role is one of cross-reactivity directed to peptides in that both peptides reflect sequences which contain MHC anchor residues and not that the T cell sees carbohydrate antigen. Cross-reactivity is therefore an important problem in tumor immunotherapy, although the problem can be overcome by in vitro immunizations (8, 9).

In contrast, glycopeptides have been shown to elicit carbohydrate-specific T-cell responses associated with MHC class I presentation (1, 30, 31). αGalNAc-reactive T cells have also been described in relation to MHC class II presentation (40). From a vaccine perspective, the construction of glycopeptide-protein immunogens is problematic. A survey of the literature suggests that the major problems that confront the syntheses of representative high-mannose glycoproteins can be divided into four categories: (i) problems associated with the assembly of the oligosaccharide moiety, (ii) problems associated with the controlled creation of the troublesome β linkage between mannose and the GlcNAc residues, (iii) problems associated with the installation of the ubiquitous asparagine residue at the anomeric center, and (iv) problems associated with the elaboration of the peptide chain.

T cells can be stimulated by nonprotein components in the context of CD1 restriction (102a). The possibility that carbohydrates can be presented to certain subsets of T cells opens a new field for investigation. It is important, then, to know the requirements for this kind of antigenic presentation. Manipulation of the conditions involved in such a process would allow the development of new strategies that can be used to overcome those disadvantages of carbohydrates as antigenic molecules. One possible strategy is the design of peptides that could mimic the antigenic response to carbohydrates being processed through CD1. Some observations support the hypothesis that peptides could be presented in a CD1-restricted fashion. It has been reported that relatively long peptides with specific hydrophobic motifs are able to bind to CD1 (14). More recently, other authors have shown CD1-restricted T cells which proliferate in the presence of ovalbumin (50).

Further evidence for CD1-processing peptides was recently reported by Schofield et al. (89), who demonstrated the appearance of an IgG immune response to glycosylphosphatidylinositol (GPI)-anchored protein antigens derived from *Plasmodium* and *Trypanosoma.* In that study they demonstrated that these IgG responses were partially driven by CD1d-restricted recognition. The identified helper population was characterized as thymus-dependent, interleukin 4-producing, $CD4^+$ natural killer (NK) T cells (89). The pathway of CD1-restricted NK T cells regulated IgG responses to the GPI-anchored surface antigens, and this might be a general mechanism for rapid MHC-unrestricted Ab responses to diverse pathogens.

The observation of such T cells and the observation that glycopeptides can be processed by MHC class I and class II molecules indicate that T-cell responses to carbohydrate antigens may be induced. T-cell-dependent IgG responses have been assumed to be exclusively restricted by class II MHC molecules. However, the expression of CD1 molecules on different subsets of human and murine B cells suggests that CD1-restricted T cells may participate in B-cell help for the production of Abs against nonconventional antigens (24). It is also possible that peptide mimotopes presented through CD1 on B cells can lend to the production of autoimmune-like responses, as evidenced by the work of Putterman and Diamond (80).

SUMMARY AND CONCLUSIONS

Nonpeptide immunogens, such as lipids and carbohydrates, not readily addressable by genetic vectors are a challenge in the design of effective vaccination approaches that target many pathogens and perhaps for antitumor immunity. Ab molecules convey a potentially unlimited molecular diversity and an antigen-binding repertoire. Similarly, phage display technology represents a powerful approach in that many millions of different peptides can be constructed into a simple system and easily selected against a given antigen. Molecular modeling and crystallographic analysis can be used to assess the molecular details that underlie the antigenic mimicry of peptides and the basis for the fine specificities of Abs for carbohydrate and peptide forms. Since peptides are simpler molecules, it is anticipated that cross-reactive immune responses would be easier to characterize and to manipulate.

Peptides have the potential of focusing the immune response on particular epitopes that may be otherwise cryptic or display tolerance. The reduction of protein structures to smaller antigenic parts, as well as the production of sets of molecular variants capable of protecting against antigenic variation, make the production of mimotopes a promising avenue for vaccine development. Peptide mimics of carbohydrates may have several advantages over the carbohydrate itself for induction of carbohydrate-specific immunity: relative ease of production and induction of long-lasting, high-affinity IgG responses. Peptides that mimic carbohydrate structures have further significant advantages as vaccines compared with carbohydrate-protein conjugates or anti-Id Abs. First, the chemical composition and purity of synthesized peptides can be precisely defined. Second, the immunogenicities of the peptides can be significantly enhanced by polymerization or by addition of relatively small carrier molecules that reduce the total amount of antigen required for immunization. Third, synthesis of peptides may be more practical than synthesis of carbohydrate-protein conjugates or production of anti-Ids. Fourth, peptide-mimicking sequences can be engineered into DNA plasmids for DNA vaccination to further manipulate T-cell responses.

The demonstration of carbohydrate-specific immunity induction by peptide-protein mimics of carbohydrates has been limited to the induction of Abs. Comparison of the functions, kinetics, and fine specificities of the immune responses elicited by carbohydrate mimics compared with those elicited by the carbohydrate itself has thus far not been performed systematically and presents an important step toward vaccination against carbohydrates, potential targets for the control of cancer, and also bacteria and viruses that cause infections.

The structural basis for peptide mimotopes of carbohydrate that induce T-cell responses, to our knowledge, has not yet been studied due to the unavailability of such mimics. An important question still remains as to whether native carbohydrate recognition by T cells differs from protein recognition by classical MHC-restricted T cells. In the induction of carbohydrate-specific, MHC-restricted T cells, the carbohydrate evidently is bound to a peptide with affinity for MHC on antigen-presenting cells. In contrast to the classical MHC class I-dependent pathway of protein presentation to classical T cells, CD1-dependent glycolipid recognition by $CD1^+$ NK T cells does not require endosomal antigen trafficking or processing. Thus, CD1-restricted NK T cells are a novel type of host defense that may be triggered by peptide mimics.

On the basis of three-dimensional analysis, the construction of an effective mimotope requires improved fitting between the bound peptide and the Ab to maximize particular interactions within the Ab heavy and light chains (116). This may be accomplished in a variety of ways, for example, (i) by comparing the topological similarities of peptide analogs relative to Ab-binding affinities, (ii) by altering the peptide to optimize the steric complementarity of its β-turn structural motif with the binding pocket of the Ab, (iii) by identifying peptides with high affinities for the Ab, and (iv) by screening libraries containing peptides constrained with cysteines at both ends. It may be that library screening processes favor peptides that can assume conformations conducive to Ab binding when they are expressed on phage but which do not achieve the same conformation when they are coupled to a carrier and are used for immunization. Consequently, peptides might be cyclized to enhance their immune responses (69).

A more precise understanding of the binding of peptides and saccharides at the molecular level is required in order to determine whether the occurrence of motifs like (W/Y)XY in mimotopes leads to molecular mimicry or simply represents an advantage provided by aromatic rings for interactions between proteins. In addition to the role that peptide mimotopes can play in exploiting the fine specificities of Abs, they may mimic PSs as immunogens and potentially elicit anti-PS responses. Not all peptides identified as antigen mimics induce PS cross-reactive immune responses (74). The problem now is to determine ways that such peptides can be modified to induce a high-level anti-PS response.

REFERENCES

1. **Abdel-Motal, U. M., L. Berg, A. Rosen, M. Bengtsson, C. J. Thorpe, J. Kihlberg, J. Dahmen, G. Magnusson, K. A. Karlsson, and M. Jondal.** 1996. Immunization with glycosylated Kb-binding peptides generates carbohydrate-specific, unrestricted cytotoxic T cells. *Eur. J. Immunol.* **26:**544–551.
2. **Adluri, S., F. Helling, S. Ogata, S. Zhang, S. H. Itzkowitz, K. O. Lloyd, and P. O. Livingston.** 1995. Immunogenecity of synthetic TF-KLH (keyhole limpet hemocyanin) and sTn-KLH conjugates in colorectal carcinoma patients. *Cancer Immunol. Immunother.* **41:**185–192.
3. **Agadjanyan, M., P. Luo, M. A. Westerink, L. A. Carey, W. Hutchins, Z. Steplewski, D. B. Weiner, and T. Kieber-Emmons** 1997. Peptide mimicry of carbohydrate epitopes on human immunodeficiency virus. *Nat. Biotechnol.* **15:**547–551.
4. **An, L. L., A. P. Hudson, R. A. Prendergast, T. P. O'Brien, E. S. Stuart, H. J. Whittum, and A. B. MacDonald.** 1997. Biochemical and functional antigenic mimicry by a polyclonal anti-idiotypic antibody for chlamydial exoglycolipid antigen. *Pathobiology* **65:**229–240.
5. **Apicella, M. A., J. M. Griffiss, and H. Schneider.** 1994. Isolation and characterization of lipopolysaccharides, lipooligosaccharides, and lipid A. *Methods Enzymol.* **235:**242–252.
6. **Apostolopoulos, V., S. A. Lofthouse, V. Popvski, G. Chelvanayagam, M. S. Sandrin, and I. F. McKenzie.** 1998. Peptide mimics of a tumor antigen induce functional cytotoxic T cells. *Nat. Biotechnol.* **16:**276–280.
7. **Apostolopoulos, V., C. Osinski, and I. F. McKenzie.** 1998. MUC1 cross-reactive Gal alpha(1,3)Gal antibodies in humans switch immune responses from cellular to humoral. *Nat. Med.* **4:**315–320.
8. **Apostolopoulos, V., M. S. Sandrin, and I. F. McKenzie.** 1999. Carbohydrate/peptide mimics: effect on MUC1 cancer immunotherapy. *J. Mol. Med.* **77:**427–436.
9. **Apostolopoulos, V., M. S. Sandrin, and I. F. McKenzie.** 1999. Mimics and cross reactions of relevance to tumour immunotherapy. *Vaccine* **18:**268–275.

10. **Appelmelk, B. J., S. I. Simoons, R. Negrini, A. P. Moran, G. O. Aspinall, J. G. Forte, V. T. De, H. Quan, T. Verboom, J. J. Maaskant, P. Ghiara, E. J. Kuipers, E. Bloemena, T. M. Tadema, R. R. Townsend, K. Tyagarajan, J. J. Crothers, M. A. Monteiro, A. Savio, and G. J. De.** 1996. Potential role of molecular mimicry between *Helicobacter pylori* lipopolysaccharide and host Lewis blood group antigens in autoimmunity. *Infect. Immun.* **64:**2031–2040.
11. **Burritt, J. B., C. W. Bond, K. W. Doss, and A. J. Jesaitis.** 1996. Filamentous phage display of oligopeptide libraries. *Anal. Biochem.* **238:**1–13.
12. **Cagas, P., and C. A. Bush.** 1992. Conformations of type 1 and type 2 oligosaccharides from ovarian cyst glycoprotein by nuclear Overhauser effect spectroscopy and T1 simulations. *Biopolymers* **32:**277–292.
13. **Cagas, P., and C. A. Bush.** 1990. Determination of the conformation of Lewis blood group oligosaccharides by simulation of two-dimensional nuclear Overhauser data. *Biopolymers* **30:**1123–1138.
14. **Castano, A. R., S. Tangri, J. E. W. Miller, H. R. Holcombe, M. R. Jackson, W. D. Huse, M. Kronenberg, and P. A. Peterson.** 1995. Peptide binding and presentation by mouse CD1. *Science* **269:**223–226.
15. **Chapman, P. B., and A. N. Houghton.** 1991. Induction of IgG antibodies against GD3 ganglioside in rabbits by an anti-idiotypic monoclonal antibody. *J. Clin. Investig.* **88:**186–192.
16. **Cheng, H. L., A. K. Sood, R. E. Ward, T. Kieber-Emmons, and H. Kohler.** 1988. Structural basis of stimulatory anti-idiotypic antibodies. *Mol. Immunol.* **25:**33–40.
17. **Cheung, N. K., A. Canete, I. Y. Cheung, J. N. Ye, and C. Liu.** 1993. Disialoganglioside GD2 anti-idiotypic monoclonal antibodies. *Int. J. Cancer* **54:**499–505.
18. **Cheung, N. K., I. Y. Cheung, A. Canete, S. J. Yeh, B. Kushner, M. A. Bonilla, G. Heller, and S. M. Larson.** 1994. Antibody response to murine anti-GD2 monoclonal antibodies: correlation with patient survival. *Cancer Res.* **54:**2228–2233.
19. **Cortese, R., P. Monaci, A. Nicosia, A. Luzzago, F. Felici, G. Galfre, A. Pessi, A. Tramontano, and M. Sollazzo.** 1995. Identification of biologically active peptides using random libraries displayed on phage. *Curr. Opin. Biotechnol.* **6:**73–80.
20. **Craig, L., P. C. Sanschagrin, A. Rozek, S. Lackie, L. A. Kuhn, and J. K. Scott.** 1998. The role of structure in antibody cross-reactivity between peptides and folded proteins. *J. Mol. Biol.* **281:**183–201.
21. **Cwirla, S. E., E. A. Peters, R. W. Barrett, and W. J. Dower.** 1990. Peptides on phage: a vast library of peptides for identifying ligands. *Proc. Natl. Acad. Sci. USA* **87:**6378–6382.
22. **Diakun, K. R., and K. L. Matta.** 1989. Synthetic antigens as immunogens. Part III. Specificity analysis of an anti-anti-idiotypic antibody to a carbohydrate tumor-associated antigen. *J. Immunol.* **142:**2037–2040.
23. **Evans, S. V., D. R. Rose, R. To, N. M. Young, and D. R. Bundle.** 1994. Exploring the mimicry of polysaccharide antigens by anti-idiotypic antibodies. The crystallization, molecular replacement, and refinement to 2.8 Å resolution of an idiotope-anti-idiotope Fab complex and of the unliganded anti-idiotope Fab. *J. Mol. Biol.* **241:**691–705.
24. **Fairhurst, R. M., C.-X. Wang, P. A. Sieling, R. L. Modlin, and J. Braun.** 1998. CD1-restricted T cells and resistance to polysaccharide-encapsulated bacteria. *Immunol. Today* **19:**257–259.
25. **Furuya, A., H. Yoshida, and N. Hanai.** 1992. Development of anti-idiotype monoclonal antibodies for sialyl Le(a) antigen. *Anticancer Res.* **12:**27–31.
26. **Ghiara, J. B., D. C. Ferguson, A. C. Satterthwait, H. J. Dyson, and I. A. Wilson.** 1997. Structure-based design of a constrained peptide mimic of the HIV-1 V3 loop neutralization site. *J. Mol. Biol.* **266:**31–39.
27. **Gulati, S., D. P. McQuillen, J. Sharon, and P. A. Rice.** 1996. Experimental immunization with a monoclonal anti-idiotope antibody that mimics the Neisseria gonorrhoeae lipooligosaccharide epitope 2C7. *J. Infect. Dis.* **174:**1238–1248.
28. **Handgretinger, R., P. Baader, R. Dopfer, T. Klingebiel, P. Reuland, J. Treuner, R. A. Reisfeld, and D. Niethammer.** 1992. A phase I study of neuroblastoma with the anti-ganglioside GD2 antibody 14.G2a. *Cancer Immunol. Immunother.* **35:**199–204.
29. **Harris, S. L., L. Craig, J. S. Mehroke, M. Rashed, M. B. Zwick, K. Kenar, E. J. Toone, N. Greenspan, F. I. Auzanneau, A. J. Marino, B. M. Pinto, and J. K. Scott.** 1997. Exploring the basis of peptide-carbohydrate crossreactivity: evidence for discrimination by peptides between closely related anti-carbohydrate antibodies. *Proc. Natl. Acad. Sci. USA* **94:**2454–2459.
30. **Haurum, J. S., G. Arsequell, A. C. Lellouch, S. Y. Wong, R. A. Dwek, A. J. McMichael, and T. Elliott.** 1994. Recognition of carbohydrate by major histocompatibility complex class I-restricted, glycopeptide-specific cytotoxic T lymphocytes. *J. Exp. Med.* **180:**739–744.

31. **Haurum, J. S., L. Tan, G. Arsequell, P. Frodsham, A. C. Lellouch, P. A. Moss, R. A. Dwek, A. J. McMichael, and T. Elliott.** 1995. Peptide anchor residue glycosylation: effect on class I major histocompatibility complex binding and cytotoxic T lymphocyte recognition. *Eur. J. Immunol.* **25:**3270–3276.
32. **Helling, F., A. Shang, M. Calves, S. Zhang, S. Ren, R. K. Yu, H. F. Oettgen, and P. O. Livingston.** 1994. GD3 vaccines for melanoma: superior immunogenicity of keyhole limpet hemocyanin conjugate vaccines. *Cancer Res.* **54:**197–203.
33. **Hiemstra, H. S., P. A. van Veelen, S. J. Willemen, W. E. Benckhuijsen, A. Geluk, R. R. de Vries, B. O. Roep, and J. W. Drijfhout.** 1999. Quantitative determination of TCR cross-reactivity using peptide libraries and protein databases. *Eur. J. Immunol.* **29:**2385–2391.
34. **Hoess, R., U. Brinkmann, T. Handel, and I. Pastan.** 1993. Identification of a peptide which binds to the carbohydrate-specific monoclonal antibody B3. *Gene* **128:**43–49.
35. **Hutchins, W., A. Adkins, T. Kieber-Emmons, and M. A. J. Westerink.** 1996. Molecular characterization of a monoclonal antibody produced in response to a group-C meningococcal polysaccharide peptide mimic. *Mol. Immunol.* **33:**503–510.
36. **Imberty, A., E. Mikros, J. Koca, R. Mollincone, R. Oriol, and S. Perez.** 1995. Computer simulation of histo-blood group oligosaccharides: energy maps of all constituting disaccharides and potential energy surfaces of 14 ABH and Lewis carbohydrate antigens. *Glycocon. J.* **12:**331–349.
37. **Imberty, A., R. Mollicone, E. Mikros, P. A. Carrupt, S. Perez, and R. Oriol.** 1996. How do antibodies and lectins recognize histo-blood group antigens? A 3D-QSAR study by comparative molecular field analysis (CoMFA). *Bioorg. Med. Chem.* **4:**1979–1988.
38. **Imberty, A., and S. Perez.** 1994. Molecular modelling of protein-carbohydrate interactions. Understanding the specificities of two legume lectins towards oligosaccharides. *Glycobiology* **4:**351–366.
39. **Jeffrey, P. D., J. Bajorath, C. Y. Chang, D. Yelton, I. Hellstrom, K. E. Hellstrom, and S. Sheriff.** 1995. The X-ray structure of an anti-tumour antibody in complex with antigen. *Nat. Struct. Biol.* **2:**466–471.
40. **Jensen, T., P. Hansen, S. L. Galli, S. Mouritsen, K. Frische, E. Meinjohanns, M. Meldal, and O. Werdelin.** 1997. Carbohydrate and peptide specificity of MHC class II-restricted T cell hybridomas raised against an O-glycosylated self peptide. *J. Immunol.* **158:**3769–3778.
41. **Kanda, S., H. Takeyama, Y. Kikumoto, S. L. Morrison, D. L. Morton, and R. F. Irie.** 1994. Both VH and VL regions contribute to the antigenicity of anti-idiotypic antibody that mimics melanoma associated ganglioside GM3. *Cell Biophys.* **2565:**65–74.
42. **Kaur, K. J., S. Khurana, and D. M. Salunke.** 1997. Topological analysis of the functional mimicry between a peptide and a carbohydrate moiety. *J. Biol. Chem.* **272:**5539–5543.
43. **Kayhty, H., and J. Eskola.** 1996. New vaccines for the prevention of pneumococcal infections. *Emerg. Infect. Dis.* **2:**289–298.
44. **Kieber-Emmons, T., B. A. Jameson, and W. J. Morrow.** 1989. The gp120-CD4 interface: structural, immunological and pathological considerations. *Biochim. Biophys. Acta* **989:**281–300.
45. **Kieber-Emmons, T., P. Luo, J. Qiu, T. Y. Chang, I. O, M. Blaszczyk-Thurin, and Z. Steplewski.** 1999. Vaccination with carbohydrate peptide mimotopes promotes anti-tumor responses. *Nat. Biotechnol.* **17:**660–669.
46. **Kieber-Emmons, T., M. M. Ward, R. E. Ward, and H. Kohler.** 1987. Structural considerations in idiotype vaccine design. *Monogr. Allergy* **22:**126–133.
47. **Klaerner, H. G., P. S. Dahlberg, R. D. Acton, R. J. Battafarano, M. E. Uknis, J. W. Johnston, and D. L. Dunn.** 1997. Immunization with antibodies that mimic LPS protects against gram negative bacterial sepsis. *J. Surg. Res.* **69:**249–254.
48. **Kudryashov, V., G. Ragupathi, I. J. Kim, M. E. Breimer, S. J. Danishefsky, P. O. Livingston, and K. O. Lloyd.** 1998. Characterization of a mouse monoclonal IgG3 antibody to the tumor-associated globo H structure produced by immunization with a synthetic glycoconjugate. *Glycoconj. J.* **15:**243–249.
49. **Lane, D. P., and C. W. Stephen.** 1993. Epitope mapping using bacteriophage peptide libraries. *Curr. Opin. Immunol.* **5:**268–271.
50. **Lee, D., A. Abeyratne, D. A. Carson, and M. Corr.** 1998. Induction of an antigen-specific, CD1-restricted cytotoxic T lymphocyte response in vivo. *J. Exp. Med.* **187:**433–438.
51. **Lescar, J., M. Pellegrini, H. Souchon, D. Tello, R. J. Poljak, N. Peterson, M. Greene, and P. M. Alzari.** 1995. Crystal structure of a cross-reaction complex between Fab F9.13.7 and guinea fowl lysozyme. *J. Biol. Chem.* **270:**18067–18076.
52. **Livingston, P. O.** 1995. Approaches to augmenting the immunogenicity of melanoma gangliosides: from whole melanoma cells to ganglioside-KLH conjugate vaccines. *Immunol. Rev.* **145**(147)**:**147–166.

53. **Longenecker, B. M., M. Reddish, R. Koganty, and G. D. MacLean.** 1993. Immune responses of mice and human breast cancer patients following immunization with synthetic sialyl-Tn conjugated to KLH plus detox adjuvant. *Ann. N. Y. Acad. Sci.* **690:**276–291.
54. **Longenecker, B. M., M. Reddish, R. Koganty, and G. D. MacLean.** 1994. Specificity of the IgG response in mice and human breast cancer patients following immunization against synthetic sialyl-Tn, an epitope with possible functional significance in metastasis. *Adv. Exp. Med. Biol.* **353:**105–124.
55. **Lou, Q., and I. Pastan.** 1999. A Lewis Y epitope mimicking peptide induces anti-Lewis Y immune responses in rabbits and mice. *J. Pept. Res.* **53:**252–260.
56. **Luo, P., M. Agadjanyan, J.-P. Qiu, M. A. J. Westerink, Z. Steplewski, and T. Kieber-Emmons.** 1998. Antigenic and immunological mimicry of peptide mimotopes of Lewis carbohydrate antigens. *Mol. Immunol.* **35:**865–879.
57. **MacLean, G. D., M. Reddish, R. R. Koganty, T. Wong, S. Gandhi, M. Smolenski, J. Samuel, J. M. Nabholtz, and B. M. Longenecker.** 1993. Immunization of breast cancer patients using a synthetic sialyl-Tn glycoconjugate plus Detox adjuvant. *Cancer Immunol. Immunother.* **36:**215–222.
58. **Mandrell, R. E., and M. A. Apicella.** 1993. Lipo-oligosaccharides (LOS) of mucosal pathogens: molecular mimicry and host-modification of LOS. *Immunobiology* **187:**382–402.
59. **McCool, T. L., C. V. Harding, N. S. Greenspan, and J. R. Schreiber.** 1999. B- and T-cell immune responses to pneumococcal conjugate vaccines: divergence between carrier- and polysaccharide-specific immunogenicity. *Infect. Immun.* **67:**4862–4869.
60. **McNamara, M. K., R. E. Ward, and H. Kohler.** 1984. Monoclonal idiotope vaccine against Streptococcus pneumoniae infection. *Science* **226:**1325–1326.
61. **Mer, G., E. Kellenberger, and J. F. Lefevre.** 1998. Alpha-helix mimicry of a beta-turn. *J. Mol. Biol.* **281:**235–240.
62. **Minasian, L. M., T. J. Yao, T. A. Steffens, D. A. Scheinberg, L. Williams, E. Riedel, A. N. Houghton, and P. B. Chapman.** 1995. A phase I study of anti-GD3 ganglioside monoclonal antibody R24 and recombinant human macrophage-colony stimulating factor in patients with metastatic melanoma. *Cancer* **75:**2251–2257.
63. **Mirkov, T. E., S. V. Evans, J. Wahlstrom, L. Gomez, N. M. Young, and M. J. Chrispeels.** 1995. Location of the active site of the bean alpha-amylase inhibitor and involvement of a Trp, Arg, Tyr triad. *Glycobiology* **5:**45–50.
64. **Monafo, W. J., N. S. Greenspan, T. J. Cebra, and J. M. Davie.** 1987. Modulation of the murine immune response to streptococcal group A carbohydrate by immunization with monoclonal anti-idiotope. *J. Immunol.* **139:**2702–2707.
65. **Mond, J. J., A. Lees, and C. M. Snapper.** 1995. T cell-independent antigens type 2. *Annu. Rev. Immunol.* **13:**655–692.
66. **Monfardini, C., T. Kieber-Emmons, J. M. VonFeldt, B. O'Malley, H. Rosenbaum, A. P. Godillot, K. Kaushansky, C. B. Brown, D. Voet, D. E. McCallus, et al.** 1995. Recombinant antibodies in bioactive peptide design. *J. Biol. Chem.* **270:**6628–6638.
67. **Monfardini, C., T. Kieber-Emmons, D. Voet, A. P. Godillot, D. B. Weiner, and W. V. Williams.** 1996. Rational design of granulocyte-macrophage colony-stimulating factor antagonist peptides. *J. Biol. Chem.* **271:**2966–2971.
68. **Moran, A. P., M. M. Prendergast, and B. J. Appelmelk.** 1996. Molecular mimicry of host structures by bacterial lipopolysaccharides and its contribution to disease. *FEMS Immunol. Med. Microbiol.* **16:**105–115.
69. **Morrow, W. J., W. M. Williams, A. S. Whalley, T. Ryskamp, R. Newman, C. Y. Kang, S. Chamat, H. Kohler, and T. Kieber-Emmons.** 1992. Synthetic peptides from a conserved region of gp120 induce broadly reactive anti-HIV responses. *Immunology* **75:**557–564.
70. **Murai, H., S. Hara, T. Ikenaka, A. Goto, M. Arai, and S. Murao.** 1985. Amino acid sequence of protein alpha-amylase inhibitor from Streptomyces griseosporeus YM-25. *J. Biochem.* **97:**1129–1133.
71. **Murali, R., and T. Kieber-Emmons.** 1997. Molecular recognition of a peptide mimic of the Lewis Y antigen by an anti-Lewis Y antibody. *J. Mol. Recognition* **10:**269–276.
72. **Oldenburg, K. R., D. Loganathan, I. J. Goldstein, P. G. Schultz, and M. A. Gallop.** 1992. Peptide ligands for a sugar-binding protein isolated from a random peptide library. *Proc. Natl. Acad. Sci. USA* **89:**5393–5397.
73. **Olsson, L.** 1987. Molecular mimicry of carbohydrate and protein structures by hybridoma antibodies. *Bioessays* **7**(3)**:**116–119.

74. **Phalipon, A., A. Folgori, J. Arondel, G. Sgaramella, P. Fortugno, R. Cortese, P. J. Sansonetti, and F. Felici.** 1997. Induction of anti-carbohydrate antibodies by phage library-selected peptide mimics. *Eur. J. Immunol.* **27:**2620–2625.
75. **Pichla, S. L., R. Murali, and R. M. Burnett.** 1997. The crystal structure of a Fab fragment to the melanoma-associated GD2 ganglioside. *J. Struct. Biol.* **119:**6–16.
76. **Pincus, S. H., M. J. Smith, H. J. Jennings, J. B. Burritt, and P. M. Glee.** 1998. Peptides that mimic the group B streptococcal type III capsular polysaccharide antigen. *J. Immunol.* **160:**293–298.
77. **Pinilla, C., J. R. Appel, G. D. Campbell, J. Buencamino, N. Benkirane, S. Muller, and N. S. Greenspan.** 1998. All-D peptides recognized by an anti-carbohydrate antibody identified from a positional scanning library. *J. Mol. Biol.* **283:**1013–1025.
78. **Pinilla, C., S. Chendra, J. R. Appel, and R. A. Houghten.** 1995. Elucidation of monoclonal antibody polyspecificity using a synthetic combinatorial library. *Pep. Res.* **8:**250–257.
79. **Prammer, K. V., J. Boyer, K. Ugen, S. J. Shattil, and T. Kieber-Emmons.** 1994. Bioactive Arg-Gly-Asp conformations in anti-integrin GPiib-iiia antibodies. *Receptor* **4:**93–108.
80. **Putterman, C., and B. Diamond.** 1998. Immunization with a peptide surrogate for double-stranded DNA (dsDNA) induces autoantibody production and renal immunoglobulin deposition. *J. Exp. Med.* **188:**29–38.
81. **Qiu, J., P. Luo, K. Wasmund, Z. Steplewski, and T. Kieber-Emmons.** 1999. Toward the development of peptide mimotopes of carbohydrate antigens as cancer vaccines. *Hybridoma* **18:**103–112.
82. **Ravindranath, M. H., and D. L. Morton.** 1991. Role of gangliosides in active immunotherapy with melanoma vaccine. *Int. Rev. Immunol.* **7:**303–329.
83. **Raychaudhuri, S., C. Y. Kang, S. V. Kaveri, T. Kieber-Emmons, and H. Kohler.** 1990. Tumor idiotype vaccines. VII. Analysis and correlation of structural, idiotypic, and biologic properties of protective and nonprotective Ab2. *J. Immunol.* **145:**760–767.
84. **Ritter, G., B. E. Ritter, R. Adluri, M. Calves, S. Ren, R. K. Yu, H. F. Oettgen, L. J. Old, and P. O. Livingston.** 1995. Analysis of the antibody response to immunization with purified O-acetyl GD3 gangliosides in patients with malignant melanoma. *Int. J. Cancer* **62:**668–672.
85. **Saleh, M. N., M. B. Khazaeli, R. H. Wheeler, L. Allen, A. B. Tilden, W. Grizzle, R. A. Reisfeld, A. L. Yu, S. D. Gillies, and A. F. LoBuglio.** 1992. Phase I trial of the chimeric anti-GD2 monoclonal antibody ch14.18 in patients with malignant melanoma. *Hum. Antibodies Hybridomas* **3:**19–24.
86. **Saleh, M. N., M. B. Khazaeli, R. H. Wheeler, E. Dropcho, T. Liu, M. Urist, D. M. Miller, S. Lawson, P. Dixon, C. H. Russell, et al.** 1992. Phase I trial of the murine monoclonal anti-GD2 antibody 14G2a in metastatic melanoma. *Cancer Res.* **52:**4342–4347.
87. **Saleh, M. N., J. D. Stapleton, M. B. Khazaeli, and A. F. LoBuglio.** 1993. Generation of a human anti-idiotypic antibody that mimics the GD2 antigen. *J. Immunol.* **151:**3390–3398.
88. **Schmolling, J., J. Reinsberg, U. Wagner, and D. Krebs.** 1995. Antiidiotypic antibodies in ovarian cancer patients treated with the monoclonal antibody B72.3. *Hybridoma* **14:**183–186.
89. **Schofield, L., M. J. McConville, D. Hansen, A. S. Campbell, B. Fraser-Reid, M. J. Grusby, and S. D. Tachado.** 1999. CD1d-restricted immunoglobulin G formation to GPI-anchored antigens mediated by NKT cells. *Science* **283:**225–229.
90. **Schreiber, J. R., K. L. Nixon, M. F. Tosi, G. B. Pier, and M. B. Patawaran.** 1991. Anti-idiotype-induced, lipopolysaccharide-specific antibody response to Pseudomonas aeruginosa. II. Isotype and functional activity of the anti-idiotype-induced antibodies. *J. Immunol.* **146:**188–193.
91. **Schreiber, J. R., M. Patawaran, M. Tosi, J. Lennon, and G. B. Pier.** 1990. Anti-idiotype-induced, lipopolysaccharide-specific antibody response to Pseudomonas aeruginosa. *J. Immunol.* **144:**1023–1029.
92. **Schreiber, J. R., G. B. Pier, M. Grout, K. Nixon, and M. Patawaran.** 1991. Induction of opsonic antibodies to Pseudomonas aeruginosa mucoid exopolysaccharide by an anti-idiotypic monoclonal antibody. *J. Infect. Dis.* **164:**507–514.
93. **Scott, J. K.** 1992. Discovering peptide ligands using epitope libraries. *Trends Biochem. Sci.* **17:**241–245.
94. **Scott, J. K., D. Loganathan, R. B. Easley, X. Gong, and I. J. Goldstein.** 1992. A family of concanavalin A-binding peptides from a hexapeptide epitope library. *Proc. Natl. Acad. Sci. USA* **89:**5398–5402.
95. **Scott, J. K., and G. P. Smith.** 1990. Searching for peptide ligands with an epitope library. *Science* **249:**386–390.
96. **Sen, G., M. Chakraborty, K. A. Foon, R. A. Reisfeld, and C. M. Bhattacharya.** 1998. Induction of IgG antibodies by an anti-idiotype antibody mimicking disialoganglioside GD2. *J. Immunother.* **21:**75–83.

97. **Shikhman, A. R., and M. W. Cunningham.** 1994. Immunological mimicry between *N*-acetyl-beta-D-glucosamine and cytokeratin peptides. Evidence for a microbially driven anti-keratin antibody response. *J. Immunol.* **152:**4375–4387.

98. **Shikhman, A. R., N. S. Greenspan, and M. W. Cunningham.** 1994. Cytokeratin peptide SFGSGFGGGY mimics *N*-acetyl-beta-D-glucosamine in reaction with antibodies and lectins, and induces in vivo anti-carbohydrate antibody response. *J. Immunol.* **153:**5593–5606.

99. **Shikhman, A. R., and M. W. Cunningham.** 1997. Trick and treat: toward peptide mimic vaccines. *Nat. Biotechnol.* **15:**512–513.

100. **Smith, G. P., and J. K. Scott.** 1993. Libraries of peptides and proteins displayed on filamentous phage. *Methods Enzymol.* **217:**228–257.

101. **Stein, K. E., and T. Soderstrom.** 1984. Neonatal administration of idiotype or antiidiotype primes for protection against Escherichia coli K13 infection in mice. *J. Exp. Med.* **160:**1001–1011.

102. **Stern, B., G. Denisova, D. Buyaner, D. Raviv, and J. M. Gershoni.** 1997. Helical epitopes determined by low-stringency antibody screening of a combinatorial peptide library. *FASEB J.* **11:**147–153.

102a. **Sugita, M., D. B. Moody, R. M. Jackman, E. P. Grant, J. P. Rosat, S. M. Behar, P. J. Peters, S. A. Porcelli, and M. B. Brenner.** 1998. CD1: a new paradigm for antigen presentation and T-cell activation. *Clin. Immunol. Immunopathol.* **87:**8–14.

103. **Taki, T., D. Ishikawa, H. Hamasaki, and S. Handa.** 1997. Preparation of peptides which mimic glycosphingolipids by using phage peptide library and their modulation on beta-galactosidase activity. *FEBS Lett.* **418:**219–223.

104. **Thurin, J., M. Thurin, Y. Kimoto, M. Herlyn, M. D. Lubeck, D. E. Elder, M. Smereczynska, K. A. Karlsson, W. M. J. Clark, and Z. Steplewski.** 1987. Monoclonal antibody-defined correlations in melanoma between levels of GD2 and GD3 antigens and antibody-mediated cytotoxicity. *Cancer Res.* **47:**1229–1233.

105. **Thurin-Blaszczyk, M., R. Murali, M. A. J. Westerink, Z. Steplewski, M.-S. Co, and T. Kieber-Emmons.** 1996. Molecular recognition of the Lewis Y antigen by monoclonal antibodies. *Protein Eng.* **9:**101–113.

106. **Transue, T. R., E. De Genst, M. A. Ghahroudi, L. Wyns, and S. Muyldermans.** 1998. Camel single-domain antibody inhibits enzyme by mimicking carbohydrate substrate. *Proteins* **32:**515–522.

107. **Tsuyuoka, K., K. Yago, K. Hirashima, S. Ando, N. Hanai, H. Saito, K. M. Yamasaki, K. Takahashi, Y. Fukuda, K. Nakao, and R. Kannagi.** 1996. Characterization of a T cell line specific to an anti-Id antibody related to the carbohydrate antigen, sialyl SSEA-1, and the immunodominant T cell antigenic site of the antibody. *J. Immunol.* **157:**661–669.

108. **Valadon, P., G. Nussbaum, L. F. Boyd, D. H. Margulies, and M. D. Scharff.** 1996. Peptide libraries define the fine specificity of anti-polysaccharide antibodies to Cryptococcus neoformans. *J. Mol. Biol.* **261:**11–22.

109. **Valadon, P., and M. D. Scharff.** 1996. Enhancement of ELISAs for screening peptides in epitope phage display libraries. *J. Immunol. Methods* **197:**171–179.

110. **Viale, G., F. Grassi, M. Pelagi, R. Alzani, S. Menard, S. Miotti, R. Buffa, A. Gini, and A. G. Siccardi.** 1987. Anti-human tumor antibodies induced in mice and rabbits by "internal image" anti-idiotypic monoclonal immunoglobulins. *J. Immunol.* **139:**4250–4255.

111. **Westerink, M. A., A. A. Campagnari, M. A. Wirth, and M. A. Apicella.** 1988. Development and characterization of an anti-idiotype antibody to the capsular polysaccharide of *Neisseria meningitidis* serogroup C. *Infect. Immun.* **56:**1120–1127.

112. **Westerink, M. A. J., P. C. Giardina, M. A. Apicella, and T. Kieber-Emmons.** 1995. Peptide mimicry of the meningococcal group C capsular polysaccharide. *Proc. Natl. Acad. Sci. USA* **92:**4021–4025.

113. **Williams, W. V., T. Kieber-Emmons, J. Von Feldt, M. I. Greene, and D. B. Weiner.** 1991. Design of bioactive peptides based on antibody hypervariable region structures. Development of conformationally constrained and dimeric peptides with enhanced affinity. *J. Biol. Chem.* **266:**5182–5190.

114. **Williams, W. V., T. Kieber-Emmons, D. B. Weiner, D. H. Rubin, and M. I. Greene.** 1991. Contact residues and predicted structure of the reovirus type 3-receptor interaction. *J. Biol. Chem.* **266:**9241–9250.

115. **Williams, W. V., D. A. Moss, T. Kieber-Emmons, J. A. Cohen, J. N. Myers, D. B. Weiner, and M. I. Greene.** 1989. Development of biologically active peptides based on antibody structure. *Proc. Natl. Acad. Sci. USA* **86:**5537–5541. (Erratum, **86:**8044.)

116. **Young, A. C., P. Valadon, A. Casadevall, M. D. Scharff, and J. C. Sacchettini.** 1997. The three-dimensional structures of a polysaccharide binding antibody to Cryptococcus neoformans and its complex with a peptide from a phage display library: implications for the identification of peptide mimotopes. *J. Mol. Biol.* **274:**622–634.
117. **Young, N. M., M. A. Gidney, B. M. Gudmundsson, C. R. MacKenzie, R. To, D. C. Watson, and D. R. Bundle.** 1999. Molecular basis for the lack of mimicry of Brucella polysaccharide antigens by Ab2gamma antibodies. *Mol. Immunol.* **36:**339–347.
118. **Zhang, H., Z. Zhong, and L. A. Pirofski.** 1997. Peptide epitopes recognized by a human anti-cryptococcal glucuronoxylomannan antibody. *Infect. Immun.* **65:**1158–1164.
119. **Zhang, S., L. A. Walberg, S. Ogata, S. H. Itzkowitz, R. R. Koganty, M. Reddish, S. S. Gandhi, B. M. Longenecker, K. O. Lloyd, and P. O. Livingston.** 1995. Immune sera and monoclonal antibodies define two configurations for the sialyl Tn tumor antigen. *Cancer Res.* **55:**3364–3368.
120. **Zhao, R., D. J. Loftus, E. Appella, and E. J. Collins.** 1999. Structural evidence of T cell xeno-reactivity in the absence of molecular mimicry. *J. Exp. Med.* **189:**359–370.

Molecular Mimicry, Microbes, and Autoimmunity
Edited by M. W. Cunningham and R. S. Fujinami

Chapter 14

Structural Basis of T-Cell Receptor Specificity and Cross-Reactivity: Implications for Pathogenesis of Human Autoimmune Diseases

Heiner Appel and Kai W. Wucherpfennig

Over the last few years, new concepts of T-cell recognition have emerged. Individual T cells were previously considered to be highly specific for single "foreign" peptides. Numerous studies have now demonstrated that the same T-cell receptor (TCR) can recognize a number of peptides that have limited primary sequence homology, indicating that degenerate recognition of major histocompatibility complex (MHC)-bound peptides is a general property of αβ TCRs (1, 2, 12, 20, 25, 43, 64, 69). "Molecular mimicry" refers to TCR cross-reactivity between a microbial peptide and a self-antigen that results in the activation of autoreactive T cells (14, 46, 69). The same TCR can also be engaged by a variety of peptides that act as agonists, partial agonists, or antagonists and that induce a wide spectrum of different T-cell responses (36). This emerging complexity of TCR recognition has important implications for self-tolerance and the pathogenesis of human autoimmune diseases.

ACTIVATION OF AUTOREACTIVE T CELLS IS REQUIRED FOR DEVELOPMENT OF AUTOIMMUNE DISEASE

Activation of autoreactive T cells is critical for the induction of autoimmunity. In animal models of autoimmunity, disease can be transferred only with activated T cells but not with resting autoreactive T cells (71, 72). This important finding has been made in several experimental models, such as the experimental autoimmune encephalomyelitis (EAE) model and the nonobese diabetic (NOD) mouse model for type I diabetes (48, 71). Resting autoreactive T cells are part of the normal immune repertoire and do not induce disease. In the EAE model, the blood-brain barrier limits access to the central nervous system (CNS) to activated T cells (29). This protective mechanism makes circulating autoreactive T cells "ignorant" of the complex set of tissue-specific self-antigens located behind the blood-brain barrier. Invasion of the CNS requires activation of autoreactive T cells in the peripheral immune system. This activation requirement raises the question as to which antigen(s) induces the activation of myelin-specific T cells outside the CNS.

Heiner Appel and Kai W. Wucherpfennig • Department of Cancer Immunology & AIDS, Dana-Farber Cancer Institute, Boston, MA 02115.

POTENTIAL MECHANISMS FOR ACTIVATION OF AUTOREACTIVE T CELLS IN AUTOIMMUNE DISEASES

Several mechanisms may be responsible for the activation of autoreactive T cells in autoimmune diseases. These include cross-reactive microbial peptides (molecular mimicry), viral or bacterial superantigens, release of autoantigen during inflammation, or bystander activation (3, 25, 40, 46). Given the complexity of human autoimmune diseases, it is unlikely that a single mechanism is responsible for these different disease processes. Also, different activation mechanisms may be operative during different stages of disease pathogenesis. For example, superantigens were shown to induce relapses of EAE when injected following an initial attack (3). However, administration of superantigen prior to disease induction resulted in deletion of naive T cells reactive with the superantigen and in resistance to subsequent disease induction by immunization with autoantigen (58). In a similar way, release of autoantigen may be important in later stages of an autoimmune process and thereby result in epitope spreading (40).

It is important to keep in mind that activation of autoreactive T cells is necessary, but not sufficient, for the development of clinical disease. Among these contributing factors, genetic susceptibility is important since genome-wide scans of patients with multiple sclerosis (MS) have demonstrated that multiple genes contribute to susceptibility (11, 21, 22). An important locus in many autoimmune diseases is the MHC on human chromosome 6 (21, 22). Also important is the induction of a sufficient degree of clonal expansion and a functional phenotype (i.e., cytokine profile) that renders such T cells pathogenic. Access to the target organ as well as the levels of expression of MHC class II and costimulatory molecules in the target organ may limit the occurrence of autoimmunity (Table 1) (25, 29).

HOW SPECIFIC IS T-CELL RECOGNITION OF MHC-PEPTIDE COMPLEXES?

The molecular mimicry hypothesis postulates that there is significant cross-reactivity between viral and bacterial T-cell epitopes and human self-peptides and that such cross-reactivity can initiate an autoimmune process (46). At first sight TCR recognition appears to be exquisitely specific since even minor substitutions in a T-cell epitope can diminish or abrogate T-cell activation (50, 67). Several lines of evidence do, however, indicate that there is a significant degree of degeneracy in peptide recognition that can result in cross-reactivity. Cross-reactivity is part of the positive selection process in the thymus that selects T cells with a low affinity for self MHC-peptide complexes and that deletes T cells with a higher affinity for these ligands (32, 33). Such cross-reactive peptides may also be important for the survival of mature T cells in the periphery (37, 63).

Table 1. Requirements for development of an autoimmune disease following an initial activation of autoreactive T cells

1. Sufficient clonal expansion of autoreactive T cells
2. Induction of a functional phenotype (i.e., cytokine profile) that makes a T-cell population pathogenic
3. Access of activated, autoreactive T cells to the target organ
4. Expression of sufficient levels of MHC molecules and costimulatory molecules on antigen-presenting cells in the target organ for reactivation of autoreactive T cells

The relatively high frequency of alloreactive T cells also indicates that T-cell recognition is not highly specific.

In addition, consideration of the structural requirements for MHC binding and TCR recognition suggests a considerable potential for cross-reactivity. Peptide binding to MHC molecules is highly degenerate, and a given MHC class II molecule binds to peptides from the majority of antigens when a set of overlapping peptides is tested. Peptide elution studies have demonstrated that a complex set of several hundred peptides is bound to a single MHC molecule (6, 31). This degeneracy is explained by the fact that for the majority of MHC class II molecules every peptide residue can be replaced by related or even unrelated amino acids (23, 24). T-cell recognition is more specific since a minor modification of critical TCR contact residues can result in a complete loss of T-cell reactivity. Such a degree of specificity is, however, limited to a single or a few residues within a T-cell epitope (50).

STRUCTURAL REQUIREMENTS FOR HLA-DR2 BINDING AND TCR RECOGNITION OF AN IMMUNODOMINANT MBP PEPTIDE

On the basis of considerations mentioned above, the specificity and cross-reactivity of myelin basic protein (MBP)-specific T-cell clones from MS patients were examined (66–69). The cause of MS is not known, but it is widely believed that MS is caused by an autoimmune process against the CNS myelin (60). Genome-wide scans of families with MS have demonstrated that multiple genes contribute to susceptibility and that the MHC is an important susceptibility locus (11, 21, 22). Within the MHC, the strongest linkage was observed with markers for the MHC class II region. Recent studies have demonstrated that only families which carry the HLA-DR2 haplotype (DRB1*1501) contribute to the observed linkage with the MHC (22). Many previous studies had demonstrated that the HLA-DR2 (DRB1*1501) haplotype is present at an increased frequency in patients with MS, in particular in patients of Scandinavian descent who carry the highest risk (30). This haplotype includes the DRB1*1501, DRB5*0101, DQA1*0102, and DQB1*0602 alleles, which are in linkage disequilibrium; these encode HLA-DR2b (DRA, DRB1*1501), HLA-DR2a (DRA, DRB5*0101), and HLA-DQ6 (DQA1*0102, DQB1*0602) (45). Disease-associated MHC alleles may confer susceptibility by affecting T-cell repertoire selection in the thymus or peripheral immune system and/or by presenting self-peptides against which an autoimmune response is directed. The observation that particular alleles of MHC genes are associated with an autoimmune disease suggests that a limited number of self-peptides are involved in early stages of the disease process (44, 70). In later stages of the disease, the autoimmune response may broaden to a larger set of antigens and peptides (epitope spreading) (40). Migration studies have also implicated environmental exposure as a risk factor for MS (39). Despite evidence that environmental exposure is important, no single pathogen has consistently been associated with the pathogenesis of MS.

Analysis of the T-cell response to MBP in MS patients and healthy subjects resulted in the identification of an immunodominant peptide that is presented by HLA-DR2 (DRA, DRB1*1501) (42, 47, 49, 56, 66, 67). This MBP peptide (residues 85 to 99) bound with a high affinity to purified HLA-DR2b (DRA, DRB1*1501) and was recognized by T-cell clones isolated from patients with MS (65, 67). The requirements for HLA-DR2 binding were defined with a large panel of single-amino-acid analogs of the MBP peptide. These experiments demonstrated that two hydrophobic amino acids, V89 and F92, were critical for DR2 binding. Replacement of these peptide residues by alanine greatly reduced the level of HLA-DR2 binding, while binding was abolished by replacement with aspartic acid. Nevertheless, these anchor residues could be replaced by other hydrophobic amino acids, such as substitution of aliphatic amino acids and phenylalanine for V89 (67). These experiments demonstrated that the HLA-DR2 binding motif was degenerate.

A limited number of peptide residues were found to be important for TCR recognition of the MBP peptide (residues 85 to 99). Three peptide residues, H90, F91, and K93, were particularly important for TCR recognition of the MBP peptide (67, 68). F91 was important for TCR recognition by every T-cell clone that was studied since replacement by alanine abrogated T-cell activation but did not affect DR2 binding. The relative contributions of other peptide residues depended on the T-cell clone. For some T-cell clones, F91 and K93 represented primary TCR contact residues, while H90 and F91 were primary TCR contact residues for other T-cell clones. Taken together, this analysis demonstrated that H90, F91, and K93 were primary TCR contact residues, while V89 and F92 were primary DR2 contact residues of the MBP peptide. The residues required for HLA-DR2 binding and TCR recognition were therefore clustered in a short segment of the MBP peptide.

IDENTIFICATION OF MICROBIAL PEPTIDES THAT ACTIVATE T CELLS SPECIFIC FOR HLA-DR2–MBP PEPTIDE COMPLEX

On the basis of the characterization of the immunodominant MBP peptide presented above, criteria were developed so that protein databases could be searched for cross-reactive viral and bacterial peptides. These search criteria focused on the core region of the peptide, residues 88 to 93, which contained MHC and TCR contact residues common to all T-cell clones. The initial search identified seven viral peptides and a bacterial peptide that stimulated at least one of seven MBP-specific T-cell clones that were studied (69). The microbial peptides were quite distinct in their primary sequence from the MBP peptide and from each other since only primary TCR contact residues were conserved.

A cross-reactive peptide from the Epstein-Barr virus (EBV) DNA polymerase was used to examine whether infected antigen-presenting cells could present a cross-reactive viral peptide to MBP-specific T cells (69). The EBV DNA polymerase gene is part of the lytic cycle and is not transcribed in EBV-transformed B cells. However, the lytic cycle and expression of the EBV DNA polymerase gene can be induced by phorbol esters (7). An HLA-DR2-positive B-cell line that had been pretreated with a phorbol ester activated an MBP-specific T-cell clone that recognized both MBP and the EBV DNA polymerase peptide. T-cell activation was blocked by a monoclonal antibody (MAb) specific for HLA-DR but not by a control MAb against HLA-DQ. These results demonstrated that the MBP-specific T-cell clone recognized not only the EBV peptide but also antigen-presenting cells that expressed the viral protein (69).

In that first study, cross-reactive microbial peptides were identified for three of seven MBP-specific T-cell clones (69). Two of the T-cell clones for which no cross-reactive peptides were initially identified were further analyzed, taking advantage of the bacterial genome sequences that have recently been reported. A total of five bacterial peptides that activated clone Ob.1A12 were identified. The second clone (clone Ob.2F3) was activated by a subset of the peptides that activated clone Ob.1A12 (27) (Table 2). The three bacterial peptides that activated both clones matched the search criteria used in the initial study, illustrating how recent progress in the sequencing of microbial genomes allowed the identification of these peptides. These peptides (derived from *Staphylococcus aureus, Mycobacterium avium,* and *Mycobacterium tuberculosis*) all had a conservative lysine-to-arginine replacement of 93K. The other two peptides did not match the original search motif because it did not consider nonconservative substitutions of 93K. These residues had not been included in the original search criteria because some of the other T-cell clones had a strong preference for a positive charge (lysine or arginine) at this position. Since many microbial genomes have not yet been sequenced, other stimulatory peptides may exist. Also, certain sequences may be overlooked by search criteria that use a motif based on single amino acid substitutions.

The search criteria also matched the requirements for an HLA-DQ1-restricted T-cell clone specific for the MBP peptide (69). This clone required the same minimal peptide segment as HLA-DR2-restricted T-cell clones (residues 87 to 97), and H90, F91, and K93 were important for T-cell recognition of the peptide. Replacement of V89 and F92 by aspartic acid greatly diminished the stimulatory capacity of the MBP peptide, while replacement of V89 and F92 hydrophobic amino acids was tolerated. These data suggested that the MBP peptide was bound in similar way to HLA-DR2 and HLA-DQ1 and that the same peptide residues were critical for TCR recognition. The DQ1-restricted T-cell clone was activated by five distinct peptides: the MBP peptide (residues 85 to 99), three viral peptides (from herpes simplex virus, adenovirus type 12, and human papillomavirus), and a bacterial peptide (from *Pseudomonas aeruginosa*). Only the peptide derived from the human papillomavirus L2 protein had obvious sequence similarity with the MBP peptide (69). For all other sequences, alignment would not have permitted their identification as cross-reactive peptides of MBP.

Taken together, the analysis of MBP-specific T-cell clones from MS patients demonstrated that microbial peptides with limited sequence homology are effective activators of autoreactive T cells. On the basis of these results, it is unlikely that a single microbial pep-

Table 2. Sequences of microbial peptides that activate human MBP-specific T-cell clones[a]

Activated T-cell clone and organism	Source protein (residues)	Sequence[b]
Ob.1A12		
Homo sapiens	Myelin basic protein	**E N P V V H F F K N I V T P R**
Staphylococcus aureus	VgaB (386–400)	V L A R L **H F** Y R **N** D **V** H K E
Mycobacterium avium	Transposase (251–265)	Q R C R **V H F** L R **N** V L A Q V
Mycobacterium tuberculosis	Transposase (229–243)	Q R C R **V H F** M R **N** L Y **T** A V
Bacillus subtilis	YqeE (193–207)	A L A **V** L **H F** Y P D K G A K N
Escherichia coli, Haemophilus influenzae	ORF[c] (359–373) HI0136 (359–373)	D F A R **V H F** I S A L H G S G
Ob.2F3		
Homo sapiens	Myelin basic protein	**E N P V V H F F K N I V T P R**
Staphylococcus aureus	VgaB (386–400)	V L A R L **H F** Y R **N** D **V** H K E
Mycobacterium avium	Transposase (251–265)	Q R C R **V H F** L R **N** V L A Q V
Mycobacterium tuberculosis	Transposase (229–243)	Q R C R **V H F** M R **N** L Y **T** A V

[a] Microbial peptides that activate two MBP-specific T cell clones (27) have been aligned with the MBP (85-99) sequence. The two T-cell clones carry the same Vα-Jα and Vβ-Jβ sequences and differ only in the N regions of TCR α and β.
[b] Boldface letters indicate identity with the MBP peptide.
[c] ORF, open reading frame.

tide is responsible for the activation of autoreactive T cells by molecular mimicry. Other cross-reactive peptides of the MBP epitope (residues 85 to 99) remain to be identified since only a small fraction of the microbial genome sequences is known.

Many studies have examined the issue of cross-reactivity by performing sequence alignments between candidate self-antigens and microbial sequences. This approach has the following related problems. (i) Frequently, it is not known whether the region of the autoantigen that was studied represents a T-cell epitope. (ii) Peptides that have visual sequence similarity may have no biological activity (lack of MHC binding, failure to activate autoreactive T cells). (iii) Peptides that are biologically active may be overlooked because the structural similarity to the self-peptide may not be obvious. The observations made in the analysis of MBP-specific T-cell clones indicate that a detailed characterization of the immunodominant epitope of an autoantigen greatly facilitates the identification of cross-reactive microbial peptides (25, 27, 69).

Peptides with minimal sequence homology have also been identified for a variety of other antigen-specific T-cell clones and T-cell hybridomas (1, 2, 12, 20, 28, 43, 64). Taken together, these results demonstrate that degenerate recognition of MHC-bound peptides is a general feature of T-cell recognition.

CRYSTAL STRUCTURE OF THE HLA-DR2–MBP PEPTIDE COMPLEX

The structure of HLA-DR2 with the bound MBP peptide (residues 85 to 99) was determined by X-ray crystallography as a step toward defining molecular mimicry at a structural level (57). For this purpose, HLA-DR2 was expressed as a soluble protein in the baculovirus system (18). The sequence of the MBP peptide (residues 85 to 99) was covalently linked to the N terminus of the mature DRβ chain, as described for murine MHC class II molecules (38). Assembly of DRα and DRβ chains was facilitated by leucine zipper dimerization domains that were cloned in frame to the 3′ end of the extracellular domains of DRα and DRβ (35). The HLA-DR2–MBP peptide complex could be crystallized at pH 3.5 to 4.5 by using polyethylene glycol as a precipitant. The structure was determined to a resolution of 2.6 Å by using a data set collected at a synchrotron radiation source (57).

Figure 1 gives an overview of the structure and illustrates features of HLA-DR2 that are important for peptide binding as well as for TCR recognition of the HLA-DR2–MBP peptide complex. The MBP peptide is bound in an extended conformation as a type II polyproline helix, as observed for other MHC class II–peptide complexes (4, 34, 61). MBP peptide side chain residues V89, F92, I95, and T97 occupy the P1, P4, P6, and P9 pockets of the binding groove, respectively (Fig. 1A). As discussed above, V89 and F92 are important anchor residues for HLA-DR2 binding of the MBP peptide, and these anchor residues occupy the hydrophobic P1 and P4 pockets, respectively (67). The P1 pocket of HLA-DR2 is identical to other DR molecules that carry a valine at DRβ 86 (5), but the P4 pocket of HLA-DR2 is distinct.

In the HLA-DR2–MBP peptide complex, the P4 pocket is occupied by an aromatic side chain (F92), which makes an important contribution to the binding of the MBP peptide (Fig. 1C). The structural basis for this side chain specificity is clear from the DR2-MBP structure, which reveals a large, predominantly hydrophobic P4 pocket. The pocket is lined with two aromatic MHC side chains, Tyr β78 and Phe β26, as well as side chains

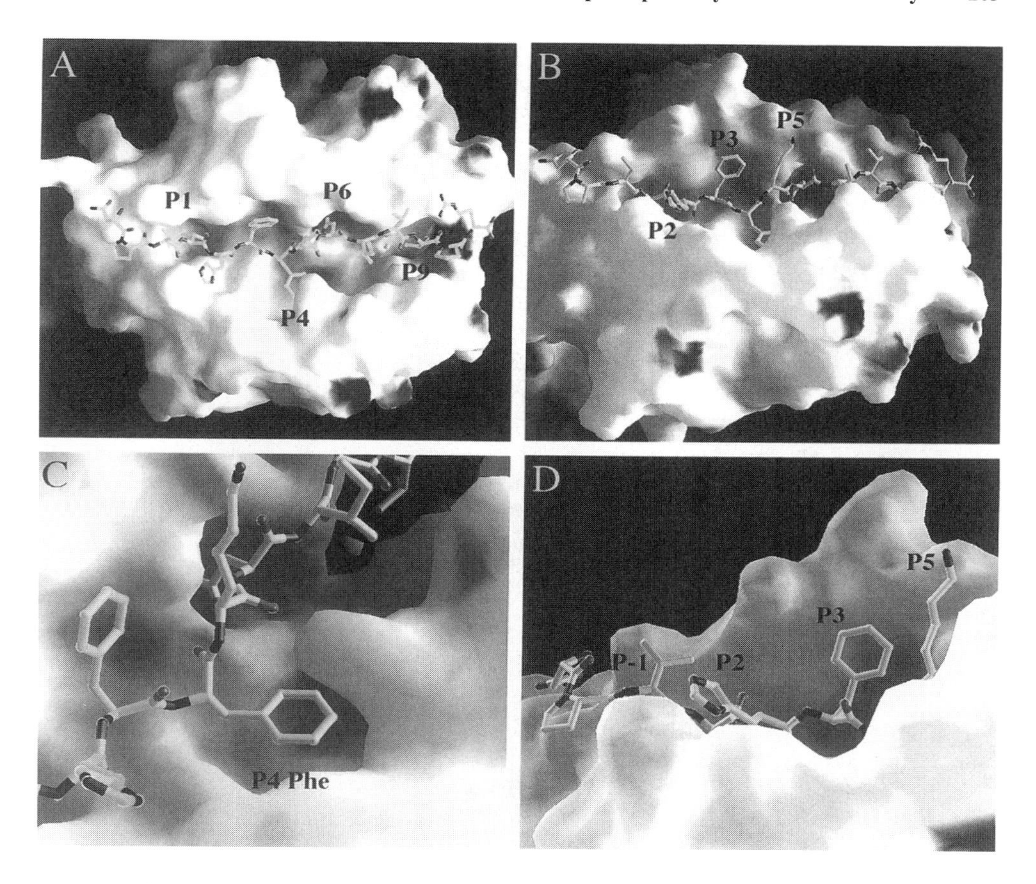

Figure 1. Crystal structure of the complex of HLA-DR2 and the MBP peptide (residues 85 to 99). (For color version of figure, see Color Plates, p. 277.) (A) Overview of the structure. MBP peptide residues V89, F92, I95, and T97 occupy the P1, P4, P6, and P9 pockets of the HLA-DR2 binding site, respectively. (B) Solvent-exposed residues that are important for TCR recognition of the MBP peptide (residues 85 to 99). MBP residues H90, F91, and K93 were identified as important TCR contact residues. These are located at the P2, P3, and P5 positions, respectively, and are available for interaction with the TCR. (C) P4 pocket of the HLA-DR2 binding site. This pocket is occupied by F92 of the MBP peptide. The necessary room for this aromatic side chain is created by the DRβ71 polymorphism. (D) Close-up view of MBP peptide residues that are important for TCR recognition. Preferences at positions P-1, P2, P3, and P5 were considered in the search criteria for cross-reactive microbial peptides. Reprinted from *The Journal of Experimental Medicine* (57) with permission of the publisher.

Ala β71, Gln β80, Asp β28, and Arg β13. The polymorphism at position 71 of DR2 (alanine) appears to be most important in creating the available space for the P4 aromatic side chain of the MBP peptide. The presence of alanine at DRβ71 is unusual among DRB1 alleles and has been observed only in DR2 alleles (DRB1*1501 to DRB1*1506) and DRB1*1309 (41). All other known DRB1 alleles carry codons for lysine, arginine, or glutamic acid at this position. In Caucasians, DRB1*1501 is the most common DR2 haplotype, and this haplotype confers susceptibility to the development of MS. It is not known if other DR2 subtypes confer susceptibility to MS since they are relatively uncommon in Europe and North America where the highest prevalence of MS is observed. Since the DRB1 gene is in linkage disequilibrium with the DRB5 gene as well as DQA-DQB genes, other MHC class II genes of the DR2 haplotype may contribute to susceptibility to MS.

The P4 pocket of HLA-DR makes an important contribution to susceptibility to other human autoimmune diseases (8, 24, 55, 70). This is explained by the fact that the P4 pocket is the most polymorphic pocket of the HLA-DR binding site. For example, an HLA-DR4 allele (DRB1*0402) that is associated with susceptibility to pemphigus vulgaris, an autoimmune disease of the skin, differs from a rheumatoid arthritis-associated allele (DRB1*0404) only at three positions, DRβ67, 70, and 71. In the pemphigus vulgaris-associated allele, DRβ70 and DRβ71 of the P4 pocket carry negative charges (aspartic acid and glutamic acid, respectively) (55). In contrast, glutamine and lysine or arginine are present at DRβ70 and DRβ71, respectively, in rheumatoid arthritis-associated alleles (DRB1*0401 and DRB1*0404). HLA-DR4 has a preference for negatively charged residues at the P4 position and does not tolerate phenylalanine or tyrosine at P4 (23, 24). A peptide from type II collagen that carries an aspartic acid residue at the P4 position has been identified in HLA-DR1 and HLA-DR4 transgenic mice and is immunodominant for murine type II collagen-specific T cells (13, 53, 54). In contrast, HLA-DR2 has a preference for hydrophobic residues at P4 and does not tolerate a negative charge at this position since replacement of F92 of the MBP peptide by aspartic acid abolishes binding (67).

Previous studies demonstrated that H90, F91, and K93 (located at positions P2, P3, and P5, respectively) are important residues for TCR recognition of the HLA-DR2–MBP peptide complex (67–69). The HLA-DR2–MBP crystal structure shows that these peptide side chains are solvent exposed (P2, P3, and P5) and are available for TCR recognition (Fig. 1B and 1D; Fig. 2). Comparison with the recently published high-resolution crystal structures of human MHC class I–peptide–TCR complexes (9, 10, 15–17) suggests that the P5 lysine would bind in a pocket formed by the complementarity-determining region 3 (CDR3) loops of the TCR α and β chains. The N-terminal segment of the MBP peptide makes a contribution to TCR recognition since truncation of P-4 and P-3 (85E, 86N) greatly diminishes TCR recognition by some DR2-MBP-specific T-cell clones without affecting DR2 binding (67). In contrast, the C-terminal segment of the peptide is not critical for recognition by these T-cell clones since almost all substitutions at P7, P8, and P9 are tolerated. The degenerate HLA-DR2 binding motif as well as the limited requirements for sequence similarity and identity at TCR contact residues accounts for the observation that microbial peptides can activate human MBP-specific T-cell clones (67–69).

COMMON FEATURES OF CROSS-REACTIVE MICROBIAL PEPTIDES

The structural information on the HLA-DR2–MBP peptide complex was used to analyze the cross-reactive peptides identified for one of the MBP-specific T-cell clones (Ob.1A12) (27). The cross-reactive peptides were aligned with the MBP peptide in order to determine residues that may be located in pockets of the HLA-DR2 binding site (Table 3). This analysis shows that the HLA-DR2 contact surface of this set of peptides is highly diverse. No sequence identity with the MBP peptide is required on the HLA-DR2 binding

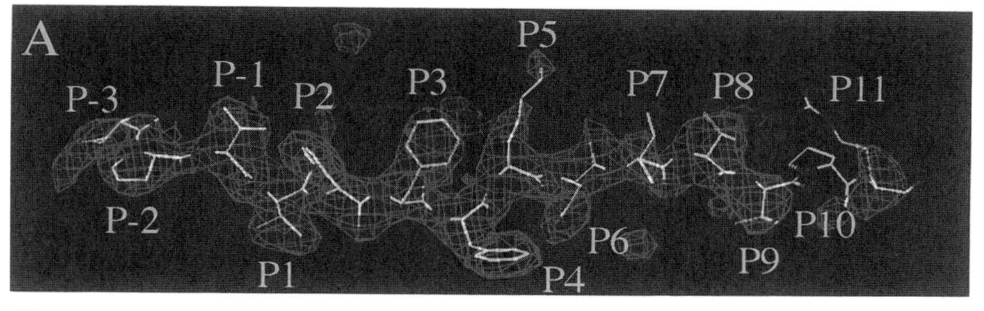

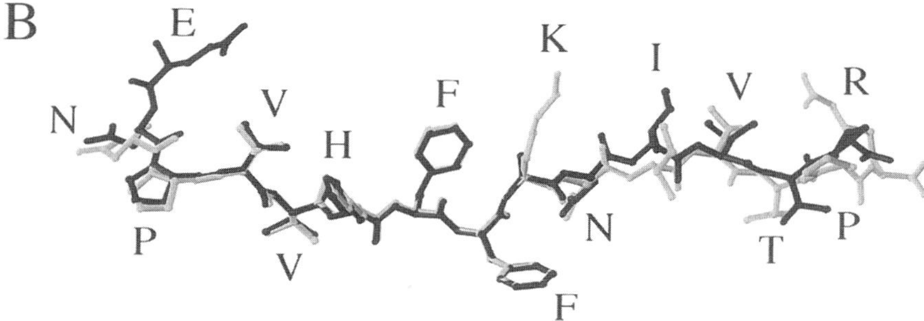

Figure 2. Electron density and model of the MBP peptide in the binding site of HLA-DR2. (For color version of figure, see Color Plates, p. 278.) (A) Electron density of the MBP peptide bound to HLA-DR2. The C terminus of the peptide (P10, P11) is partially disordered. (B) Superposition of the two MBP peptides in the asymmetric unit. The DR2-MBP peptide complex crystallized as a dimer of dimers, like other HLA-DR molecules (4, 61). The model for the MBP peptide includes residues P-3 to P11 and P-4 to P10 for the two copies in the asymmetric unit, yellow and blue, respectively. The peptide backbones superimpose in the P-1 to P4 segment and are more divergent in the C-terminal segment due to different crystal contacts. A crystal contact between peptide residue P-3 in one molecule and P5 from a symmetrically related molecule stabilizes the N terminus of one peptide, enabling P-4 to be included in the model for this peptide and P5 Lys to be included in the model for the other peptide. Reprinted from *The Journal of Experimental Medicine* (57) with permission of the publisher.

surface, as illustrated by comparison of the MBP and *Bacillus subtilis* peptides. Common features of putative HLA-DR2 contact residues are the presence of an aliphatic residue (V, L) in the P1 pocket and of a large hydrophobic residue (F, Y, L, I, M) in the P4 pocket. At P6, a preference for asparagine is observed, while the residues that occupy the P7 and P9 positions are diverse. These data are in agreement with HLA-DR2 binding studies, which indicated that only two positions of the peptide (V89 and F92) are critical for binding (67). These anchor residues could be replaced by other aliphatic residues or phenylalamine (V89) or by other hydrophobic residues (F92).

Analysis of the residues that may be solvent exposed and that could interact with the TCR shows a higher degree of sequence similarity or identity, in particular in the center of the epitope (Table 3). All peptides that activate this T-cell clone carry the two primary TCR

Table 3. Alignment of microbial peptides based on crystal structure of HLA-DR2–MBP peptide complex

Type of residues and organism	Source protein (residues)	Sequence[a]
Residues located in HLA-DR2 binding pockets		
Homo sapiens	Myelin basic protein	**- - - - V - - F - N I - T - -**
Staphylococcus aureus	VgaB (386–400)	- - - - L - - Y - **N** D - H - -
Mycobacterium avium	Transposase (251–265)	- - - - **V** - - L - **N** V - A - -
Mycobacterium tuberculosis	Transposase (229–243)	- - - - **V** - - M - **N** L - **T** - -
Bacillus subtilis	YqeE (193–207)	- - - - L - - Y - D K - A - -
Escherichia coli, Haemophilus influenzae	ORF[b] (359–373) HI0136 (359–373)	- - - - **V** - - I - A L - G - -
Solvent-exposed residues		
Homo sapiens	Myelin basic protein	**E N P V - H F - K - - V - P R**
Staphylococcus aureus	VgaB (386–400)	V L A R - **H F** - R - - **V** - K E
Mycobacterium avium	Transposase (251–265)	Q R C R - **H F** - R - - L - Q V
Mycobacterium tuberculosis	Transposase (229—243)	Q R C R - **H F** - R - - Y - A V
Bacillus subtilis	YqeE (193–207)	A L A **V** - **H F** - P - - G - K N
Escherichia coli, Haemophilus influenzae	ORF (359–373) HI0136 (359–373)	D F A R - **H F** - S - - H - S G

[a] Boldface indicates identity with the MBP peptide.

[b] ORF, open reading frame.

contact residues of the MBP peptide (H90 and F91). Also, a preference for a positively charged residue (K or R) is observed at P5. Nevertheless, a great deal of sequence diversity is observed in the N- and C-terminal flanking segments, even though these segments are required for efficient T-cell stimulation. Analysis of MBP peptides with N- or C-terminal truncations demonstrated that removal of residues 85E and 86N abolished T-cell activation (67). While the MBP (85-99) peptide had full activity, only high concentrations of MBP (86–98) activated this T-cell clone. Taken together, these data indicate that the flanking segments are required to stabilize the TCR-MHC-peptide complex, but that they contribute relatively little to the specificity of recognition.

Comparison of the fine specificity of the MBP-reactive T-cell clones demonstrated obvious differences in the degree of specificity and cross-reactivity. For some of the T-cell clones, such as Ob1.A12, amino acid identity was required at two TCR contact residues of the MBP peptide, while every TCR contact residue of the MBP peptide could be replaced by at least one structurally related amino acid for other T-cell clones (67–69). A particularly striking example of a highly degenerate T-cell clone has recently been described (28). This MBP (85-99)-specific T-cell clone was stimulated by synthetic random peptide libraries, in which every peptide position was synthesized with a mixture of amino acids. These libraries represented large numbers of individual peptides (2×10^{14} different sequences in an X11 library). It is not known what fraction of T-cell clones have such a degree of degeneracy since other MBP (85-99)-specific T-cell clones were not stimulated by such random peptide libraries. Notably, the random peptide libraries did not stimulate the T-cell clones that we have studied, indicating that such a level of degeneracy is not required for cross-reactivity with microbial peptides.

The data also indicate that combinatorial effects shape the peptide surface that can be recognized by a TCR. Algorithms that are based on single amino acid substitutions can be used to predict peptides that will bind with a high affinity to particular MHC class II molecules. In contrast, particular amino acid combinations can be important in creating peptide surfaces that can be recognized by a TCR (1). This notion is supported by the observation that the majority of microbial peptides that perfectly matched the MHC binding and TCR recognition motif did not stimulate the MBP-specific T-cell clones (27, 69). Identification of a complete set of peptide sequences that represent agonists for a TCR will require analysis of complex peptide libraries. At present, such analyses represent a technical challenge since a large number of peptides may need to be sequenced from phage display libraries or peptide libraries on beads. However, the complexity of the peptide repertoire that is recognized by an individual TCR may be greatly underestimated unless the combinatorial nature of peptide recognition by the TCR is taken into consideration.

STRUCTURAL FEATURES OF AUTOREACTIVE TCRs THAT CONTRIBUTE TO THE DEGREE OF CROSS-REACTIVITY

Structural aspects of human TCRs that allow activation of autoreactive T cells by diverse microbial peptides were examined for two MBP-specific T-cell clones (Ob.1A12 and Ob.2F3) that had identical $V\alpha$-$J\alpha$ and $V\beta$-$J\beta$ rearrangements and that differed only in the N-regions of TCR α and β (27). These two clones had similar fine specificities for the MBP peptide except for the P5 position of the peptide (K93). In the crystal structure of the HLA-DR2–MBP peptide complex, P5 lysine was a prominent, solvent-exposed residue in the center of the DR2-MBP peptide surface (Fig. 1) (57). The peptide recognition motif of both T-cell clones was very similar except for 93K. Three analogs of 93K (P, A, R) strongly stimulated clone Ob.2F3, while seven analogs (P, A, M, R, Q, S, T) stimulated clone Ob.1A12 (where the underscores indicate analogs that only stimulate clone Ob.1A12). For both clones, replacement of 93K by aromatic or negatively charged residues resulted in a major reduction or complete loss of T-cell activation (27).

As described above, five microbial peptides that activated clone Ob.1A12 were identified; three of these also activated clone Ob.2F3 (Table 2). The microbial peptides that activated clone Ob.1A12 had conservative or nonconservative changes at P5 (lysine to arginine, serine, or proline). In contrast, clone Ob.2F3 was activated only by those three peptides that had a conservative lysine-to-arginine substitution. The degree of specificity or degeneracy in recognition of the P5 side chain was the key difference between these TCRs since the *Escherichia coli-Haemophilus influenzae* peptide stimulated both clones when the P5 position was changed from serine to arginine. These results demonstrate that the CDR3 loops contribute to the degree of degeneracy in peptide recognition by human MBP-specific TCRs.

In the crystal structures of MHC class I–peptide complexes (9, 10, 15–17), the P5 side chain occupies a central pocket created by the CDR3 loops of $\alpha\beta$ TCRs. This suggests that the two MBP-specific TCRs differ in terms of the size and shape of the TCR pocket that accommodates 93K of the MBP peptide. The P5 pocket of the Ob.1A12 TCR may be more degenerate because a larger variety of amino acids can be accommodated.

A striking similarity was observed for the two human T-cell leukemia virus type 1 (HTLV-1) Tax-specific T-cell clones (A6 and B7) for which the crystal structure of the MHC-peptide-TCR complex has been determined (9, 15, 26). The B7 TCR was exquisitely specific for the P5 tyrosine of Tax (residues 11 to 19) since only aromatic substitutions were tolerated. Only the Y5F analog showed a dose-response similar to that of the wild-type peptide, while the Y5W analog showed an approximately 100-fold reduction in activity. In contrast, the A6 TCR was much more degenerate for this peptide position since 10 of 17 analog peptides induced lysis at low peptide concentrations (26). The absolute requirement for an aromatic side chain by the B7 T-cell clone could be explained by the structure of the P5 pocket (9, 15). In the B7 structure, the P5 tyrosine represented a tight fit for this TCR pocket, and the aromatic ring of the P5 tyrosine stacked against an aromatic residue of the CDR3 loop of the TCR β chain (Y104 β). In contrast, the A6 TCR had a larger P5 pocket than the B7 TCR, and the P5 tyrosine did not interact with an aromatic TCR residue. These data demonstrate that the TCR CDR3 loops can determine the degree of specificity and degeneracy of the central TCR pocket. Taken together, these examples illustrate structural features of TCRs that can contribute to the degree of specificity or cross-reactivity in peptide recognition.

GENERAL STRUCTURAL FEATURES OF TCR RECOGNITION THAT DETERMINE SPECIFICITY AND CROSS-REACTIVITY

The crystal structures of the human and murine MHC-peptide-TCR complexes that have been determined provide important general insights into the structural basis of TCR specificity and cross-reactivity. Crystal structures have been determined for two human TCRs that are specific for the HLA-A2-bound Tax peptide (residues 11 to 19) of HTLV-1, as well as for the murine 2C TCR that is specific for *H-2K*^{b} and the dEV8 self-peptide (9, 10, 15–17). These structures demonstrate several common features of MHC-peptide recognition by αβ TCRs. The first common feature that has been observed is a diagonal binding mode that buries most of the peptide in the MHC class I–TCR complex and that allows the TCR surface to interact with the peptide by fitting down between the highest points of the MHC helices. In all cases, the TCR made substantial contacts with the MHC helices. TCR contacts with conserved MHC residues may therefore account for the common diagonal orientation. The position of the Vα domains is very similar among these structures, suggesting that the Vα domain is important in defining the common diagonal orientation. In contrast, there is substantial variation in the relative position of the Vβ domain among these structures. Mutagenesis experiments indicate that such a diagonal binding mode is general (62).

A second common feature is the positioning of the TCR α and β chains over the N- and C-terminal segments of the peptide, respectively. The CDR3 loops of TCR α and β are located in the center of the TCR contact surface and create a deep central cavity. Apart from this central cavity, the contact surfaces of these TCRs are remarkably flat. An important general feature that accounts for TCR cross-reactivity is the fact that the bound peptide represents only a relatively small fraction of the total MHC-peptide contact surface with the TCR. In the HLA-A2–Tax–TCR structure, the peptide represented ~33% of the contact surface (15). In the murine 2C TCR complex, it represented only 22% of the MHC-peptide contact surface with the TCR (17).

The dEV8 self-peptide represents a ligand with a low affinity for the 2C TCR. This peptide allows positive selection of the 2C TCR in transgenic mice and is a weak agonist of the 2C TCR. The crystal structure demonstrates a poor fit of the 2C TCR with the *H-2K*^{b}–dEV8 complex. The interface between the peptide and the TCR exhibits extremely poor shape complementarity, and there are large empty spaces (17). Importantly, the hydrophobic central cavity created by the CDR3 loops of α and β is empty. This TCR pocket is filled with the P5 tyrosine of the HTLV-1 Tax peptide in the HLA-A2–Tax peptide–TCR structures. In addition, the CDR3 loop of the β chain shows only a minimal interaction with the dEV8 peptide. These results demonstrate that the fit with the peptide could be greatly improved. An interesting comparison of the bound and unbound 2C TCR structures was also made. Large conformational changes were observed in three of the TCR CDR loops as the result of binding to the MHC-peptide complex (17). An induced fit provides an additional mechanism for the observed structural plasticity of TCRs. An induced fit has also been observed in crystal structures of antibody-antigen complexes (52).

Recently, three additional structures of the HLA-A2–Tax-specific A6 TCR were reported in which the Tax peptide was replaced by three single-amino-acid analog peptides that represented partial agonists or antagonists (10). The three-dimensional structures of these A6-TCR–peptide–MHC complexes are remarkably similar to each other and to the wild-type agonist complex, with minor adjustments at the interface. Very small differences at the interface can therefore result in large differences in the biological response and in the half-life of the complex. Cross-reactivity is achieved by small structural adjustments in which the TCR, peptide, and MHC residues rearrange to better accommodate the altered peptide. In the V7R complex, the valine at position 7 has been replaced by arginine. This results in displacement of a six-amino-acid segment of the CDR3 loop of the β chain. Three glycine residues are present in this segment (at positions 97, 100, and 101), which may facilitate movement of this loop. In the Y8A complex, removal of the tyrosine ring and replacement by alanine removes the interaction of the Y8 hydroxyl group with the CDR2β residue Glu-30. This also results in a displacement of the CDR3β loop, as observed in the V7R structure (Gly-101 has moved 3.5 Å rather than 4.4 Å). In each complex, a new or enlarged cavity is found as a result of a less perfect fit of the TCR to the MHC-peptide complex. In the Y8A complex, such a cavity is found adjacent to the alanine that replaced Y8. These results demonstrate that the TCR CDR loops can be flexible and can thereby accommodate changes in the bound peptide.

IMPLICATIONS OF DEGENERATE PEPTIDE RECOGNITION FOR PATHOGENESIS OF HUMAN AUTOIMMUNE DISEASES

The results presented here demonstrate that degenerate peptide recognition is an inherent feature of TCR recognition and that TCR cross-reactivity is not a rare phenomenon. The diverse nature of the viral and bacterial peptides that stimulate MBP (85-99)-specific T-cell clones makes it unlikely that a single infectious agent is responsible for the initiation of autoimmunity by molecular mimicry. Since there are several potential mechanisms for the activation of autoreactive T cells, it is also unlikely that a single mechanism is solely responsible for the induction of human autoimmune diseases.

The diverse nature of the molecular mimicry peptides and the ubiquitous presence of some of these pathogens may make it difficult to establish a direct epidemiological link between infectious agents and the occurrence of certain autoimmune diseases. In particular, the temporal relationship between an infection and the development of an autoimmune process may in many cases not be clear because of the time that it can take until clinical symptoms become obvious and a diagnosis is made. In MS, a clinical diagnosis is frequently made at a time when magnetic resonance imaging scans demonstrate a number of old lesions, indicating that the disease process has already been active for months or even years (59).

However, recent data have demonstrated that certain autoimmune diseases are clearly associated with preceding infections. For example, it has been clearly demonstrated that infection with *Campylobacter jejuni* precedes the occurrence of Guillain-Barré syndrome, an acute demyelinating disease of the peripheral nervous system. Infection with *Campylobacter* was demonstrated in approximately a third of new cases and in 2% of the appropriate household controls (51). Chapter 12 in this volume, by Gross and Huber, reviews recent data on the pathogenesis of chronic Lyme arthritis that is caused by *Borrelia burgdorferi* (19). These new observations are beginning to define the epidemiology as well as the molecular mechanisms responsible for the development of human T-cell-mediated inflammatory diseases.

Acknowledgments. We acknowledge the important contributions that our colleagues and collaborators have made toward this research. In particular, we acknowledge the contributions of Don C. Wiley, Jack L. Strominger, David A. Hafler, Katherine Smith, Stefan Hausmann, Laurent Gauthier, and Jason Pyrdol. This work was supported by grants from the National Multiple Sclerosis Society and the National Institutes of Health.

REFERENCES

1. **Ausubel, L. J., C. K. Kwan, A. Sette, V. Kuchroo, and D. A. Hafler.** 1996. Complementary mutations in an antigenic peptide allow for cross-reactivity of autoreactive T-cell clones. *Proc. Natl. Acad. Sci. USA* **93:**15317–15322.
2. **Bhardwaj, V., V. Kumar, H. M. Geysen, and E. E. Sercarz.** 1993. Degenerate recognition of a dissimilar antigenic peptide by myelin basic protein-reactive T cells. Implications for thymic education and autoimmunity. *J. Immunol.* **151:**5000–5010.
3. **Brocke, S., A. Gaur, C. Piercy, A. Gautam, K. Gijbels, C. G. Fathman, and L. Steinman.** 1993. Induction of relapsing paralysis in experimental autoimmune encephalomyelitis by bacterial superantigen. *Nature* **365:**642–644.
4. **Brown, J. H., T. S. Jardetzky, J. C. Gorga, L. J. Stern, R. G. Urban, J. L. Strominger, and D. C. Wiley.** 1993. Three-dimensional structure of the human class II histocompatibility antigen HLA-DR1. *Nature* **364:**33–39.
5. **Busch, R., C. M. Hill, J. D. Hayball, J. R. Lamb, and J. B. Rothbard.** 1991. Effect of natural polymorphism at residue 86 of the HLA-DRβ chain on peptide binding. *J. Immunol.* **147:**1292–1298.
6. **Chicz, R. M., R. G. Urban, J. C. Gorga, D. A. Vignali, W. S. Lane, and J. L. Strominger.** 1993. Specificity and promiscuity among naturally processed peptides bound to HLA-DR alleles. *J. Exp. Med.* **178:**27–47.
7. **Datta, A. K., R. J. Feighny, and J. S. Pagano.** 1980. Induction of Epstein-Barr virus-associated DNA polymerase by 12-O-tetradecanoylphorbol-13-acetate. Purification and characterization. *J. Biol. Chem.* **255:**5120–5125.
8. **Dessen, A., C. M. Lawrence, S. Cupo, D. M. Zaller, and D. C. Wiley.** 1997. X-ray crystal structure of HLA-DR4 (DRA*0101, DRB1*0401) complexed with a peptide from human collagen II. *Immunity* **7:**473–481.
9. **Ding, Y. H., K. J. Smith, D. N. Garboczi, U. Utz, W. E. Biddison, and D. C. Wiley.** 1998. Two human T cell receptors bind in a similar diagonal mode to the HLA-A2/Tax peptide complex using different TCR amino acids. *Immunity* **8:**403–411.
10. **Ding, Y. H., B. M. Baker, D. N. Garboczi, W. E. Biddison, and D. C. Wiley.** 1999. Four A6-TCR/peptide/HLA-A2 structures that generate very different T cell signals are nearly identical. *Immunity* **11:**45–56.
11. **Ebers, G. C., K. Kukay, D. E. Bulman, A. D. Sadovnick, G. Rice, C. Anderson, H. Armstrong, K. Cousin, R. B. Bell, W. Hader, D. W. Paty, S. Hashimoto, J. Oger, P. Duquette, S. Warren, T. Gray, P. O'Connor, A. Nath, A. Auty, L. Metz, G. Francis, J. E. Paulseth, T. J. Murray, W. Pryse-Phillips, R. Nelson, M. Freedman, D. Brunet, J.-P. Bouchard, D. Hinds, and N. Risch.** 1996. A full genome search in multiple sclerosis. *Nat. Genet.* **13:**472–476.
12. **Evavold, B. D., J. Sloan-Lancaster, K. J. Wilson, J. B. Rothbard, and P. M. Allen.** 1995. Specific T cell

recognition of minimally homologous peptides: evidence for multiple endogenous ligands. *Immunity* **2:**655–663.
13. **Fugger, L., J. B. Rothbard, and G. Sonderstrup-McDevitt.** 1996. Specificity of an HLA-DRB1*0401-restricted T cell response to type II collagen. *Eur. J. Immunol.* **26:**928–933.
14. **Fujinami, R. S., and M. B. Oldstone.** 1985. Amino acid homology between the encephalitogenic site of myelin basic protein and virus: mechanism for autoimmunity. *Science* **230:**1043–1045.
15. **Garboczi, D. N., P. Ghosh, U. Utz, Q. R. Fan, W. E. Biddison, and D. C. Wiley.** 1996. Structure of the complex between human T-cell receptor, viral peptide and HLA-A2. *Nature* **384:**134–141.
16. **Garcia, K. C., M. Degano, R. L. Stanfield, A. Brunmark, M. R. Jackson, P. A. Peterson, L. Teyton, and I. A. Wilson.** 1996. An αβ T cell receptor structure at 2.5 Å and its orientation in the TCR-MHC complex. *Science* **274:**209–219.
17. **Garcia, K. C., M. Degano, L. R. Pease, M. Huang, P. A. Peterson, L. Teyton, and I. A. Wilson.** 1998. Structural basis of plasticity in T cell receptor recognition of a self peptide-MHC antigen. *Science* **279:**1166–1172.
18. **Gauthier, L., K. J. Smith, J. Pyrdol, A. Kalandadze, J. L. Strominger, D. C. Wiley, and K. W. Wucherpfennig.** 1998. Expression and crystallization of the complex of HLA-DR2 (DRA, DRB1*1501) and an immunodominant peptide of human myelin basic protein. *Proc. Natl. Acad. Sci. USA* **95:**11828–11833.
19. **Gross, D. M., T. Forsthuber, M. Tary-Lehmann, C. Etling, K. Ito, Z. A. Nagy, J. A. Field, A. C. Steere, and B. T. Huber.** 1998. Identification of LFA-1 as a candidate autoantigen in treatment-resistant Lyme arthritis. *Science* **281:**703–706.
20. **Hagerty, D. T., and P. M. Allen.** 1995. Intramolecular mimicry. Identification and analysis of two cross-reactive T cell epitopes within a single protein. *J. Immunol.* **155:**2993–3001.
21. **Haines, J. L., M. Ter-Minassian, A. Bazyk, J. F. Gusella, D. J. Kim, H. Terwedow, M. A. Pericak-Vance, J. B. Rimmler, C. S. Haynes, A. D. Roses, A. Lee, B. Shaner, M. Menold, E. Seboun, R. P. Fitoussi, C. Gartioux, C. Reyes, F. Ribierre, G. Gyapay, J. Weissenbach, S. L. Hauser, D. E. Goodkin, R. Lincoln, K. Usuku, and J. R. Oksenberg.** 1996. A complete genomic screen for multiple sclerosis underscores a role for the major histocompatibility complex. The Multiple Sclerosis Genetics Group. *Nat. Genet.* **13:**469–471.
22. **Haines, J. L., H. A. Terwedow, K. Burgess, M. A. Pericak-Vance, J. B. Rimmler, E. R. Martin, J. R. Oksenberg, R. Lincoln, D. Y. Zhang, D. R. Banatao, N. Gatto, D. E. Goodkin, and S. L. Hauser.** 1998. Linkage of the MHC to familial multiple sclerosis suggests genetic heterogeneity. The Multiple Sclerosis Genetics Group. *Hum. Mol. Genet.* **7:**1229–1234.
23. **Hammer, J., E. Bono, F. Gallazzi, C. Belunis, Z. Nagy, and F. Sinigaglia.** 1994. Precise prediction of major histocompatibility complex class II-peptide interaction based on peptide side chain scanning. *J. Exp. Med.* **180:**2353–2358.
24. **Hammer, J., F. Gallazzi, E. Bono, R. W. Karr, J. Guenot, P. Valsasnini, Z. A. Nagy, and F. Sinigaglia.** 1995. Peptide binding specificity of HLA-DR4 molecules: correlation with rheumatoid arthritis association. *J. Exp. Med.* **181:**1847–1855.
25. **Hausmann, S., and K. W. Wucherpfennig.** 1997. Activation of autoreactive T cells by peptides from human pathogens. *Curr. Opin. Immunol.* **9:**831–838.
26. **Hausmann, S., W. E. Biddison, K. J. Smith, Y. H. Ding, D. N. Garboczi, U. Utz, D. C. Wiley, and K. W. Wucherpfennig.** 1999. Peptide recognition by two HLA-A2/Tax11-19-specific T cell clones in relationship to their MHC/peptide/TCR crystal structures. *J. Immunol.* **162:**5389–5397.
27. **Hausmann, S., M. Martin, L. Gauthier, and K. W. Wucherpfennig.** 1999. Structural features of autoreactive TCR that determine the degree of degeneracy in peptide recognition. *J. Immunol.* **162:**338–344.
28. **Hemmer, B., B. T. Fleckenstein, M. Vergelli, G. Jung, H. McFarland, R. Martin, and K. H. Wiesmuller.** 1997. Identification of high potency microbial and self ligands for a human autoreactive class II-restricted T cell clone. *J. Exp. Med.* **185:**1651–1659.
29. **Hickey, W. F.** 1991. Migration of hematogenous cells through the blood-brain barrier and the initiation of CNS inflammation. *Brain Pathol.* **1:**97–105.
30. **Hillert, J., T. Kall, M. Vrethem, S. Fredrikson, M. Ohlson, and O. Olerup.** 1994. The HLA-Dw2 haplotype segregates closely with multiple sclerosis in multiplex families. *J. Neuroimmunol.* **50:**95–100.
31. **Hunt, D. F., R. A. Henderson, J. Shabanowitz, K. Sakaguchi, H. Michel, N. Sevilir, A. L. Cox, E. Appella, and V. H. Engelhard.** 1992. Characterization of peptides bound to the class I MHC molecule HLA-A2.1 by mass spectrometry. *Science* **255:**1261–1263.

32. **Ignatowicz, L., J. Kappler, and P. Marrack.** 1996. The repertoire of T cells shaped by a single MHC/peptide ligand. *Cell* **84:**521–529.
33. **Ignatowicz, L., W. Rees, R. Pacholczyk, H. Ignatowicz, E. Kushnir, J. Kappler, and P. Marrack.** 1997. T cells can be activated by peptides that are unrelated in sequence to their selecting peptide. *Immunity* **7:**179–186.
34. **Jardetzky, T. S., J. H. Brown, J. C. Gorga, L. J. Stern, R. G. Urban, J. L. Strominger, and D. C. Wiley.** 1996. Crystallographic analysis of endogenous peptides associated with HLA-DR1 suggests a common, polyproline II-like conformation for bound peptides. *Proc. Natl. Acad. Sci. USA* **93:**734–738.
35. **Kalandadze, A., M. Galleno, L. Foncerrada, J. L. Strominger, and K. W. Wucherpfennig.** 1996. Expression of recombinant HLA-DR2 molecules. Replacement of the hydrophobic transmembrane region by a leucine zipper dimerization motif allows the assembly and secretion of soluble DR αβ heterodimers. *J. Biol. Chem.* **271:**20156–20162.
36. **Kersh, G. J., and P. M. Allen.** 1996. Essential flexibility in the T-cell recognition of antigen. *Nature* **380:**495–498.
37. **Kirberg, J., A. Berns, and H. von Boehmer.** 1997. Peripheral T cell survival requires continual ligation of the T cell receptor to major histocompatibility complex-encoded molecules. *J. Exp. Med.* **186:**1269–1275.
38. **Kozono, H., J. White, J. Clements, P. Marrack, and J. Kappler.** 1994. Production of soluble MHC class II proteins with covalently bound single peptides. *Nature* **369:**151–154.
39. **Kurtzke, J. F.** 1980. Epidemiologic contributions to multiple sclerosis: an overview. *Neurology* **30:**61–79.
40. **Lehmann, P. V., T. Forsthuber, A. Miller, and E. E. Sercarz.** 1992. Spreading of T-cell autoimmunity to cryptic determinants of an autoantigen. *Nature* **358:**155–157.
41. **Marsh, S. G., and J. G. Bodmer.** 1995. HLA class II region nucleotide sequences. *Tissue Antigens* **46:**258–280.
42. **Martin, R., D. Jaraquemada, M. Flerlage, J. Richert, J. Whitaker, E. O. Long, D. E. McFarlin, and H. F. McFarland.** 1990. Fine specificity and HLA restriction of myelin basic protein-specific cytotoxic T cell lines from multiple sclerosis patients and healthy individuals. *J. Immunol.* **145:**540–548.
43. **Nanda, N. K., K. K. Arzoo, H. M. Geysen, A. Sette, and E. E. Sercarz.** 1995. Recognition of multiple peptide cores by a single T cell receptor. *J. Exp. Med.* **182:**531–539.
44. **Nepom, G. T., and H. Erlich.** 1991. MHC class-II molecules and autoimmunity. *Annu. Rev. Immunol.* **9:**493–525.
45. **Oksenberg, J. R., E. Seboun, and S. L. Hauser.** 1996. Genetics of demyelinating diseases. *Brain Pathol.* **6:**289–302.
46. **Oldstone, M. B. A.** 1998. Molecular mimicry and immune-mediated diseases. *FASEB J.* **12:**1255–1265.
47. **Ota, K., M. Matsui, E. L. Milford, G. A. Mackin, H. L. Weiner, and D. A. Hafler.** 1990. T-cell recognition of an immunodominant myelin basic protein epitope in multiple sclerosis. *Nature* **346:**183–187.
48. **Peterson, J. D., B. Pike, M. McDuffie, and K. Haskins.** 1994. Islet-specific T cell clones transfer diabetes to nonobese diabetic (NOD) F1 mice. *J. Immunol.* **153:**2800–2806.
49. **Pette, M., K. Fujita, D. Wilkinson, D. M. Altmann, J. Trowsdale, G. Giegerich, A. Hinkkanen, J. T. Epplen, L. Kappos, and H. Wekerle.** 1990. Myelin autoreactivity in multiple sclerosis: recognition of myelin basic protein in the context of HLA-DR2 products by T lymphocytes of multiple-sclerosis patients and healthy donors. *Proc. Natl. Acad. Sci. USA* **87:**7968–7972.
50. **Reay, P. A., R. M. Kantor, and M. M. Davis.** 1994. Use of global amino acid replacements to define the requirements for MHC binding and T cell recognition of moth cytochrome *c* (93–103). *J. Immunol.* **152:**3946–3957.
51. **Rees, J. H., S. E. Soudain, N. A. Gregson, and R. A. Hughes.** 1995. Campylobacter jejuni infection and Guillain-Barré syndrome. *N. Engl. J. Med.* **333:**1374–1379.
52. **Rini, J. M., U. Schulze-Gahmen, and I. A. Wilson.** 1992. Structural evidence for induced fit as a mechanism for antibody-antigen recognition. *Science* **255:**959–965.
53. **Rosloniec, E. F., D. D. Brand, L. K. Myers, K. B. Whittington, M. Gumanovskaya, D. M. Zaller, A. Woods, D. M. Altmann, J. M. Stuart, and A. H. Kang.** 1997. An HLA-DR1 transgene confers susceptibility to collagen-induced arthritis elicited with human type II collagen. *J. Exp. Med.* **185:**1113–1122.
54. **Rosloniec, E. F., D. D. Brand, L. K. Myers, Y. Esaki, K. B. Whittington, D. M. Zaller, A. Woods, J. M. Stuart, and A. H. Kang.** 1998. Induction of autoimmune arthritis in HLA-DR4 (DRB1*0401) transgenic mice by immunization with human and bovine type II collagen. *J. Immunol.* **160:**2573–2578.
55. **Scharf, S. J., A. Friedmann, C. Brautbar, F. Szafer, L. Steinman, G. Horn, U. Gyllensten, and H. A.**

Erlich. 1988. HLA class II allelic variation and susceptibility to pemphigus vulgaris. *Proc. Natl. Acad. Sci. USA* **85:**3504–3508.

56. **Scholz, C., K. T. Patton, D. E. Anderson, G. J. Freeman, and D. A. Hafler.** 1998. Expansion of autoreactive T cells in multiple sclerosis is independent of exogenous B7 costimulation. *J. Immunol.* **160:**1532–1538.
57. **Smith, K. J., J. Pyrdol, L. Gauthier, D. C. Wiley, and K. W. Wucherpfennig.** 1998. Crystal structure of HLA-DR2 (DRA*0101, DRB1*1501) complexed with a peptide from human myelin basic protein. *J. Exp. Med.* **188:**1511–1520.
58. **Soos, J. M., J. Schiffenbauer, and H. M. Johnson.** 1993. Treatment of PL/J mice with the superantigen, staphylococcal enterotoxin B, prevents development of experimental allergic encephalomyelitis. *J. Neuroimmunol.* **43:**39–43.
59. **Stadt, D., L. Kappos, E. Rohrbach, R. Heun, and M. Ratzka.** 1990. Occurrence of MRI abnormalities in patients with isolated optic neuritis. *Eur. Neurol.* **30:**305–309.
60. **Steinman, L.** 1996. Multiple sclerosis: a coordinated immunological attack against myelin in the central nervous system. *Cell* **85:**299–302.
61. **Stern, L. J., J. H. Brown, T. S. Jardetzky, J. C. Gorga, R. G. Urban, J. L. Strominger, and D. C. Wiley.** 1994. Crystal structure of the human class II MHC protein HLA-DR1 complexed with an influenza virus peptide. *Nature* **368:**215–221.
62. **Sun, R., S. E. Shepherd, S. S. Geier, C. T. Thomson, J. M. Sheil, and S. G. Nathenson.** 1995. Evidence that the antigen receptors of cytotoxic T lymphocytes interact with a common recognition pattern on the H-2K^b molecule. *Immunity* **3:**573–582.
63. **Tanchot, C., F. A. Lemonnier, B. Perarnau, A. A. Freitas, and B. Rocha.** 1997. Differential requirements for survival and proliferation of CD8 naive or memory T cells. *Science* **276:**2057–2062.
64. **Ufret-Vincenty, R. L., L. Quigley, N. Tresser, S. H. Pak, A. Gado, S. Hausmann, K. W. Wucherpfennig, and S. Brocke.** 1998. In vivo survival of viral antigen-specific T cells that induce experimental autoimmune encephalomyelitis. *J. Exp. Med.* **188:**1725–1738.
65. **Vogt, A. B., H. Kropshofer, H. Kalbacher, M. Kalbus, H. G. Rammensee, J. E. Coligan, and R. Martin.** 1994. Ligand motifs of HLA-DRB5*0101 and DRB1*1501 molecules delineated from self-peptides. *J. Immunol.* **153:**1665–1673.
66. **Wucherpfennig, K. W., J. Zhang, C. Witek, M. Matsui, Y. Modabber, K. Ota, and D. A. Hafler.** 1994. Clonal expansion and persistence of human T cells specific for an immunodominant myelin basic protein peptide. *J. Immunol.* **152:**5581–5592.
67. **Wucherpfennig, K. W., A. Sette, S. Southwood, C. Oseroff, M. Matsui, J. L. Strominger, and D. A. Hafler.** 1994. Structural requirements for binding of an immunodominant myelin basic protein peptide to DR2 isotypes and for its recognition by human T cell clones. *J. Exp. Med.* **179:**279–290.
68. **Wucherpfennig, K. W., D. A. Hafler, and J. L. Strominger.** 1995. Structure of human T-cell receptors specific for an immunodominant myelin basic protein peptide: positioning of T-cell receptors on HLA-DR2/peptide complexes. *Proc. Natl. Acad. Sci. USA* **92:**8896–8900.
69. **Wucherpfennig, K. W., and J. L. Strominger.** 1995. Molecular mimicry in T cell-mediated autoimmunity: viral peptides activate human T cell clones specific for myelin basic protein. *Cell* **80:**695–705.
70. **Wucherpfennig, K. W., and J. L. Strominger.** 1995. Selective binding of self peptides to disease-associated major histocompatibility complex (MHC) molecules: a mechanism for MHC-linked susceptibility to human autoimmune diseases. *J. Exp. Med.* **181:**1597–1601.
71. **Zamvil, S., P. Nelson, J. Trotter, D. Mitchell, R. Knobler, R. Fritz, and L. Steinman.** 1985. T-cell clones specific for myelin basic protein induce chronic relapsing paralysis and demyelination. *Nature* **317:**355–358.
72. **Zamvil, S. S., and L. Steinman.** 1990. The T lymphocyte in experimental allergic encephalomyelitis. *Annu. Rev. Immunol.* **8:**579–621.

Molecular Mimicry, Microbes, and Autoimmunity
Edited by M. W. Cunningham and R. S. Fujinami

Chapter 15

Molecular Mimicry and Determinant Spreading

Anthony Quinn and Eli E. Sercarz

The question of how autoimmunity is initiated and maintained has been argued for a considerable time in the literature. Whether there is a microbial instigator which induces a lymphocyte response cross-reactive with a self-determinant (molecular mimicry) (8, 24) or whether the prime mover relates mainly to an environmental inflammatory event, or both, is still under debate. In both cases, propagation of disease then depends on determinant spreading (14). Certainly, strong arguments can be made for both positions. For example, acute rheumatic fever, which represents one of the best examples of molecular mimicry, is an autoimmune disease that is precipitated by exposure to an infectious agent which colonizes an anatomical site distinct from the afflicted organ. The onset of rheumatic carditis is preceded by a pharyngeal infection with certain serotypes of group A streptococcus; those serotypes associated with skin infections apparently lack rheumatogenic properties. Strikingly, while antibodies and T cells responsive to streptococcal antigens are detected in the blood and hearts of patients with rheumatic carditis, the bacterium itself has not been shown to be cardiotropic, and in many instances, the organism cannot be isolated from affected patients (7). Therefore, it seems unlikely that the observed myocarditis and valvulitis are collateral damage resulting from the initial bacterial insult. On the other hand, coxsackievirus B3, which is also associated with immune-mediated myocarditis (11), has been shown to be cardiotropic, and viral RNA can be detected in the myocardium (6, 34). Although the precise mechanisms responsible for myocyte death in coxsackievirus-induced myocarditis are not completely resolved, the presence of the virus in the myocardium may be sufficient to provoke an immune response that could progress into autoimmunity and morbidity. Probably just as important as the foregoing factors is whether regulation of the self-reactive response supervenes, preventing the further propagation of autoimmunity and pathogenesis. Clearly, the progression from infection to autoimmunity to autoimmune disease is a rare event such that the majority of individuals exposed to group A streptococcus, including those who may develop autoreactive antibodies, do not develop rheumatic heart disease. The evidence to date suggests that the autoimmune response must surpass some threshold (quantitative or qualitative) to sufficiently perturb immunological homeostasis and create a state of dysregulation and autoimmune disease (16).

Anthony Quinn and Eli E. Sercarz • Division of Immune Regulation, La Jolla Institute for Allergy and Immunology, 10355 Science Center Drive, San Diego, CA 92121.

REGULATORY T CELLS AND TRANSIENT AUTOIMMUNITY

Transience of autoimmune symptoms might be commonplace in the acquisition of resistance to certain autoimmune diseases (17). In the regulation of experimental autoimmune encephalomyelitis in the B10.PL mouse, the animal injected with myelin basic protein or its amino-terminal peptide becomes paralyzed for 1 to 2 weeks and then manages a complete recovery (17). Furthermore, the mouse gains resistance to subsequent attempts to reintroduce the disease! This resistance has proved to be dependent on raising endogenous regulatory CD4 and CD8 T cells, which are specific for Vβ8.2 T-cell receptor (TCR) determinants on the aggressive autoreactive T cell (17). In this mouse model, the regulatory TCR-reactive T cells appear spontaneously without the necessity to introduce the TCR as an experimental immunogen. It is not known whether this type of scenario occurs with all Vβ TCRs, but if it is frequent, then a crucial aspect of the induction of autoimmunity would be the failure to engage the appropriate regulatory populations.

The response to foreign heat shock proteins such as hsp65 of *Mycobacterium tuberculosis* (Bhsp) in the Lewis rat is extremely broad (21). The initial response to Bhsp diversifies from the N terminus and central portion of the molecule to the C-terminal determinants (CTDs), owing to enhanced processing of both Bhsp and the ever-present, cross-reactive rat protein Rhsp65. Intramolecular spreading permits the emergence of new T cells reactive against bacterial CTD (BCTD) as well as rat CTD (RCTD). Since earlier experiments demonstrated the regulatory nature of these C-terminal determinants (21), spreading in this case leads to a tremendous diversification as well as a turning off of the whole response to the heat shock protein. Interestingly, WKY rats that are resistant to adjuvant arthritis enlist these CTD-specific regulatory cells in the early phase of their response to bacterial hsp65. Thus, the susceptibility to autoimmune disease on the part of Lewis rats may not hinge solely on the presence of pathogenic immune cells but on the late enlistment of regulatory cells. In this instance, it is likely that Lewis rats fail to render regulation-inducing determinants available by antigen processing in a timely manner.

REMOVAL OF REGULATORY CELLS

Evidence for this failure has recently been described in the removal of a regulatory population of $CD25^+$ T cells (12). When this is accomplished by treatment of the autoimmune-susceptible individual with anti-CD25, there is a spontaneous appearance of a variety of autoimmune conditions, as well as the competence to raise antitumor responsiveness (26). These data strongly imply that a suppressive force continually monitors the activation of a variety of self-reactive responses. Fitting this same category is the control exerted by the B7 coreceptor, CTLA-4, which has recently been shown (3) to downregulate responses or, through its inhibition, to potentiate antitumor responses.

SPONTANEOUS AUTOIMMUNITY IN THE NOD MOUSE

To return to preregulatory events, however, one of the most intensively studied animal models of autoimmunity is insulin-dependent diabetes mellitus (IDDM) in the nonobese diabetic (NOD) mouse. It is rather similar to IDDM in humans in that the disease is often contracted before the end of adolescence and there is a response directed against several self-molecules found in the pancreas. In the NOD mouse, these different responses seem to arise sequentially, and the first response appears to be directed against the glutamic acid decarboxylase (GAD) molecule, followed by responses to determinants of insulin, heat shock protein 60, carboxypeptidase H, and several others (13). Accordingly, the question of whether there was a single initiator in the response, followed by determinant spreading to other epitopes on the initiating antigen as well as intermolecular spreading to other antigens, was raised (18–20, 28, 29). This question not only is pertinent in terms of spreading, but it also relates to the means of induction of disease (14). This is because if a mimicry event is involved, it would likely involve a single set of T cells directed to the mimicked determinant on a single protein, followed by spreading to other determinants and other molecules. However, if there was a cataclysmic inflammatory event or a microbial incursion which initiated responsiveness to a variety of newly expressed immunogenic determinants, mimicry need not be involved at all. Any proinflammatory stimulus that upregulates expression of costimulators, major histocompatibility complex (MHC) class I and II molecules, as well as enhanced antigen processing, thereby revealing new immunogenic self-determinants, would be sufficient to induce autoimmunity. Cytokines found in the inflammatory site, such as interleukin 6, may actually alter antigen processing and yield an altered pattern of determinant display on activated antigen-presenting cells (APCs) (5). In the NOD mouse model, even at 3 weeks of life, evidence exists for spontaneous reactivity to GAD, a suggestion that some unusual event in GAD processing could be at the crux of disease initiation. Several viral agents are suspected in the induction of diabetes, including coxsackievirus, which shares a region of sequence similarity with GAD65 (1). In a case of molecular mimicry it would usually be necessary for a dominant antigenic determinant associated with a microbial agent to cross-react with a cryptic determinant in the host; otherwise, these recipient T cells should have been rendered tolerant during development. Alternatively, the microbial mimic could behave as an altered-peptide ligand, possibly providing stimulation strong enough to deviate benign Th2 responses into proinflammatory Th1 cells (15).

LOCAL INFLAMMATION FOLLOWED BY DETERMINANT SPREADING

It is evident that a scenario need not involve a molecular mimicry event which initiates the autoimmunity. Local inflammation, followed by determinant spreading, is sufficient to propagate disease, so long as regulatory cells do not interfere to abruptly end the outbreak. One excellent example has been worked out in the Theiler's virus (Theiler's murine encephalomyelitis virus [TMEV]) infection system in SJL mice (4, 30, 32, 33): the viral infection causes local disease and subsequent inflammation, where there is no evidence for a mimicking cross-reactivity between viral and host antigenic determinants. Infection with TMEV leads to a demyelinating disease whose onset is slow and which does not peak until 6 months after infection (4). TMEV determinants do not cross-react with determinants on any of the known myelin antigens. When minimal clinical and histological signs of infection are first seen at about day 40, the response is directed solely against TMEV epitopes and cannot be raised against proteolipid protein determinants. Responsiveness against myelin determinants occurs only after virally induced local myelin destruction is evident, so that at 80 to 95 days after TMEV infection, when there is severe disease, a response to many of the dominant and subdominant central nervous system determinants can be raised without the necessity for antigen pulsing. Another rather similar instance has been reported by Horwitz et al. (10), in which pancreatic autoantigens are released following the occurrence of damage caused by infection with coxsackievirus, and IDDM follows. These illustrative examples suggest that some care and analysis need be applied to situations which resemble mimicry, followed by determinant spreading and an absence of regulation efficient enough to permit chronic disease to persist. Nevertheless, owing to TCR degeneracy, to be discussed below, it may often be difficult to exclude cross-reactivity without extensive testing.

TRUE DEGENERACY IN TCR RECOGNITION

Degeneracy, redundancy, and pleiotropy are characteristics that are widespread in the immune system (e.g., see Cohen [3a]). Apparently, for the system to function optimally, it is necessary to avoid the establishment of only one way of accomplishing its ends. In many cases, apparent redundancy occurs and there are pleiotropic effector molecules that act on different targets. It can be imagined that there has been selection for maintenance of a variety of approaches, and therefore, with regard to survival and to economy of function, the diversity of tactics usable in the system still fit the requirements of Occam's razor. At an earlier time, the molecular mimicry hypothesis might have been considered less likely than it is today, simply because it was not fully accepted that the TCR displays a degeneracy of specificity. Thus, each TCR definitely has discrete requirements for a particular topography of interaction, which ordinarily includes some residues from the MHC molecule and others from the peptide, especially those pointing upward from the MHC binding groove. However, that topography is not unique to one molecular complex and can be simulated accurately enough to permit binding to the same TCR. In our laboratory, having encountered an early example of this degeneracy, we were aware of the phenomenon and therefore came across many examples. In our first paper that described degeneracy (2), we sought to discriminate between "molecular mimicry" and "degeneracy of TCR recognition." In the few examples up until that time, mimicry in the T-cell compartment involved the finding of peptide determinants from diverse organisms which shared several (usually four to six) amino acid residues and immunological cross-reactivity. In the case that we were examining, in the response to the amino-terminal determinant on myelin basic protein (2), there appeared to be a sharing of only one to two residues, and even then they were not identical but only homologous: such mimicry was of a different order and had to involve "shape mimicry" and aspects of degenerate recognition. Apparently, many different ligands were redundantly able to bind to the same TCR and activate the T cell. Moreover, this did not necessarily imply any flexibility of recognition by the T cell, although it was not excluded.

One of the most surprising cases of degenerate recognition appeared in our laboratory at this same time. It was observed by Hongkui Deng in the course of his experiments with a hen lysozyme (HEL)-specific T-cell hybridoma, activated by p20-35/A^d (unpublished data). The hybridoma also could be turned on by A^k APCs in the presence of a component of the culture medium, which contained fetal bovine serum. It was later shown that this peptide was a single 15-mer from a set of overlapping peptides that covered the amino-terminal half of fetal bovine serum albumin residues which no doubt was cross-reactive with fetal mouse serum albumin. A third ligand for this T cell included a still unknown peptide from the M12 APC, recognized in the context of E^d. (Several peaks eluted from E^d could stimulate the hybridoma, but the analysis could not discriminate the peptides within these peaks.) Thus, three different peptides within three different MHC class II molecules could serve as alternative ligands for this T cell, an indication of a degree of redundancy that exceeded our expectations: (MHC[A] + AgX = MHC[B] + AgY = MHC[C] + AgZ). Recently, a similar level of degeneracy has been described for the 2c receptor (27). As summarized in Table 1, the concept of degeneracy in recognition is already infused into many areas of immunology.

Approaching a Description of the Bounds of Specificity in Degenerate Recognition

We particularly focused on two areas that might yield to experimental analysis in arriving at an understanding of the degeneracy of T-cell recognition. In the first, positive selection, we asked whether it was possible to select for an E^d-HEL-specific set of clones with a non-E^d-binding peptide in the thymic epithelium. This was accomplished with an epithelial knockout of E^d, requiring that any positive selection would occur on the A^d molecule. Nevertheless, even with this nonhomologous MHC class restriction molecule, about 20% of the usual response to E^d-HEL arose, indicating that it was possible to find this level of degeneracy within a heterogeneous population of T cells (23). This was supporting evidence for the $A + X = B + Y = C + Z$ experiments, again indicating a perceived degenerate recognition of MHC molecules of different haplotypes.

Flexibility of T-Cell Recognition

A second question was whether it was possible to affect the recognition by the TCR even by changing the residues that flank the site of MHC binding of the peptide within the groove. This proved to be the case, demonstrating a surprising level of flexibility by the TCR. In addition to the ability to recognize substitutions of residues within the same register of a determinant region, the TCR also possesses a quality that could be considered "flexibility of core recognition." The core of an antigenic determinant had been defined as the essential portion of the determinant required by all T cells that respond to that determinant envelope (9). We sought to stretch this definition with respect to the core amino acids of a sperm whale myoglobin determinant, A^d-110-118, in the BALB/c mouse. We asked whether the peptide from residues 105 to 116, which was not recognized by the tested I-A^d-restricted T-cell hybridoma, could be altered to induce a response. Indeed, among 180

Table 1. Examples of degeneracy, redundancy, and pleiotropy

- Positive selection takes place by low-level recognition by a T cell of an unrelated self-peptide in an MHC groove
- Cross-reactivity of TCRs for a dominant viral peptide and a cryptic self-peptide can lead to autoimmune sequelae
- Finding of T-cell hybridomas which can specifically respond to several MHCs, each with a different peptide attached
- Combinatorial peptide libraries reveal peptide superagonists with complete amino acid differences from a known TCR ligand
- B1 antibodies can bind to a large variety of ligands
- Public "regulatory" idiotopes appear on antigen-specific receptors of B cells with different specificities
- Heterodimer receptors for different cytokines of the same family often share one receptor chain

variants of this 12-mer, eight different peptides were stimulatory to the hybridoma, at rather high levels. Further extension down toward residues 102 to 113 with other mutations revealed more agonist peptide cores for this hybridoma. Eventually, five distinct core determinants had been identified, demonstrating a remarkable flexibility of T-cell recognition (22). This so-called flexibility appears to be a separate aspect of degeneracy and might also be considered a type of epitope shift (31) that would contribute to maximization of determinant spreading under conditions of inflammation.

This state of affairs suggests that it is not necessary to postulate an enormous set of T-cell rearrangements to permit the T-cell repertoire to enter into all of its involvements. Recent work with combinatorial peptide libraries indicates that there are many mimics that are entirely different from the starting material and yet that activate the T cell (25). Altered peptide ligands, which usually differ in a small number of residues from the original ligand, have been shown to exert quite a variety of effects, from partial agonism to antagonism, so it is obvious that completely different ligands might exert a much wider spectrum of effects.

SUMMARY

In summary, we have considered how molecular mimicry, when viewed in the broader context of degeneracy of T-cell recognition specificity, provides a rationale for the existence of frequent autoimmunity. Whether the initiating event is a viral infection or inflammation at a site where APCs become ready to present and process self-antigens or foreign antigens efficiently, the commonality is the presentation of unusual, generally nondisplayed antigenic determinants to T cells which have evaded the mechanisms of self-tolerance induction. Autoimmune disease does not automatically follow such an initiation, since a variety of conditions that permit a self-reactive response must exist. Determinant spreading must take place to perpetuate this response, as the initiator effector T cell must produce a Th1 cytokine array that will lead to enhanced MHC display, heightened levels of antigen processing, and the appearance of adhesion and accessory molecules and costimulators on the surface of interacting cells, all to fan the flames of inflammation and to recruit more self-reactive effectors. However, provided that regulation is not impaired, these events are usually brought under control. For autoimmunity to occur, it is absolutely necessary for the initiating response to expand. Accordingly, since so many T cells potentially can be addressed by a large diversity of ligands, it is clear that many, apparently redundant regulatory mechanisms must have evolved to ensure the sanctity of the self from errant breakaway clones. In the first decades of the 21st century investigators will need to define and learn to mobilize and maintain these regulators.

Acknowledgments. This work was supported in part by grants from the National Multiple Sclerosis Society, the National Institutes of Health, and the Juvenile Diabetes Foundation International.

REFERENCES

1. **Atkinson, M. A., M. A. Bowman, L. Campbell, B. L. Darrow, D. L. Kaufman, and N. K. Maclaren.** 1994. Cellular immunity to a determinant common to glutamate decarboxylase and coxsackie virus in insulin-dependent diabetes. *J. Clin. Investig.* **94:**2125–2129.

2. **Bhardwaj, V., V. Kumar, H. M. Geysen, and E. E. Sercarz.** 1993. Degenerate recognition of a dissimilar antigenic peptide by myelin basic protein-reactive T cells. Implications for thymic education and autoimmunity. *J. Immunol.* **151:**5000–5010.
3. **Chambers, C. A., and J. P. Allison.** 1999. Costimulatory regulation of T cell function. *Curr. Opin. Cell Biol.* **11:**203–210.
3a. **Cohen, I.R.** 2000. *Tending Adam's Garden: Evolving the Immune Cognitive Self.* Academic Press, San Diego, Calif.
4. **Dal Canto, M. C., R. W. Melvold, B. S. Kim, and S. D. Miller.** 1995. Two models of multiple sclerosis: experimental allergic encephalomyelitis (EAE) and Theiler's murine encephalomyelitis virus (TMEV) infection. A pathological and immunological comparison. *Microsc. Res. Tech.* **32:**215–229.
5. **Drakesmith, H., D. O'Neil, S. C. Schneider, M. Binks, P. Medd, E. Sercarz, P. Beverley, and B. Chain.** 1998. In vivo priming of T cells against cryptic determinants by dendritic cells exposed to interleukin 6 and native antigen. *Proc. Natl. Acad. Sci. USA* **95:**14903–14908.
6. **Easton, A. J., and R. P. Eglin.** 1988. The detection of coxsackievirus RNA in cardiac tissue by in situ hybridization. *J. Gen. Virol.* **69:**285–291.
7. **Feuer, J., and H. Spiera.** 1997. Acute rheumatic fever in adults: a resurgence in the Hasidic Jewish community. *J. Rheumatol.* **24:**337–340.
8. **Fujinami, R. S.** 1988. Virus-induced autoimmunity through molecular mimicry. *Ann. N. Y. Acad. Sci.* **540:**210–217.
9. **Gammon, G., H. M. Geysen, R. J. Apple, E. Pickett, M. Palmer, A. Ametani, and E. E. Sercarz.** 1991. T cell determinant structure: cores and determinant envelopes in three mouse major histocompatibility complex haplotypes. *J. Exp. Med.* **173:**609–617.
10. **Horwitz, M. S., L. M. Bradley, J. Harbertson, T. Krahl, J. Lee, and N. Sarvetnick.** 1998. Diabetes induced by coxsackie virus: initiation by bystander damage and not molecular mimicry. *Nat. Med.* **4:**781–785.
11. **Huber, S. A., and L. P. Job.** 1983. Cellular immune mechanisms in coxsackievirus group B, type 3 induced myocarditis in BALB/c mice. *Adv. Exp. Med. Biol.* **161:**491–508.
12. **Itoh, M., T. Takahashi, N. Sakaguchi, Y. Kuniyasu, J. Shimizu, F. Otsuka, and S. Sakaguchi.** 1999. Thymus and autoimmunity: production of $CD25^{+}CD4^{+}$ naturally anergic and suppressive T cells as a key function of the thymus in maintaining immunologic self-tolerance. *J. Immunol.* **162:**5317–5326.
13. **Kaufman, D. L., M. Clare-Salzler, J. Tian, T. Forsthuber, G. S. Ting, P. Robinson, M. A. Atkinson, E. E. Sercarz, A. J. Tobin, and P. V. Lehmann.** 1993. Spontaneous loss of T-cell tolerance to glutamic acid decarboxylase in murine insulin-dependent diabetes. *Nature* **366:**69–72.
14. **Kumar, V.** 1998. Determinant spreading during experimental autoimmune encephalomyelitis: is it potentiating, protecting or participating in the disease? *Immunol. Rev.* **164:**73–80.
15. **Kumar, V., V. Bhardwaj, L. Soares, J. Alexander, A. Sette, and E. Sercarz.** 1995. Major histocompatibility complex binding affinity of an antigenic determinant is crucial for the differential secretion of interleukin 4/5 or interferon gamma by T cells. *Proc. Natl. Acad. Sci. USA* **92:**9510–9514.
16. **Kumar, V., and E. Sercarz.** 1996. Dysregulation of potentially pathogenic self reactivity is crucial for the manifestation of clinical autoimmunity. *J. Neurosci. Res.* **45:**334–339.
17. **Kumar, V., and E. E. Sercarz.** 1993. The involvement of T cell receptor peptide-specific regulatory $CD4^{+}$ T cells in recovery from antigen-induced autoimmune disease. *J. Exp. Med.* **178:**909–916.
18. **Lehmann, P. V., T. Forsthuber, A. Miller, and E. E. Sercarz.** 1992. Spreading of T-cell autoimmunity to cryptic determinants of an autoantigen. *Nature* **358:**155–157.
19. **Lehmann, P. V., E. E. Sercarz, T. Forsthuber, C. M. Dayan, and G. Gammon.** 1993. Determinant spreading and the dynamics of the autoimmune T-cell repertoire. *Immunol. Today* **14:**203–208.
20. **McRae, B. L., C. L. Vanderlugt, M. C. Dal Canto, and S. D. Miller.** 1995. Functional evidence for epitope spreading in the relapsing pathology of experimental autoimmune encephalomyelitis. *J. Exp. Med.* **182:**75–85.
21. **Moudgil, K. D., T. T. Chang, H. Eradat, A. M. Chen, R. S. Gupta, E. Brahn, and E. E. Sercarz.** 1997. Diversification of T cell responses to carboxy-terminal determinants within the 65-kD heat-shock protein is involved in regulation of autoimmune arthritis. *J. Exp. Med.* **185:**1307–1316.
22. **Nanda, N. K., K. K. Arzoo, H. M. Geysen, A. Sette, and E. E. Sercarz.** 1995. Recognition of multiple peptide cores by a single T cell receptor. *J. Exp. Med.* **182:**531–539.
23. **Nanda, N. K., and E. E. Sercarz.** 1995. The positively selected T cell repertoire: is it exclusively restricted to the selecting MHC? *Int. Immunol.* **7:**353–358.

24. **Oldstone, M. B.** 1989. Molecular mimicry as a mechanism for the cause and a probe uncovering etiologic agent(s) of autoimmune disease. *Curr. Top. Microbiol. Immunol.* **145:**127–135.
25. **Pinilla, C., R. Martin, B. Gran, J. R. Appel, C. Boggiano, D. B. Wilson, and R. A. Houghten.** 1999. Exploring immunological specificity using synthetic peptide combinatorial libraries. *Curr. Opin. Immunol.* **11:**193–202.
26. **Shimizu, J., S. Yamazaki, and S. Sakaguchi.** 1999. Induction of tumor immunity by removing $CD25^+CD4^+$ T cells: a common basis between tumor immunity and autoimmunity. *J. Immunol.* **163:**5211–5218.
27. **Speir, J. A., K. C. Garcia, A. Brunmark, M. Degano, P. A. Peterson, L. Teyton, and I. A. Wilson.** 1998. Structural basis of 2C TCR allorecognition of H-2Ld peptide complexes. *Immunity* **8:**553–562.
28. **Takacs, K., and D. M. Altmann.** 1998. The case against epitope spread in experimental allergic encephalomyelitis. *Immunol. Rev.* **164:**101–110.
29. **Takacs, K., P. Chandler, and D. M. Altmann.** 1997. Relapsing and remitting experimental allergic encephalomyelitis: a focused response to the encephalitogenic peptide rather than epitope spread. *Eur. J. Immunol.* **27:**2927–2934.
30. **Tsunoda, I., and R. S. Fujinami.** 1996. Two models for multiple sclerosis: experimental allergic encephalomyelitis and Theiler's murine encephalomyelitis virus. *J. Neuropathol. Exp. Neurol.* **55:**673–686.
31. **Tuohy, V. K., M. Yu, B. Weinstock-Guttman, and R. P. Kinkel.** 1997. Diversity and plasticity of self recognition during the development of multiple sclerosis. *J. Clin. Investig.* **99:**1682–1690.
32. **Tuohy, V. K., M. Yu, L. Yin, J. A. Kawczak, J. M. Johnson, P. M. Mathisen, B. Weinstock-Guttman, and R. P. Kinkel.** 1998. The epitope spreading cascade during progression of experimental autoimmune encephalomyelitis and multiple sclerosis. *Immunol. Rev.* **164:**93–100.
33. **Vanderlugt, C. L., K. L. Neville, K. M. Nikcevich, T. N. Eagar, J. A. Bluestone, and S. D. Miller.** 2000. Pathologic role and temporal appearance of newly emerging autoepitopes in relapsing experimental autoimmune encephalomyelitis. *J. Immunol.* **164:**670–678.
34. **Weiss, L. M., L. A. Movahed, M. E. Billingham, and M. L. Cleary.** 1991. Detection of coxsackievirus B3 RNA in myocardial tissues by the polymerase chain reaction. *Am. J. Pathol.* **138:**497–503.

Molecular Mimicry, Microbes, and Autoimmunity
Edited by M. W. Cunningham and R. S. Fujinami

Chapter 16

Molecular Mimicry: Lessons from Experimental Models of Systemic Lupus Erythematosus and Antiphospholipid Syndrome

Miri Blank, Ilan Krause, and Yehuda Shoenfeld

The healthy immune system is tolerant to the molecules of which the body is composed. However, one can find that among the major antigens recognized during a wide variety of bacterial, viral, and parasitic diseases, many belong to conserved protein families, and they share extensive sequence identity or conformational fits with the host's molecules, namely, *molecular mimicry.*

There is a general consensus that autoimmune diseases have a multifactorial etiology, depending on both genetic and environmental factors (58). Various examples of molecular mimicry in different autoimmune systems were described, and the major ones are summarized in Table 1. Viruses, microbes, and parasites may brake peripheral self-tolerance and induce and maintain autoimmunity via several overlapping mechanisms (1, 4, 7, 14, 16). The shared epitopes between the pathogens and autoantigen may induce chronic inflammation, which is mandatory for the establishment of all other mechanisms that contribute to autoimmunity: unveiling of "hidden" self-epitopes (4), cross-reactive peptide presentation; determinant spreading, upregulation of major histocompatibility complex (MHC), adhesion, and costimulatory molecules on antigen-presenting cells (APCs); upregulation of cellular and extracellular processing; apoptosis; infection of professional APCs; autoantibody production (37, 45, 46, 53, 66); and subversion of T-cell responses and the immunological homunculus network (14).

MOLECULAR MIMICRY BETWEEN PATHOGENS AND TARGET EPITOPES FOR AUTOANTIBODIES IN EXPERIMENTAL SLE

The immunogenicity of self-like microbial molecules is strikingly illustrated in several cases of murine and human systemic lupus erythematosus (SLE). The etiology of the autoimmune status associated with SLE and the nature of the triggering antigen for the anti-DNA response are still mysteries. It is not clear whether the anti-double-stranded DNA (anti-dsDNA) response represents a response to immunogenic DNA (such as bacter-

Miri Blank, Ilan Krause, and Yehuda Shoenfeld • Research Unit of Autoimmune Diseases, Internal Medicine 'B,' Sheba Medical Center, Tel-Hashomer 52621, Israel.

Table 1. Major examples of molecular mimicry in association with autoimmune diseases

Disease[a]	Homologous antigens or pathogens	Selected reference no(s).
EAE	Myelin basic protein and hepatitis B virus polymerase	20
Myasthenia gravis	Acetylcholine receptor, neurofilaments, herpes viruses	42, 55
Multiple sclerosis	Myelin proteins and peptides from coronavirus, measles virus, mumps, Epstein-Barr virus, herpes viruses, etc.	59, 63, 71
Guillain-Barré syndrome	Peripheral nerve, gangliosides, and *Campylobacter jejuni*	73
IDDM	Islet antigens (GAD 65, proinsulin carboxypeptidase H) and coxsackie B virus, rotaviruses, herpesviruses, hepatitis C virus, rhinovirus, hantavirus, faviviruses, retroviruses	6, 33, 36

[a] EAE, experimental autoimmune encephalomyelitis; IDDM, insulin-dependent diabetes mellitus.

ial, viral, or oxidized DNA), the nucleosome, apoptotic cells, or some different cross-reactive self-antigen or is a by-product of a response to a foreign, non-DNA antigen. Perhaps the best-known instance occurs during evaluation of the New Zealand mouse model (65). In this system of genetically defined mice that spontaneously develop autoimmune disease, persistent infection with a DNA virus like polyomavirus or an RNA virus like lymphocytic choriomeningitis virus speeds up the time when antibodies (Abs) to DNA, red blood cells, and other autoantigens appear. Thus, the kinetics and the severity of the resulting autoimmune disease are accelerated. In the New Zealand White (NZW) mouse, which carries the gene(s) necessary for the development of anti-DNA and other autoimmune responses and diseases, the kinetics and the severity of autoimmune disease are accelerated upon viral infection. Molecular mimicry of host structures by pathogens was raised as an explanation for the initial activation of autoreactive B cells that does not require priming by autoantigen (58). This was supported by the presence of an anti-DNA idiotype (16/6 Id) in individuals with microbial infections such as *Klebsiella* infection or pulmonary tuberculosis (19, 56). Furthermore, monoclonal antibacterial Waldenstrom's macroglobulins had immunochemical similarities with anti-DNA from a lupus patient with 16/6 Id (2, 44). Cross-reactivity between dsDNA and bacterial antigen has been documented in studies of both human (19, 26, 60) and mouse (38, 52) Abs. Furthermore, TB-68, a mouse monoclonal anti-tuberculosis glycolipid which binds to dsDNA and which carries the 16/6 Id, as well as a T-cell line directed to this monoclonal antibody (MAb), were able to induce experimental SLE in naive mice (8, 10).

Abs to phosphorylcholine, a component of pneumococcal cell wall polysaccharide, were found to be protective against a lethal pneumococcal infection in mice (38). In vitro substitution of glutamic acid to alanine at residue 35 in the complementarity-determining region (CDR) 1 (CDR1) of the heavy chain of antiphosphorylcholine resulted in a shift to dsDNA binding (17). Some of these MAbs led to extensive glomerular and tubular deposition similar to that caused by anti-DNA Abs from lupus-prone mice. These data led to the suggestion that autoreactive B cells may be similarly generated in vivo during the autoimmune response to bacterial antigen (17).

In experiments with mice that constitutively express the *bcl-2* gene, which inhibits apoptosis and, hence, clonal deletion, Abs to both dsDNA and phosphorylcholine were generated after immunization with a pneumococcal cell wall hapten (52). Furthermore, transgenic mice that constitutively express *bcl-2* in the B-cell compartment display a marked increase in B-cell number and, in certain genetic backgrounds, display autoantibody production as well (62). To elucidate the potential role of molecular mimicry in SLE patients, a study was carried out by creating a combinatorial library for a lupus patient given a pneumococcal polysaccharide vaccine 10 days prior to splenectomy (35). Fab clones that bore the 3I Id, which is expressed on both pathogenic anti-DNA Abs and anti-pneumococcal Abs, were analyzed. These Fabs displayed reactivity with bacterial polysaccharides, phosphorylcholine and the nonpolymorphic cell wall polysaccharide also present in the pneumococcal vaccine, DNA, or some combination of these antigens. On the basis of their data the investigators suggest that bacterial antigen might activate an anti-DNA response. These studies provide support at the molecular level for a potential role of molecular mimicry in the generation of anti-DNA Abs.

As for anti-DNA and antibacterial molecular mimicry, autoantibodies from patients with SLE bind to a shared sequence of SmD and Epstein-Barr virus-encoded nuclear antigen I (EBNA-I) (54). SmD is one of the small nuclear ribonucleoproteins frequently targeted by autoantibodies in SLE. Abs isolated from sera from patients with lupus have been found to bind specifically to the C-terminal region of SmD (sequence positions 95 to 119). This region is highly homologous to sequence positions 35 to 58 of the EBNA-I antigen, one of the nuclear antigens induced by infection with Epstein-Barr virus. Abs affinity purified over a column with the peptide from positions 95 to 119 were able to recognize this sequence in the context of the whole SmD molecule. Anti-SmD (positions 95 to 119) Abs also bound to the EBNA-I peptide from positions 35 to 58 and detected the EBNA-I molecule in a total cell extract from Epstein-Barr virus-infected lines. A population of anti-SmD Abs is therefore able to bind to an epitope shared by the autoantigen and the viral antigen EBNA-I. Moreover, sera from mice immunized with the EBNA-I peptide from positions 35 to 58 displayed the same pattern of reactivity as spontaneously produced anti-SmD Abs. These data suggest that molecular mimicry may play a role in the induction of anti-SmD autoantibodies. This work was later confirmed by PCR and computer analyses (31).

MOLECULAR MIMICRY BETWEEN PATHOGENS AND TARGET EPITOPES FOR ANTIPHOSPHOLIPID ANTIBODIES IN EXPERIMENTAL APS

Hughes syndrome, also known as antiphospholipid syndrome (APS), is characterized by a wide variety of hemocytopenic and vaso-occlusive manifestations and is associated with antibodies directed against negatively charged phospholipids. Features of APS include hemolytic anemia, thrombocytopenia, venous and arterial occlusions, livedo reticularis, pulmonary manifestations, recurrent fetal loss, neurological manifestations (stroke, transverse myelitis, Guillain-Barré syndrome), and a positive Coombs' test, positivity for anticardiolipin antibodies, or lupus anticoagulant activity (27, 30, 41). The factor(s) that causes production of the antiphospholipid antibodies in APS remains unidentified.

It is widely accepted that anticardiolipin Abs purified from patients with primary or secondary APS bind to anionic phospholipid through the β_2-glycoprotein -I (β_2GPI) molecule (21, 30, 40, 41). β_2GPI (50 kDa) is a member of the complement control protein, is the target antigen for many of the autoimmune anti-β_2GPI Abs, and is composed of five respective consensus repeats (30, 41). β_2GPI exhibits several properties in vitro which define it as an anticoagulant (e.g., inhibition of prothrombinase activity, ADP-induced platelet aggregation, and platelet factor IX production) (34, 41).

It has been postulated that anti-β_2GPI Abs exert a direct pathogenic effect by interfering with the homeostatic reactions that occur on the surfaces of platelets or vascular endothelial cells (9, 24, 47, 57, 70). Passive transfer of these Abs into naive mice results in the induction of experimental APS (9, 24).

Several indirect arguments support the idea that microbial agents influence the course of APS. An association between APS and pathogens was documented, such as hepatitis C virus (48), *Salmonella* lipopolysaccharide (28), and *Mycoplasma penetrans,* a rare bacterium that has so far only been found in human immunodeficiency virus (HIV)-infected persons and that was isolated from the blood and throat of a non-HIV-infected patient with primary APS (whose etiology and pathogenesis are unknown) (72).

Furthermore, using the shotgun phage display technique, Zhang et al. (74) identified a *Staphylococcus aureus* protein, Sbi, which bound to β_2GPI and which serves as the target molecule for immunoglobulin G (IgG) binding. It was also shown that protein Sbi, and thus the β_2GPI binding potential, is expressed on the staphylococcal cell surface at levels that vary between strains (74).

Synthetic peptides that share a putative phospholipid binding region of the β_2GPI molecule and that share a high degree of homology with human cytomegalovirus induce the generation of antiphospholipid and anti-β_2GPI antibodies in NIH/Swiss mice (25). These findings demonstrated that some phospholipid-binding viral and bacterial proteins function like β_2GPI in inducing antiphospholipid (αPL) and anti-β_2GPI production, and this induction function is consistent with a role for such viral and bacterial proteins in the induction of αPL antibody production in humans.

Using a hexapeptide phage display library, we identified three hexapeptides that react specifically with the anti-β_2GPI MAbs, which cause endothelial cell activation and induce experimental APS (12). All three peptides specifically inhibit both in vitro and in vivo the biological functions of the corresponding anti-β_2GPI MAbs. Exposure of endothelial cells to anti-β_2GPI MAbs and their corresponding peptides led to the inhibition of endothelial cell activation, as shown by decreased levels of expression of adhesion molecules (E-selectin, ICAM-1, VCAM-1) and monocyte adhesion. In vivo infusion of each of the anti-β_2GPI MAbs into BALB/c mice, followed by administration of the corresponding specific peptides, prevented the peptide-treated mice from developing experimental APS. The use of synthetic peptides that focus on neutralization of pathogenic anti-β_2GPI Abs represents a possible new therapeutic approach to APS (12).

Use of the sequences in the Swiss protein database revealed high degrees of homology between the hexapeptide LKTPRV and TLRVYK and different bacteria and viruses (summarized in Table 2). We prepared relevant particles from *Pseudomonas aeruginosa, Haemophilus influenzae, Streptococcus pneumoniae, Shigella dysenteriae,* and *Neisseria gonorrhoeae* and from the yeast *Candida albicans.* Naive mice were immunized in the hind footpads with the particles from the pathogens (10 μg/mouse in complete Freund's adjuvant), and 3 weeks later a booster injection in phosphate-buffered saline was given. As shown in Table 3, all the immunized mice developed anticardiolipin antibodies. However, the most significant levels of mouse anti-β_2GPI Abs were detected in the mice immunized with *H. influenzae* or *N. gonorrhoeae* and bound to β_2GPI in a dose-dependent manner. Mouse IgG specific to the relevant peptide (equivalent to β_2GPI) was prepared by passing the sera from mice immunized with *P. aeruginosa, H. influenzae, S. pneumoniae, S. dysenteriae,* or *N. gonorrhoeae* on a peptide-specific column. The peptide-specific column elut-

Table 2. Homology between the studied peptides and bacteria, viruses, and yeast

Type of pathogen	Organism with homology to the following peptide:		
	LKTPRV	KDKATFG	TLRVYK
Bacteria	*Pseudomonas aeruginosa* *Yersinia pseudotuberculosis*		*Streptococcus pneumoniae* *Shigella dysenteriae* *Haemophilus influenzae* *Neisseria gonorrhoeae*
Virus	Cytomegalovirus Polyomavirus jc Adenovirus type 40		Epstein-Barr virus
Yeast	*Streptomyces lividans* *Saccharomyces cerevisiae*	*Saccharomyces cerevisiae*	*Candida albicans*
Parasite	*Schistosoma mansoni*		

Table 3. Presence of autoantibodies in the sera of mice immunized with bacteria or yeast homologous to the relevant peptides[a]

Bacterium	Antibody to:	Direct binding[b]	Competition[c]
P. aeruginosa	Cardiolipin	0.774 ± 0.213	0.645 ± 0.218
	β_2GPI	0.377 ± 0.108	0.116 ± 0.078
	Phosphatidylcholine	0.352 ± 0.198	ND[d]
	dsDNA	0.225 ± 0.071	ND
S. pneumoniae	Cardiolipin	0.879 ± 0.105	0.816 ± 0.109
	β_2GPI	0.434 ± 0.174	0.162 ± 0.112
	Phosphatidylcholine	0.453 ± 0.117	0.408 ± 0.077
	dsDNA	0.697 ± 0.091	0.613 ± 0.105
S. dysenteriae	Cardiolipin	0.367 ± 0.214	0.861 ± 0.117
	β_2GPI	0.263 ± 0.231	0.103 ± 0.081
	Phosphatidylcholine	0.355 ± 0.112	ND
	dsDNA	0.118 ± 0.111	ND
H. influenzae	Cardiolipin	0.798 ± 0.097	0.756 ± 0.142
	β_2GPI	0.836 ± 0.216	0.251 ± 0.118
	Phosphatidylcholine	0.421 ± 0.117	ND
	dsDNA	0.372 ± 0.124	ND
N. gonorrhoeae	Cardiolipin	0.765 ± 0.237	0.672 ± 0.241
	β_2GPI	0.628 ± 0.194	0.102 ± 0.116
	Phosphatidylcholine	0.182 ± 0.114	ND
	dsDNA	0.235 ± 0.121	ND
C. albicans	Cardiolipin	0.835 ± 0.121	0.767 ± 0.118
	β_2GPI	0.445 ± 0.117	0.131 ± 0.096
	Phosphatidylcholine	0.328 ± 0.126	ND
	dsDNA	0.251 ± 0.111	ND

[a] Mice immunized with irrelevant bacteria such as *Escherichia coli* and *Proteus mirabilis* (gram-negative organisms) or *Klebsiella pneumoniae* (a gram-positive organism), did not develop mouse anti-β_2GPI antibodies (optical densities at 405 nm, 0.086 ± 0.064 to 0.243 ± 0.097); however, the titers of anticardiolipin antibodies remained high (optical densities at 405 nm, 0.854 ± 0.116 to 1.217 ± 0.134).

[b] Data are presented as the mean ± standard deviation optical density at 405 nm for a serum dilution of 1:200.

[c] A relevant peptide was used as a competitor (50 μM) in the presence of serum at a concentration that permitted 50% binding.

[d] ND, not determined.

ed immunoglobulin found to bind to β_2GPI. These specific immunoglobulins were infused intravenously into BALB/c mice, and the development of clinical manifestations of APS was studied. As described in Table 4, only in the mice which were infused with mouse Abs that were derived from mice immunized with *H. influenzae* or *N. gonorrhoeae* and that were directed to the peptide TLRVYK did the peptides have the potential to induce clinical manifestations which resemble experimental APS (e.g., thrombocytopenia, prolonged activated partial thromboplastin time [aPTT], and an elevated percentage of fetal loss). We hypothesize, in the current case, that the mechanism of pathogenic anti-β_2GPI generation is induced by epitope mimicry. B cells present the mimicking epitopes of bacteria or viruses to T cells via an MHC class II pathway. These B cells produce Abs that have specificity for the instigating epitope and that cross-react with host β_2GPI as a native molecule or in a complex with platelets or endothelial cells that express the mimicked self-epitope. Depending on the appropriate secondary signaling, a pathogenic autoreactivity develops.

MOLECULAR MIMICRY IN EXPERIMENTAL LUPUS INDUCED BY IDIOTYPIC MANIPULATION

The molecular basis of antigen mimicry by anti-idiotypic antibodies was studied extensively. On the basis of Jerne's theory (32), after immunization with an autoantibody that carries a specific idiotype (Ab1), naive mice develop an antiautoantibody (anti-Id; which is also known as Ab2) and then generate anti-anti-Id (Ab3) a few weeks later. The Ab3 Abs elicited by anti-Id Abs resemble the Ab1 that originated with the anti-Id. There are two major possibilities for the induction of Ab3. The first is that an anti-Id will mimic Ab1 so closely as to produce an Ab response equivalent to the Ab response produced by the original Ab1. The second possibility is that the anti-Id, when used as an immunogen, will back-elicit a clone similar to that from which it originated, thus stimulating lymphocytes that display anti-anti-Id and that should be very closely related to Ab1. Immunization with anti-DNA by us (8, 10, 43, 58) and others (15, 29, 64) resulted in the generation of mouse anti-DNA Abs associated with the clinical manifestations of lupus (e.g., leukopenia, elevated erythrocyte sedimentation rate, and kidney damage). Inoculation of the 16/6 Id (Ab1) into naive mice was followed by generation of anti-Id, which led to the generation of high titers of mouse Ab3 which had the same binding characteristics and in vivo patho-

Table 4. Clinical manifestations in mice infused with mouse Abs directed to the peptide homologous to the bacterium or yeast[a]

Parameter	Anti-β_2GPI from mice immunized with:					
	P. aeruginosa (n = 8)	*S. pneumoniae* (n = 7)	*S. dysenteriae* (n = 7)	*H. influenzae* (n = 12)	*N. gonorrhoeae* (n = 10)	*C. albicans* (n = 8)
Platelet count (cells/mm^3 [10^3])	976 ± 152	918 ± 216	1,012 ± 214	527 ± 126	603 ± 142	994 ± 221
aPTT (s)	3 ± 2	7 ± 2	3 ± 1	12 ± 2	9 ± 3	4 ± 1
Resorption (%)	8 ± 2	6 ± 1	4 ± 2	28 ± 3	17 ± 2	7 ± 2

[a] BALB/c mice were infused with 40 μg of affinity-purified anti-β_2GPI antibodies from mice immunized with bacteria or yeast intravenously on day 0 of pregnancy. Mice infused with irrelevant mouse IgG had platelet counts of $1,037 \times 10^3 \pm 217 \times 10^3$ cells/mm^3, an aPTT of 3 ± 1 s, and resorption of 7% ± 2%.

genicity as the human Ab1 (16/6 Id). Although Ab3 mimicked the same binding epitopes on DNA as the original Ab1, sequence analysis did not reveal high degrees of homology between the CDRs of the Ab1 and the Ab3 molecules (67, 68). Thus, human Ab1 (16/6 Id) and Ab3 (mouse 16/6 Id) may have conformational similarities at their binding sites and may target similar epitopes. A mechanism of epitope spreading was observed in mice immunized with either the human or the mouse 16/6 Id, which led to autoantibody spread and which resulted in high titers of antibodies to DNA, histones, nuclear proteins, and negatively charged phospholipids (8, 10, 43, 58, 67). Furthermore, immunization with anti-DNA (16/6 Id) CDR-based peptides has been shown to be involved in both the induction of experimental lupus and the inhibition of the autoimmune response (69). These peptides prevented autoantibody production in neonatal mice that were later immunized either with the peptide or with the pathogenic autoantibody and inhibited the specific proliferation of lymph node cells of mice immunized with the same peptide or with the human 16/6 Id. In other words, CDR-based peptides mimic the pathogenic Id located on the CDR of the pathogenic mouse 16/6 Id and mimic the human 16/6 Id target epitope as well. In addition, immunization of mice with synthetic peptides that mimic the target epitope of anti-DNA R4A Id (23) or anti-DNA Id (p64) (18) resulted in lupus nephritis in mice by mimicking the biological function of the original anti-DNA Id epitope on the MAb (18, 50). Moreover, a peptide that mimics the target epitope of anti-DNA R4A Id was shown to inhibit glomerular deposition of the pathogenic anti-DNA R4A Id Ab by neutralizing its binding properties (23).

MOLECULAR MIMICRY IN EXPERIMENTAL APS INDUCED BY IDIOTYPIC MANIPULATION

Immunization of naive mice with anticardiolipin β_2GPI-dependent MAbs and polyclonal antibodies or their corresponding scFv, such as Ab1, resulted in the production in the inoculated mice of autoantibody directed to cardiolipin and to the cardiolipin β_2GPI-dependent antibody Ab3. The immunized mice developed a clinical picture of experimental APS: thrombocytopenia, a high percentage of fetal resorptions, and a prolonged coagulation time (3, 11, 13, 61).

Using a hexapeptide phage display library, we identified three hexapeptides which reacted specifically with anti-β_2GPI mAbs, MAbs ILA-1, ILA-3, and H-3, which cause endothelial cell activation and which induce experimental APS (12, 24). These synthetic peptides are molecular mimics of the target epitopes on the β_2GPI molecule for anti-β_2GPI MAbs ILA-1, ILA-3, and H-3. The β_2GPI-mimicking peptides specifically inhibited the in vitro and in vivo biological functions of the corresponding anti-β_2GPI MAbs (12).

SUMMARY AND CONCLUSION

Molecular mimicry is one mechanism by which experimental lupus or experimental APS can occur in association with pathogens or with dysregulation of the idiotypic network. The concept of molecular mimicry remains a viable hypothesis that can be used to frame questions and approaches to uncovering the initiating infectious agent or idiotypic network as well as to recognize the self-determinant, understand the pathogenic mechanisms involved, and design strategies for the treatment and prevention of lupus or APS disorders. Studies on experimental lupus and experimental APS prove the existence of molecular mimicry between pathogens and the autoantigens involved in experimental lupus and APS. We speculate that the mimicking antigen that is similar to the pathogen at only one epitope may initiate a primary cross-reactive response to that epitope that subsequently results in recognition of numerous epitopes in the mimicked host β_2GPI or DNA. Recognition of multiple antigens and epitopes is evident in insulin-dependent diabetes mellitus, SLE, APS, rheumatoid arthritis, PBC, and probably most autoimmune diseases, with spreading of the named epitope leading to autoantibody spread. Mimicry may be one mechanism by which the balance of tolerance can be broken and the autoimmune response can be triggered, but the mere presence of self-determinants in a virus or a bacterium need not result in pathogenesis. Tolerance to self-antigens is normally maintained in an environment of natural autoantibodies and T cells that will cross-react with self-molecules, allowing the immune system to cope with the peripheral presence of autoreactive B and T cells.

The use of molecular mimicry between synthetic peptides that correspond to the autoantigen on the basis of either the CDR sequence or peptide phage display libraries is a potent tool for the treatment of experimental lupus and experimental APS.

REFERENCES

1. **Abu-Shakra, M., D. Buskila, and Y. Shoenfeld.** 1999. Molecular mimicry between host and pathogen: examples from parasites and implications. *Immunol. Lett.* **67:**147–152.
2. **Atkinson, P. M., G. W. Lampman, B. C. Furie, Y. Naparstek, R. S. Schwartz, B. D. Stollar, and B. Furie.** 1985. Homology of the NH_2-terminal amino acid sequences of the heavy and light chains of human monoclonal lupus autoantibodies containing the dominant 16/6 idiotype. *J. Clin. Investig.* **75:**1138–1143.
3. **Bakimer, R., P. Fishman, M. Biann, B. Sredni, M. Djaldetti, and Y. Shoenfeld.** 1992. Induction of primary anti-phospholipid syndrome in mice by immunization with human monoclonal anti-cardiolipin antibody (H-3). *J. Clin. Investig.* **89:**1558–1563.
4. **Barnaba, V.** 1996. Hidden self-epitopes and autoimmunity. *Immunol. Rev.* **152:**47–66.
5. **Barnett, L. A., and R. S. Fujinami.** 1992. Molecular mimicry: a mechanism for autoimmune injury. *FASEB J.* **6:**840–844.
6. **Baum, H., V. Brusik, K. Choudhuri, P. Cunningham, D. Vergani, and M. Peakman.** 1995. MHC molecular mimicry in diabetes. *Nat. Med.* **1:**388.
7. **Behar, S. M., and S. A. Porcelli.** 1995. Mechanisms of autoimmune disease induction: the role of the immune response to microbial pathogens. *Arthritis Rheum.* **38:**458–476.
8. **Blank, M., M. Krup, S. Mendlovic, H. Fricke, E. Mozes, N. Talal, A. R. Coates, and Y. Shoenfeld.** 1990. The importance of the pathogenic 16/6 idiotype in the induction of SLE in naive mice. *Scand. J. Immunol.* **31:**45–52.
9. **Blank, M., J. Cohen, V. Toder, and Y. Shoenfeld.** 1991. Induction of primary anti-phospholipid syndrome in mice by passive transfer of anti-cardiolipin antibodies. *Proc. Natl. Acad. Sci. USA* **88:**3069–3073.
10. **Blank, M., S. Mendlovic, E. Mozes, A. R. Coates, and Y. Shoenfeld.** 1991. Induction of systemic lupus erythematosus in naive mice with T-cell lines specific for human anti-DNA antibody SA-1 (16/6 Id+) and for mouse tuberculosis antibody TB/68 (16/6 Id+). *Clin. Immunol. Immunopathol.* **60:**471–483.
11. **Blank, M., I. Krause, M. Ben-Bassat, and Y. Shoenfeld.** 1992. Induction of experimental anti-phospholipid syndrome associated with SLE following immunization with human monoclonal pathogenic anti-DNA idiotype. *J. Autoimmun.* **5:**495–509.
12. **Blank, M., Y. Shoenfeld, S. Cabilly, Y. Heldman, M. Fridkin, and E. Katchalski-Katzir.** 1999. Prevention of experimental antiphospholipid syndrome and endothelial cell activation by synthetic peptides. *Proc. Natl. Acad. Sci. USA* **96:**5164–5198.
13. **Blank, M., A. Waisman, E. Mozes, T. Koike, and Y. Shoenfeld.** 1999. Characteristics and pathogenic role of anti-beta-2-glycoprotein-I (β_2GPI) single chain Fv domains: induction of experimental antiphospholipid syndrome. *Int. Immunol.* **11:**1917–1926.

14. **Cohen, I. R., and D. B. Young.** 1991. Autoimmunity, microbial immunity and the immunological homunculus. *Immunol. Today* **12:**105–110.
15. **Dang, H., A. G. Geiser, J. J. Letterio, T. Nakabayashi, L. Kong, G. Fernandes, and N. Talal.** 1995. SLE-like autoantibodies and Sjögren's syndrome-like lymphoproliferation in TGF-beta knockout mice. *J. Immunol.* **155:**3205–3212.
16. **Davis, J. M.** 1997. Molecular mimicry: can epitope mimicry induce autoimmune disease? *Immunol. Cell. Biol.* **75:**113–126.
17. **Diamond, B., and M. D. Scharff.** 1984. Somatic mutation of the T15 heavy chain gives rise to an antibody with autoantibody specificity. *Proc. Natl. Acad. Sci. USA* **81:**5841–5844.
18. **Eivazova, E. R., J. M. McDonnel, B. J. Sulton, and N. A. Staines.** 1999. Anti-idiotype antibodies with dual specificity for DNA and their potential pathogenicity in SLE. *Arthritis Rheum.* **42:**19.
19. **El-Roey, A., W. L. Gross, J. Luedemann, D. A. Isenberg, and Y. Shoenfeld.** 1986. Preferential secretion of a common anti-DNA idiotype (16/6 Id) and anti-polynucleotide antibodies by normal mononuclear cells following stimulation with Klebsiella pneumoniae. *Immunol. Lett.* **12:**313–319.
20. **Fujinami, R. S., and M. B. A. Oldstone.** 1985. Amino acid homology between the encephalogenic site of myelin basic protein and virus: mechanism for autoimmunity. *Science* **230:**1043–1045.
21. **Galli, M., P. Comfurius, C. Maassen, H. C. Hemker, M. H. de Baets, P. J. van Breda-Vriesman, T. Barbui, R. F. Zwaal, and E. M. Bevers.** 1990. Anti-cardiolipin antibodies (ACA) directed not to cardiolipin but to a plasma protein cofactor. *Lancet* **335:**1544–1547.
22. **Gauntt, C. J., H. M. Arizpe, A. L. Higdon, H. J. Wood, D. F. Bowers, M. M. Rozek, and R. Crawley.** 1995. Molecular mimicry, anti-coxsackievirus B3 neutralizing monoclonal antibodies, and myocarditis. *J. Immunol.* **154:**2983–2995.
23. **Gaynor, B., C. Putterman, P. Valadon, L. Spatz, M. D. Scharff, and B. Diamond.** 1997. Peptide inhibition of glomerular deposition of an anti-DNA antibody. *Proc. Natl. Acad. Sci. USA* **94:**1955–1960.
24. **George, J., M. Blank, Y. Levy, E. Grinbaum, S. Cohen, M. Damianovich, A. Tincani, and Y. Shoenfeld.** 1998. Differential effects of anti-beta2-glycoprotein I antibodies on endothelial cells and on the manifestations of experimental antiphospholipid syndrome. *Circulation* **97:**900–907.
25. **Gharavi, E. E., H. Chaimovich, E. Cucurull, C. M. Celli, H. Tang, W. A. Wilson, and A. E. Gharavi.** 1999. Induction of antiphospholipid antibodies by immunization with synthetic viral and bacterial peptides. *Lupus* **8:**449–455.
26. **Grayzel, A., A. Solomon, C. Aranow, and B. Diamond.** 1991. Antibodies elicited by pneumococcal antigens bear an anti-DNA associated idiotype. *J. Clin. Investig.* **87:**842–846.
27. **Harris, E. N., A. E. Gharavi, M. L. Boey, B. M. Patel, C. G. Mackworth-Young, S. Loizou, and G. R. V. Hughes.** 1983. Anticardiolipin antibodies: detection by radioimmunoassay and association with thrombosis in systemic lupus erythematosus. *Lancet* **ii:**1211–1214.
28. **Hayem, G., N. Kassis, P. Nicaise, P. Bouvet, A. Andremont, C. Labarre, M. F. Kahn, and O. Meyer.** 1999. Systemic lupus erythematosus-associated catastrophic antiphospholipid syndrome occurring after typhoid fever: a possible role of Salmonella lipopolysaccharide in the occurrence of diffuse vasculopathy-coagulopathy. *Arthritis Rheum.* **42:**1056–1061.
29. **Howe, C. A., B. Hartley, D. G. Williams., S. Muller, and N. A. Staines.** 1997. Active immunization with anti-DNA autoantibody idiopeptide 88H.64-80 is nephritogenic in (NZB × NZW)F_1 and BALB/c mice. *Biochem. Soc. Trans.* **25:**316S.
30. **Hughes, G. R., E. N. Harris, and A. E. Gharavi.** 1986. The anti-cardiolipin syndrome. *J. Rheumatol.* **13:**486–489.
31. **Incaprera, M., L. Rindi, A. Bazzichi, and C. Garzelli.** 1998. Potential role of the Epstein-Barr virus in systemic lupus erythematosus autoimmunity. *Clin. Exp. Rheumatol.* **16:**289–294.
32. **Jerne, N. K.** 1984. Idiotypic networks and other preconceived ideas. *Immunol. Rev.* **79:**5–24.
33. **Jones, D. B., and N. W. Armstrong.** 1995. Coxsackie virus and diabetes revisited. *Nat. Med.* **1:**284.
34. **Kandiah, D. A., and S. A. Krilis.** 1994. Beta 2-glycoprotein-I *Lupus* **3:**207–212.
35. **Kowal, C., A. Weinstein, and B. Diamond.** 1999. Molecular mimicry between bacterial and self antigen in a patient with systemic lupus erythematosus. *Eur. J. Immunol.* **29:**1901–1911.
36. **Krauss, D. S., M. D. Aronson, D. W. Gump, and D. S. Newcombe.** 1974. Haemophilus influenzae septic arthritis. A mimicker of gonococcal arthritis. *Arthritis Rheum.* **17:**267–271.
37. **Kuo, P., C. Kowal, B. Tadmor, and B. Diamond.** 1997. Microbial antigens can elicit autoantibody production: a potential pathway to autoimmune diseases. *Ann. N. Y. Acad. Sci.* **815:**230–236.

38. **Limpanisithikul, W., S. Ray, and B. Diamond.** 1995. Cross-reactive antibodies have both protective and pathogenic potential. *J. Immunol.* **155:**967–973.
39. **MacLaren, N. K., and M. A. Alkinson.** 1997. Insulin dependent diabetes mellitus: the hypothesis of molecular mimicry between islet cell antigens and microorganisms. *Mol. Med. Today* **3:**76–83.
40. **Marx, A., A. Wilisch, A. Schultz, S. Gattenlohner, R. Nenninger, and H. K. Muller-Hermelink.** 1997. Pathogenesis of myasthenia gravis. *Virchows Arch.* **430:**355–364.
41. **Matsuura, E., Y. Igarashi, T. Yasuda, D. A. Triplett, and T. Koike.** 1994. Anti-cardiolipin antibodies recognize β2-glycoprotein-I structure altered by interacting with oxygen-modified solid phase surface. *J. Exp. Med.* **179:**457–462.
42. **McNeil, H. P., C. N. Chesterman, and S. A. Krilis.** 1991. Immunological and clinical importance of antiphospholipid antibodies. *Adv. Immunol.* **49:**193–280.
43. **Mendlovic, S., S. Brocke, Y. Shoenfeld, M. Ben-Bassat, A. Meshorer, R. Bakimer, and E. Mozes.** 1988. Induction of a systemic lupus erythematosus-like disease in mice by a common human anti-DNA idiotype. *Proc. Natl. Acad. Sci. USA* **85:**2260–2264.
44. **Naparstek, Y., D. Duggan, A. Schattner, M. P. Madaio, F. Goni, B. Frangione, B. D. Stollar, E. A. Kabat, and R. S. Schwartz.** 1985. Immunochemical similarities between monoclonal antibacterial Waldenstrom's macroglobulins and monoclonal anti-DNA lupus autoantibodies. *J. Exp. Med.* **161:**1525–1538.
45. **Oldstone, M. B. A.** 1989. Molecular mimicry as a mechanism for the cause and as a probe uncovering etiologic agent(s) of autoimmune disease. *Curr. Top. Microbiol. Immunol.* **145:**127–135.
46. **Oldstone, M. B. A.** 1998. Molecular mimicry and immune-mediated diseases. *FASEB J.* **12:**1255–1265.
47. **Pierangeli, S. S., X. W. Liu, G. Anderson, J. H. Barker, and E. N. Harris.** 1996. Thrombogenic properties of murine anti-cardiolipin antibodies induced by beta 2 glycoprotein 1 and human immunoglobulin G antiphospholipid antibodies. *Circulation* **94:**1746–1751.
48. **Puri, V., A. Bookman, E. Yeo, R. Cameron, and E. J. Heathcote.** 1999. Antiphospholipid antibody syndrome associated with hepatitis C infection. *J. Rheumatol.* **26:**509–510.
49. **Putterman, C., W. Limpanisithikul, and B. Diamond.** 1996. The double-edged sword of the immune response: mutational analysis of a murine anti-pneumococcal, anti-DNA antibody. *J. Clin. Investig.* **97:**2251–2259.
50. **Putterman, C., and B. Diamond.** 1998. Immunization with a peptide surrogate for double-stranded DNA (dsDNA) induces autoantibody production and renal immunoglobulin deposition. *J. Exp. Med.* **188:**29–38.
51. **Radway-Bright, E. L., M. Inanc, and D. A. Isenberg.** 1999. Animal models of the antiphospholipid syndrome. *Rheumatology* **38:**591–601.
52. **Ray, S. K., C. Putterman, and B. Diamond.** 1996. Pathogenic autoantibodies are routinely generated during the response to foreign antigen: a paradigm for autoimmune disease. *Proc. Natl. Acad. Sci. USA* **93:**2019–2024.
53. **Rosa, F. D., and V. Barnaba.** 1998. Persisting viruses and chronic inflammation: understanding their relation to autoimmunity. *Immunol. Rev.* **164:**17–27.
54. **Sabbatini, A., S. Bombardieri, and P. Migliorini.** 1993. Autoantibodies from patients with systemic lupus erythematosus bind a shared sequence of SmD and Epstein-Barr virus-encoded nuclear antigen EBNA I. *Eur. J. Immunol.* **23:**1146–1152.
55. **Schwimmbeck, P. L., T. Dyrberg, D. B. Drachman, and M. B. A. Oldstone.** 1989. Molecular mimicry and myasthenia gravis. *J. Clin. Investig.* **84:**1174–1180.
56. **Sela, O., A. El-Roeiy, D. A. Isenberg, R. C. Kennedy, C. B. Colaco, J. Pinkhas, and Y. Shoenfeld.** 1987. A common anti-DNA idiotype in sera of patients with active pulmonary tuberculosis. *Arthritis Rheum.* **30:**50–56.
57. **Shi, W., B. H. Chong, and C. N. Chesterman.** 1993. Beta 2-glycoprotein I is a requirement for anticardiolipin antibodies binding to activated platelets: differences with lupus anticoagulants. *Blood* **81:**1255–1262.
58. **Shoenfeld, Y.** 1994. Idiotypic induction of autoimmunity: a new aspect of the idiotypic network. *FASEB J.* **8:**1296–1301.
59. **Soldan, S. S., R. Berti, Nsalem, P. Secchiero, L. Flamand, P. A. Calabresi, M. B. Brennan, H. W. Malon, H. F. NcFarland, H. C. Lin, et al.** 1997. Association of human herpes virus 6 (HHV-6) with multiple sclerosis: increased IgM response to HHV-6 early antigen and detection of serum HHV-6 DNA. *Nat. Med.* **3:**1394–1397.
60. **Spellerberg, M. B., C. J. Chapman, C. I. Mockridge, D. A. Isenberg, and D. K. Stevenson.** 1995. Dual recognition of lipid A and DNA by human antibodies encoded by the VH-21 gene: a possible link between infection and lupus. *Hum. Antibodies Hybridomas* **6:**52–56.

61. **Sthoeger, Z. M., E. Mozes, and B. Tartakovsky.** 1993. Anti-cardiolipin antibodies induce pregnancy failure by impairing embryonic implantation. *Proc. Natl. Acad. Sci. USA* **190:**6464–6467.

62. **Strasser, A., S. Whittingham, D. L. Vaux, E. Webb, M. L. Bath, J. M. Adams, and A. W. Harris.** 1990. Enforced BCL2 expression in B-lymphoid cells prolongs antibody responses and elicits autoimmune disease. *Proc. Natl. Acad. Sci. USA* **88:**8661–8665.

63. **Talbot, P. J.** 1997. Virus-induced autoimmunity in multiple sclerosis: the coronavirus paradigm. *Adv. Clin. Neurosci.* **7:**215–233.

64. **Tincani, A., G. Balestrieri, F. Allegri, R. Cattaneo, A. Fornasieri, M. Li, A. Sinico, and G. D'Amico.** 1993. Induction of experimental SLE in naive mice by immunization with human polyclonal anti-DNA antibody carrying the 16/6 idiotype. *Clin. Exp. Rheumatol.* **11:**129–134.

65. **Tonietti, G., M. B. A. Oldstone, and F. J. Dixon.** 1970. The effect of induced chronic viral infections on the immunologic diseases of New Zealand mice. *J. Exp. Med.* **132:**89–109.

66. **Von Herrath, M., and M. B. A. Oldstone.** 1996. Virus-induced autoimmune disease. *Curr. Top. Microbiol. Immunol.* **8:**878–885.

67. **Waisman, A., and E. Mozes.** 1993. Variable region sequences of autoantibodies from mice with experimental systemic lupus erythematosus. *Eur. J. Immunol.* **23:**1566–1573.

68. **Waisman, A., Y. Shoenfeld, M. Blank, R. J. Ruiz, and E. Mozes.** 1995. The pathogenic human monoclonal anti-DNA that induces experimental systemic lupus erythematosus in mice is encoded by a VH4 gene segment. *Int. Immunol.* **7:**689–696.

69. **Waisman, A., P. L. Ruiz, E. Israeli, E. Eilat, S. Konen-Waisman, H. Zinger, M. Dayan, and E. Mozes.** 1997. Modulation of murine systemic lupus erythematosus with peptides based on complementarity determining regions of a pathogenic anti-DNA monoclonal antibody. *Proc. Natl. Acad. Sci. USA* **94:**4620–4625.

70. **Wang, M. X., D. A. Kandiah, K. Ichikawa, M. Khamashta, G. Hughes, T. Koike, R. Roubey, and S. A. Krilis.** 1995. Epitope specificity of monoclonal anti-beta 2-glycoprotein I antibodies derived from patients with the antiphospholipid syndrome. *J. Immunol.* **155:**1629–1636.

71. **Wucherpfennig, K. W., and J. L. Strominger.** 1995. Molecular mimicry in T cell-mediated autoimmunity: viral peptides activate human T cell clones specific for myelin basic protein. *Cell* **80:**695–705.

72. **Yanez, A., L. Cedillo, O. Neyrolles, E. Alonso, M. C. Prevost, J. Rojas, H. L. Watson, A. Blanchard, and G. H. Cassell.** 1999. Mycoplasma penetrans bacteremia and primary antiphospholipid syndrome. *Emerg. Infect. Dis.* **5:**164–167.

73. **Yuki, N., Y. Tagawa, and S. Handa.** 1996. Autoantibodies to peripheral nerve glycosphingolipids SPG, SLPG and SGPG in Gillain-Barré syndrome and chronic inflamatory demyelinating polyneuropathy. *J. Neuroimmunol.* **70:**1–6.

74. **Zhang, L., K. Jacobsson, K. Strom, M. Lindberg, and L. Frykberg.** 1999. Staphylococcus aureus expresses a cell surface protein that binds both IgG and beta2-glycoprotein I. *Microbiology* **145:**177–183.

Molecular Mimicry, Microbes, and Autoimmunity
Edited by M. W. Cunningham and R. S. Fujinami

Chapter 17

Contributions of Viruses and Immunity as Causes of Diabetes and Development of Strategies for Treatment and Prevention of Autoimmune Disease

Matthias G. von Herrath

What is the evidence that viruses or other microbes can precipitate autoimmune diabetes? First, viruses have been found in organs affected by the disease, such as the pancreatic islets of Langerhans (61, 103). In this situation, β-cell destruction can result from direct viral lysis (33, 70) or, alternatively, can occur as a consequence of bystander killing mediated by antiviral cytokines or by antiviral T lymphocytes that cross-react with islet cell antigens (mimicry). Additionally, a nonlytic virus can persist and replicate in β cells of the islets and then alter their insulin-producing function without killing them (54, 68). To define molecular mimicry with an example, coxsackie B4 virus, which shares sequence homology with glutamic acid decarboxylase (GAD), an islet cell self-antigen (6, 7, 37, 38), can cross-react with that islet cell antigen. In contrast, by using another model system, cytokine-mediated bystander killing in the absence of mimicry has been suggested to play a role in β-cell destruction after infection with coxsackie B4 virus (31, 32). Second, the infectious agent may not need to be present in the target organ or cell per se, if molecular mimicry and the resulting activation of autoreactive lymphocytes are the only factors required to cause disease (64, 65, 67). Many such examples of associations between infectious agents, their sequences, and self-determinants and diseases are discussed in detail in the introductory section of this book (see chapter 1) and in its other chapters (5, 77, 101).

Why, then, despite the high degree of circumstantial evidence, has it been difficult to obtain definite proof for the involvement of mimicry in causing autoimmune diseases in vivo? One reason is that conformational rather than linear sequence comparisons are required to precisely predict the outcome of major histocompatibility complex (MHC) and peptide T-cell interactions (102). At present, it is still difficult to model such interactions or to know the important recognition residues needed to compare self with microbial ligands. Another reason is the recent observation that cross-reactivity with certain altered peptide ligands (APLs) (APL effects) can turn off autoreactive T cells and ameliorate autoimmunity (73). The difference between an agonist and an APL effect can frequently be determined by a change in only one amino acid (69). Additionally, a minimal number of autoreactive lymphocytes must be activated to cause disease (86, 88, 92, 94), a finding that has not

Matthias G. von Herrath • Division of Virology, IMM6, Department of Neuropharmacology, The Scripps Research Institute, 10550 N. Torrey Pines Road, La Jolla, CA 92037.

always been taken into account in studies that involve transfers of in vitro peptide-stimulated lymphocytes. In such experiments, the number of autoreactive cells may be artificially augmented in vitro and not reflect in vivo events. All these issues illustrate the complexity of associating viral infections with autoimmunity.

Because of these uncertainties, a transgenic animal model was designed to study how virus induces autoimmune diabetes and the role played by molecular mimicry. Here I introduce and explain the rat insulin promoter (RIP)-lymphocytic choriomeningitis virus (LCMV) model of virally induced autoimmune diabetes. Special emphasis is given to the issues of molecular mimicry, the potential involvement of several viral infections in inducing or abrogating insulin-dependent diabetes mellitus (IDDM), new therapeutic perspectives, and pathogenically important findings that stem from the use of this model. In this chapter, I will discuss these issues, as well as other factors that precipitate or prevent IDDM.

THE RIP-LCMV TRANSGENIC MOUSE MODEL OF VIRUS-INDUCED DIABETES

Description of the Model and Rationale

The RIP-LCMV transgenic mouse model was developed in 1991 simultaneously in the laboratories of Michael Oldstone and colleagues (66) and Rolf Zinkernagel and colleagues (63). The pancreatic β cells of these mice express, under control of RIP, either the nucleoprotein (NP) or glycoprotein (GP) of LCMV as a target antigen (self-antigen) (88). Expression of the viral antigen per se does not lead to β-cell dysfunction, islet cell infiltration, hyperglycemia, or spontaneous activation of autoreactive (anti-LCMV) lymphocytes (88). However, infection with LCMV induces autoimmune diabetes (IDDM) in 95 to 100% of such transgenic mice, whereas their nontransgenic littermates never develop IDDM or insulitis after LCMV infection (Fig. 1). IDDM is characterized by infiltration of mononuclear cells into the islets, an infiltrate that contains equal amounts of $CD4^+$ and $CD8^+$ T cells and B lymphocytes; in addition, significant amounts of dendritic cells and macrophages are present (Fig. 2) (88, 95). Thus, RIP-LCMV transgenic mice are not tolerant to the viral self-antigen, since autoreactive lymphocytes escape thymic selection and are present in the periphery. However, these cells are not activated, unless LCMV is administered and infects professional antigen-presenting cells (APCs), which then activate naive lymphocytes. Upon activation of LCMV-reactive (anti-self) lymphocytes, their unresponsiveness toward the viral transgene in the β cells is broken and islet cell death results (90). Clearly, MHC class I-restricted, LCMV-specific autoreactive $CD8^+$ cytotoxic lymphocytes (CTLs) (LCMV-CTLs) are a crucial factor for the development of IDDM, since neither diabetes nor islet cell infiltration occurs in their absence (88). However, they are probably not the only prerequisite for islet cell destruction, since adoptive transfer of such cells most often leads to insulitis but very rarely to diabetes (91), a provocative finding that is discussed in more detail later in this chapter (section on costimulation and professional APCs). It is important to understand that mice of all RIP-LCMV transgenic lines clear LCMV infection with precisely the same kinetics as healthy nontransgenic mice of the same MHC background (66). This removal of LCMV is entirely dependent on the effector function of $CD8^+$ CTLs (10, 11, 66). Establishment of this transgenic model demonstrates that unresponsive, naive autoreactive lymphocytes can emerge from the thymus and occu-

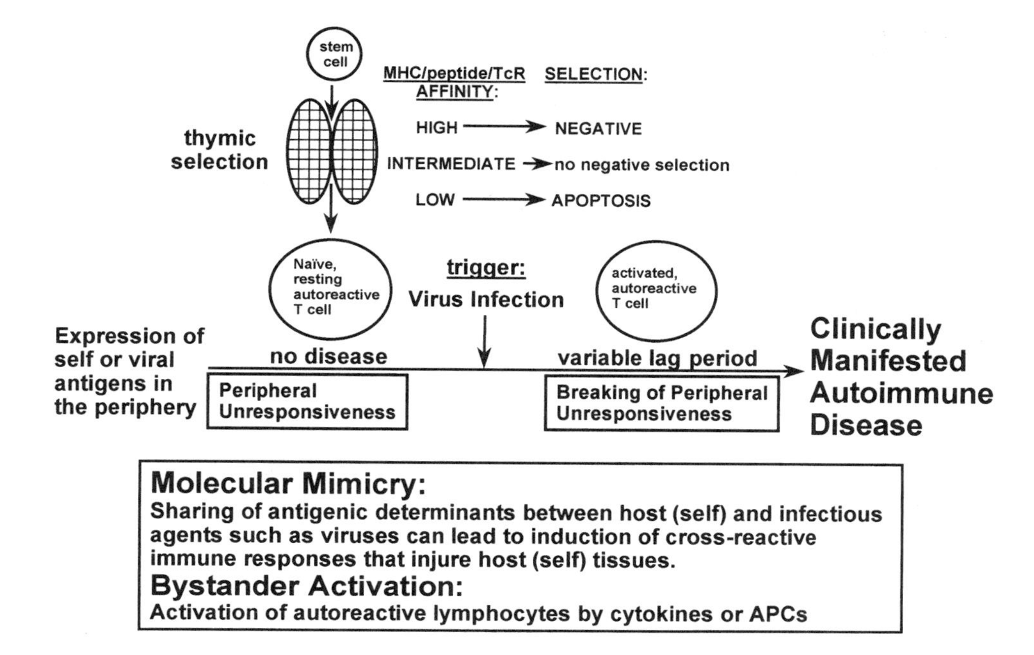

Figure 1. Principle of RIP-LCMV transgenic mouse model. Pancreatic β cells of RIP-LCMV transgenic (tg) mice express the NP or GP of LCMV under control of RIP. As the viral gene is integrated into the host's genome and passed on to progeny mice, it becomes essentially a host self-antigen. These mice are not tolerant to the LCMV (self) transgene but are unresponsive since naive autoreactive (LCMV-specific) lymphocytes escape thymic selection and are present in the periphery. Such cells are not activated under normal circumstances. However, infection with LCMV or expression of certain cytokines (IFN-γ) or activation molecules (B7.1) breaks this unresponsiveness, and as a consequence, activated LCMV (anti-self)-specific lymphocytes attack the β cells that express the appropriate LCMV proteins. The result is that autoimmune diabetes develops in 95 to 100% of RIP-LCMV transgenic mice after LCMV infection.

py the periphery. Their activation follows infection with a virus that shares antigens with the self (viral) transgene expressed in β cells. Since the virus protein expressed in these β cells is the product of a gene integrated in the host's chromosome that is passed on to progeny, the viral transgene can essentially be considered a "self-antigen." The subsequent virus infection involves antigen presentation by APCs and in this way breaks peripheral unresponsiveness. The activated autoreactive CTLs are instrumental in initiating β-cell destruction, which results in disease (IDDM).

Advantages of the RIP-LCMV Transgenic Model

The RIP-LCMV transgenic mouse model differs in some important aspects from other established antigen-specific models of autoimmunity as well as from the nonobese diabetic (NOD) mouse model of spontaneous IDDM (4). Below I discuss the advantages of the RIP-LCMV model and the questions best suited to be answered by using this system.

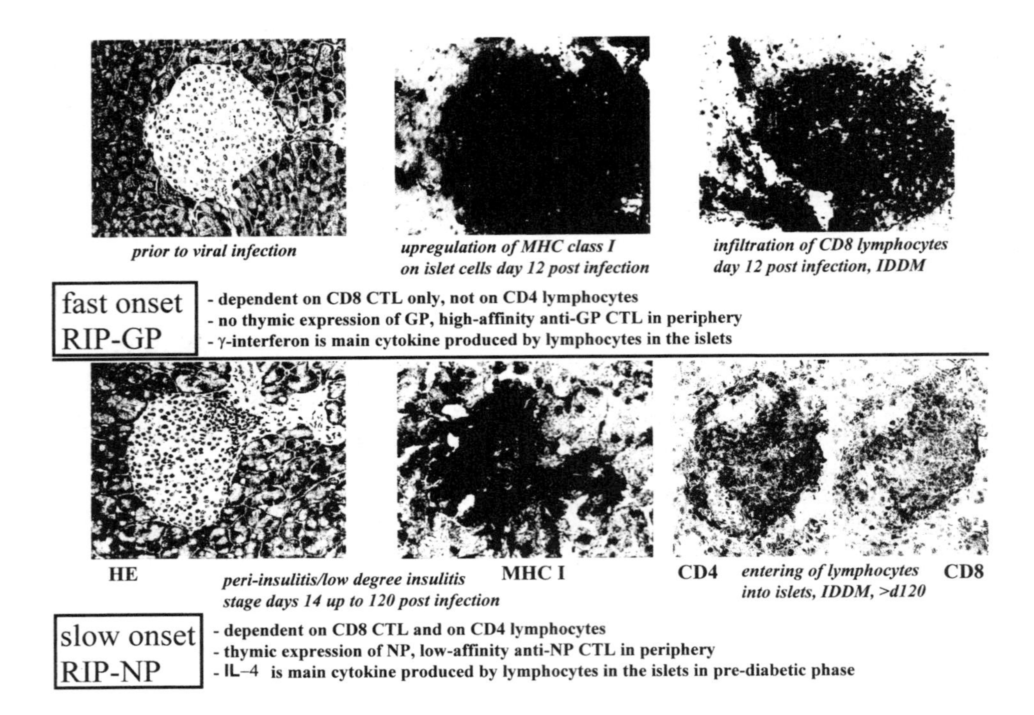

Figure 2. Histological findings in the RIP-LCMV transgenic models of slow- and rapid-onset IDDM. RIP-LCMV transgenic mice without thymic expression of the viral self-antigens expressed in their β cells develop $CD4^+$ independent, rapid-onset IDDM. A predominance of INF-γ-producing lymphocytes enters the islets. In contrast, the slow-onset IDDM that develops in mice with thymic expression is $CD4^+$ dependent and has a lag phase before IDDM occurs. This is a nondiabetic period during which lymphocytes mainly appear around, but not in, islets. Of the lymphocytes found within islets, the majority express or secrete IL-4 during this stage. He, hematoxylineosin staining.

In contrast to RIP-LCMV transgenic mice, NOD mice develop spontaneous diabetes (4). Neither the initiating cause(s) nor the trigger(s) is known (4). However, a multitude of genetic factors has been discovered, indicating that a complex network of genes for susceptibility and resistance to diabetes is involved (43, 99). Clearly, the initial anticipation of mapping autoimmune diabetes to a few cardinal genetic factors is overly optimistic (99). Furthermore, after more than 15 years of research with this model, the islet cell autoantigens initially recognized by autoreactive lymphocytes are still unclear and controversial, although recent evidence (using antisense technology) has implied that GAD (glutamic acid decarboxylase) is the primary self-antigen in IDDM of NOD mice (104)—which is difficult to reconcile with the fact that high numbers of insulin-B-specific CTL are present in NOD islets very early (100a). Insulin and other self-antigens have also been discovered (47), but the T- and B-cell responses to these autoantigens have been difficult to standardize among separate laboratories (unpublished report from the 2nd IDS Congress, Canberra, Australia, December 1996). Thus, negligible to modest knowledge about the location, activation, trafficking, and numbers of autoreactive T-lymphocytes in NOD mice is available, despite their massive usage in multiple laboratories. This dilemma has partly been addressed by generating T-cell receptor (TCR) transgenic NOD (BDC2.5) mice and using such cells for adoptive transfers (12, 27). However, again, the islet cell antigen recognized by these T cells is unknown. Table 1 illustrates the advantages offered by the RIP-LCMV transgenic mouse model which can be used to question several aspects that are not easily addressed by studying NOD mice. First, and of particular importance, in the RIP-LCMV model, the numbers and the activities of autoreactive CTLs can be defined and monitored over time (90). Second, with the RIP-LCMV model, the precise time point for induction of autoimmunity can be chosen experimentally, allowing clear comparison of immunity before and after the onset of disease (88). Third, RIP-LCMV transgenic mice, unlike NOD mice, do not have a general predisposition for autoimmunity and do not exhibit autoimmune disease at multiple sites, e.g., sialoadenitis or thyroiditis (34). Thus, RIP-LCMV transgenic mice have strengths not provided by the NOD model and, as such, nicely complement studies with the latter strain.

How do RIP-LCMV transgenic mice compare to other models of transgenic mice that express target antigens in β cells? In one, the hemagglutinin (HA) of influenza virus was expressed under control of the RIP (59). These RIP-HA transgenic mice developed a low degree of spontaneous IDDM. Most notably, however, infection with influenza virus (PR8) provided negligible or very modest enhancement of IDDM (72). When such mice were crossed with TCR transgenic mice in which the TCR was made specific for a determinant of HA, the incidence of spontaneous IDDM in their offspring increased markedly (59). These studies demonstrated that higher numbers of autoreactive lymphocytes can achieve the threshold level required for autoimmunity to develop. Reasons for the low incidence of PR8-induced IDDM in these mice compared to the high incidence of LCMV-induced IDDM in RIP-LCMV transgenic mice might be the different replication kinetics of influenza virus compared to those of LCMV, the viruses' unequal abilities to infect the pancreas (LCMV has a greater ability than influenza virus to infect the pancreas), as well as the discrepancy in the levels of expression of HA and LCMV proteins in β cells. In a similar model of RIP–simian virus 40 (SV40)–T-antigen transgenic mice, both tolerance and autoimmune phenotypes were also observed (1, 17, 24, 78).

In summary, the RIP-LCMV model is unique. It provides a reproducibly high incidence of IDDM that is not altered by diet, handling, etc., as is usual in NOD mice. Fur-

Table 1. Why the RIP-LCMV model?[a]

Characteristic	NOD mice	RIP-LCMV transgenic mice
Known initiating self-antigen	No	Yes
Experimentally defined prediabetic lag or "honeymoon" phase	No	Yes
Trackability of autoreactive cells	Yes and no	Yes
Environmental defined trigger allows direct comparison of before and after	No	Yes
CD4 and CD8 dependence	Yes	Yes
Genetic factors (MHC and non-MHC)	Yes	Yes

[a] Since human diabetes only shows a limited concordance in monocygotic twins, this model demonstrates that an external trigger (virus) can induce autoimmune disease.

thermore, RIP-LCMV transgenic mice do not manifest peripheral tolerance. Perhaps the fact that LCMV infects the pancreas, but influenza virus and SV40 do not, accounts for some of the differences between these RIP models. Also, unlike SV40 and influenza virus, which have a nuclear phase of replication, LCMV does not. Because the expression of LCMV mRNA in the nucleus (as a transgene) is less stable than that in other viruses, just the correct amount of LCMV transgene may be provided to promote IDDM.

IMMUNOPATHOGENESIS OF IDDM IN RIP-LCMV MICE

How Virus Can Induce Rapid- and Slow-Onset IDDM: Role of the Thymus in Determining Affinities and Numbers of Autoreactive CTLs

Analysis of several different lines of RIP-LCMV transgenic mice developed by Michael Oldstone and colleagues (66) resulted in the discovery of an interesting difference: some lines developed rapid-onset IDDM that occurred 10 to 18 days postinfection with LCMV. This corresponded to the lines developed independently by Rolf Zinkernagel, Pam Ohashi, and colleagues (63). In contrast, some lines developed in the Oldstone laboratory displayed slow-onset IDDM that occurred 2 to 6 months after LCMV infection (88). Comparison of the RIP-LCMV-GP rapid-onset line and the RIP-LCMV-NP slow-onset line showed that whereas both lines expressed equivalent levels of the transgene in the pancreas (87, 88), only the slow-onset RIP-NP lines also expressed the viral transgene in the thymus. Further analysis showed that thymic expression of the transgene led to deletion of most LCMV-specific (anti-self) high-affinity CTLs, but lower-affinity CTLs were detectable in the peripheries of these mice (88). Although these low-affinity CTLs were able to trigger IDDM, its onset was delayed; in addition, IDDM did not develop in the absence of $CD4^+$ helper lymphocytes. In contrast, rapid-onset IDDM in RIP-GP mice did not depend on the participation of $CD4^+$ (88). Thus, thymic expression of a self-antigen does not necessarily eliminate all autoreactive lymphocytes (see also Fig. 1 and 2). Those low-affinity CTLs that remain can cause slowly progressive self-destruction and autoimmune disease, a process dependent on the presence of $CD4^+$ help. Thus, the slow-onset RIP-LCMV model reproduces a scenario similar to that of human IDDM. Furthermore, because of the prolonged lag phase that precedes IDDM, the slowly developing events offer a suitable window for the testing of therapeutic interventions.

Other evidence (T. Dyrberg and M. Oldstone, unpublished data) suggests that the autoimmune attack initially induced against the LCMV transgene involves other islet cell antigens before the onset of IDDM (Fig. 3). This is reflected in the generation of antibodies to GAD in RIP-LCMV transgenic mice with slow-onset IDDM and parallels the findings that the occurrence of islet cell antibodies precedes IDDM in humans (6, 7, 41). Furthermore, the anti-GAD antibody serves as a marker in both RIP-LCMV transgenic animals and patients for initiation of therapies before the onset of clinical diabetes.

Role of MHC Genes in Virus-Induced IDDM

To test the influence of the transgenic MHC background on the development of virus-induced IDDM in RIP-LCMV transgenic mice, RIP-NP mice (slow-onset IDDM) were backcrossed into mice with the MHC backgrounds *H-2*^{d}, *H-2*^{d}, and *H-2*^{k}. IDDM occurred with midrange timing (within 2 to 6 months) in the *H-2*^{b} mice, more rapidly in *H-2*^{d} mice

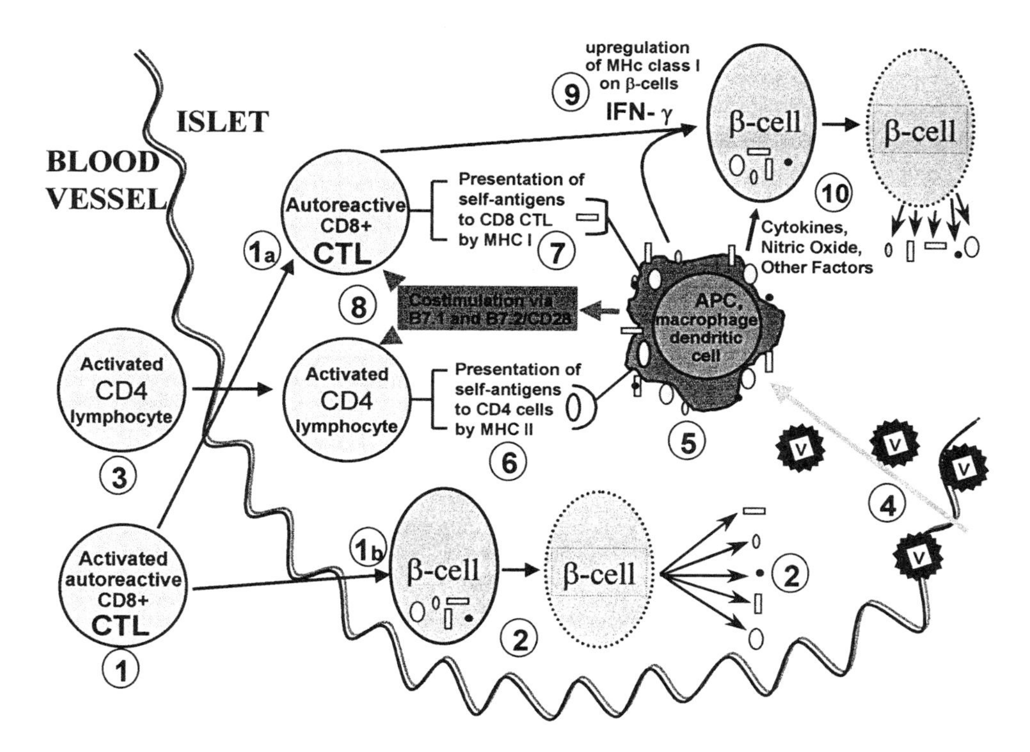

Figure 3. Important events (numbered 1 through 10) in the pathogenesis of IDDM. Autoreactive lymphocytes (events 1 and 3), once activated, enter the islets in large numbers. There, $CD8^+$ CTLs kill (event 1b) some β cells using perforin, but not a sufficient amount to cause clinical IDDM. Antigens released by the dying β cells (event 2) are taken up in the islets by professional APCs (event 5), such as dendritic cells or macrophages. Additionally, an external noxious factor like viral infection (event 4) of the pancreas or another inflammation is likely needed to recruit and activate enough APCs into the area. These APCs (event 5) can then present β-cell antigens to infiltrating CD4 and CD8 lymphocytes (events 6 and 7) and provide crucial costimulatory signals, for example, through B7.1 and B7.2 CD28 interactions (event 8). As a result, the infiltrating lymphocyte population is expanded and can attack more β cells (event 9). Inflammatory cytokines are also secreted and may directly contribute to β-cell destruction (event 10). In this way, the autoimmune process is perpetuated locally until all islet cells are destroyed so that IDDM results. The three main players in this model, autoreactive lymphocytes, activated APCs, and regulatory factors, are all targets with immunotherapeutic potential.

(within 1 to 2 months), and slowly (>8 months) or not at all (0 to 40%) in $H\text{-}2^k$ mice (85, 88). The incidence of IDDM with respect to the different MHC backgrounds correlated well with the magnitude of the LCMV (NP)-specific CTL response when the numbers and/or affinities of the autoreactive CTLs generated were measured. RIP-NP transgenic $H\text{-}2^d$ mice generated greater numbers of low-affinity CTLs than RIP-NP transgenic $H\text{-}2^b$ mice, which in turn generated more CTLs than $H\text{-}2^k$ mice did. Our studies showed that the excess of CTLs in RIP-NP transgenic *H-2*^{d} mice compared to the number in RIP-NP transgenic *H-2*^{b} mice stemmed from the higher affinity of the NP epitope in *H-2*^{b} mice that resulted from more efficient negative selection in the thymus. In many studies, MHC genes were involved in the susceptibility of humans to IDDM (79). Thus, the RIP-LCMV transgenic mouse model illustrates and incorporates this issue and additionally offers a potential mechanism of its development. Since most responses to pathogens, particularly viruses, are controlled in an MHC-restricted manner, the linkage of IDDM to certain MHC class I and II alleles serves as a widely applicable experimental model.

Role of Non-MHC-Linked Polymorphism in Susceptibility to IDDM

Non-MHC-linked genes can also influence the incidence of IDDM in animal models. For example, a multitude of IDDM susceptibility genes has been identified in the NOD mouse (99). Furthermore, studies with RIP-influenza virus HA transgenic mice have shown that the incidence of IDDM varies significantly between transgenic mice bred to mice that have different genetic backgrounds but that are alike in MHC type (75). Similarly, colleagues and I observed recently that the incidence of rapid-onset as well as slow-onset IDDM was dramatically reduced (90%) in RIP-LCMV transgenic *H-2*^{b} mice that were crossed to mice of the Sv129 *H-2*^{b} strain for one generation (M. von Herrath, unpublished results). This reduction occurred even though these two strains had identical immune responses to LCMV and identical rates of viral clearance. Currently, colleages and I are analyzing whether the 129 *H-2*^{b} background provides genes protective against IDDM or whether the two strains differ in β-cell regeneration.

Numbers of Autoreactive Lymphocytes Correlate with Incidence and Severity of Autoimmune Disease

The RIP-LCMV transgenic mouse model has also proved to be useful for testing of whether the number of self-reactive CTLs generated correlates with the severity of an autoimmune process and incidence of IDDM and, if so, how well they correspond. Evidence was obtained from four different experiments. In the first, RIP-LCMV transgenic mice were infected with vaccinia virus recombinants that expressed the LCMV NP or GP. Infection with these recombinants induced 10 to 50 fewer LCMV NP- or GP-specific CTLs, according to limiting dilution analysis, and lower levels of intracellular cytokine (gamma interferon [INF-γ]) expression than those induced by LCMV infection alone (66, 90, 91). None of the mice infected with vaccinia virus-LCMV recombinants developed IDDM, but insulitis was observed. The second line of evidence emerged from studies with RIP-LCMV × RIP-B7.1 double transgenic mice. B7.1 is an important costimulatory molecule that is expressed on professional APCs such as macrophages and dendritic cells (22, 23, 25, 50). Our results showed that B7.1, when expressed on β cells, led to local activation and amplification of CTLs. IDDM occurred spontaneously without LCMV infection in RIP-B7.1 × RIP-GP mice (no thymic expression of the GP), and the virally induced, slow-onset IDDM normally observed in RIP-LCMV-NP transgenic mice with thymic expression of the transgene was markedly accelerated (90). These differences in the incidence of IDDM directly correlated with the numbers of CTL precursors (pCTLs) isolated locally from the pancreas or the spleen. Local expression of B7.1 led to a 10-fold increase in the number of pCTLs in the pancreata of RIP-LCMV thymic expressor transgenic mice, an increase that correlated with the accelerated incidence of IDDM. The findings were similar for the spontaneous IDDM in RIP-LCMV nonthymic expressor transgenic mice: expression of B7.1 induced local activation and amplification of CTLs in the pancreas, resulting in numbers of CTLs comparable to those found in RIP-LCMV single transgenic mice with IDDM (90). The third set of experiments used several strains of LCMV, each characterized by the induction of large or small numbers of anti-self (LCMV-NP) pCTLs in RIP-NP transgenic mice. Again, the incidence of IDDM correlated directly with the amount of anti-self (viral) pCTLs generated. For example, IDDM was clearly evident in 90 to 100% of transgenic mice infected with our regular strain, LCMV Armstrong, which induces 1/50 to 1/300 LCMV-NP pCTLs per overall splenocytes 7 days postinfection. In contrast, LCMV Pasteur induced pCTL levels of about 1/20,000, and fewer than 15% of RIP-NP mice manifested IDDM. In general, when less than 1/6,000 pCTLs were generated, no diabetes developed. Lastly, this type of cutoff for the number of autoreactive pCTLs required for disease development was also reflected by comparing the incidence of IDDM in RIP-LCMV thymic expressors and nonexpressors with the $H\text{-}2^b$ background. Nonexpressors developed IDDM within 2 weeks (1/100 pCTLs), whereas thymic expressor lines had IDDM at 2 to 6 months postinfection ($<$1/6,000 pCTLs).

Clearly, a threshold effect is present, in that the number of autoreactive lymphocytes available determines whether disease develops. Several studies with animal models and humans have documented low numbers of naive, nonactivated autoreactive cells in the peripheral blood (20, 80). Their quantity or degree of activation is presumably too low to cause disease. However, if external factors (such as virus-induced cellular activation [cross-activation]) or predisposing genetic factors fall together, this fragile equilibrium can be broken so that a pathogenic process begins the cascade of autoimmunity (80). The fact that the number of autoreactive lymphocytes correlates with the incidence and severity of IDDM has been shown in several animal models other than those described above (57), an outcome that can be exploited therapeutically.

As shown recently, MHC class I blocking peptides that bind with a high affinity to the MHC D^b allele but that do not engage the TCR of LCMV $H\text{-}2D^b$-restricted lymphocytes can, in vivo, prevent the expansion of autoreactive lymphocytes and the onset of IDDM in RIP-LCMV transgenic mice (86). LCMV-CTL levels dropped below 1/6,000 splenocytes 7 days postinfection in blocking peptide-treated mice (86). The design and development of peptide analogs that do not undergo rapid in vivo degradation will be necessary to bring such an approach closer to use for the clinical treatment of patients.

Role of Professional APCs and Costimulation

An important question for the pathogenesis of autoimmune diseases is, are large quantities of activated, autoreactive lymphocytes alone sufficient to cause disease or are other factors required? Recent evidence suggests that factors from the islets are also necessary to drive the local inflammatory process that leads to IDDM (30). In RIP-LCMV transgenic mice, colleagues and I observed that adoptive transfer of large numbers of highly activated LCMV (self-transgene)-specific CTLs into RIP-LCMV transgenic uninfected recipients rarely led to IDDM. Although insulitis routinely followed, β cells were not usually destroyed. This prompted us to consider whether unique pathogenic changes in islets of LCMV-infected transgenic mice preceded infiltration and accumulation of autoreactive

LCMV-specific CTLs. Indeed, investigation showed that upregulation of MHC class II and activation of macrophages occurred around the islets as soon as 4 days after infection, clearly before the first CTLs and CD4 lymphocytes were seen in the pancreas (day 7) (91). Possibly, these changes resulted, at least partially, from the presence of infectious virus in the pancreas. Once activated lymphocytes reached the target organ, professional presentation of viral antigens initiated the expansion of an activated lymphocyte population. During this process, the viral transgene expressed on β cells was also recognized, which likely resulted in the direct or indirect killing of β cells by LCMV CTLs. Still not clear is how much of a role cross-presentation (41a) of β-cell antigens and LCMV transgenes in the pancreatic draining node plays in the continuation of the autoimmune process. At later stages, at about the time when clinical IDDM develops and most β cells are destroyed, islet infiltrates resembled a "mini-lymph node" around a network of dendritic cells, and activated macrophages were found only at the periphery of islets (88). Similar findings have been reported with the RIP-HA transgenic mouse model. The importance of costimulation in the development of autoimmune diabetes is also underlined by our findings with RIP-B7.1 transgenic mice (see previous paragraph), which undergo markedly enhanced IDDM. The need for professional APCs and costimulation for the activation and sustenance of autoreactive lymphocytes can be exploited therapeutically. That is, systemic treatment with antibodies to the B7.1, B7.2, or CD40 ligand given during a susceptible phase after LCMV infection but preceding clinical IDDM can completely abrogate the development of autoimmune disease (M. von Herrath, unpublished observations).

Role of Cytokines

Cytokines are important messenger and regulatory molecules that constitute a major part of the "cross-talk" network that connects lymphocytes, APCs, and other cells of the immune system. In the past, the so-called Th1/Th2 paradigm gained increasing importance but also stimulated controversy in characterizing the regulation of immune processes (28, 36, 48, 81). This scheme divides cytokines into two main groups. The first, the Th1 or Tc1 cytokines, includes INF-γ, tumor necrosis factor alpha (TNFα), interleukin 12 (IL-12), and others, all of which are classified as inflammatory cytokines. Secretion of these cytokines results in upregulation of MHC molecules on APCs, enhancement of CTL killing, increased activation-induced cell death, proliferation, and recruitment of other cells that secrete Th1 cytokines. The main immunoglobulin subclass induced by a Th1 response is immunoglobulin G2a (IgG2a). In contrast, the second or Th2/Tc2 group of cytokines including IL-4, IL-5, IL-6, and IL-13 is believed to exert mainly a regulatory function (58). First and foremost of these is the dampening of Th1 processes, but other tasks are performed, such as activation of B cells to secrete IgE and IgG1, enhancement of cellular proliferation, and prolongation of naive lymphocytes' lives. Yet, the foregoing findings make it quite clear that the Th1/Th2 paradigm is not an entirely "black-and-white" situation and that, within a "gray zone," Th2 cytokines might enhance or maintain an inflammatory process and Th1 cytokines might conversely terminate an inflammation through induction of more activation-dependent cell death. For example, P. Marrack and colleagues have noted that IL-4 can significantly increase the life spans of naive lymphocytes in vitro (84). Second, increasing the level of IL-2 can lower the incidence of IDDM in NOD mice by increasing the level of apoptosis of autoreactive lymphocytes (71). Therefore, evaluation of cytokine-dependent effects is required, because these effects vary among autoimmune diseases and depend on the organ affected. In the NOD mouse, Th1 and Th2 clones have been isolated from the pancreas, and in some instances, Th2 clones conferred protection (36), whereas in other cases, they were diabetogenic (28; D. Wegmann, unpublished results).

With this background, the role of cytokines in the RIP-LCMV model of virus-induced IDDM was explored via two routes. First, several cytokines were overexpressed, i.e., were expressed above basal levels, locally in β cells by generating RIP-cytokine transgenic mice that were then crossed with RIP-LCMV transgenic mice. Second, cytokine knockout mice were backcrossed to the RIP-LCMV transgenic mice. The incidences of spontaneous and LCMV-induced IDDM were then monitored in both groups of mice and compared.

When INF-γ was overexpressed in β-cells, IDDM occurred spontaneously without LCMV infection in RIP-LCMV × RIP–IFN-γ double transgenic mice. Massive inflammation and upregulation of MHC class I molecules pervaded the pancreas, and anti-LCMV CTLs were activated spontaneously without LCMV infection. Such effector CTLs were found in the islets (45, 74). In contrast, when RIP-LCMV transgenic INF-γ-deficient mice were infected with LCMV, IDDM did not develop (95). Our studies showed that this was the case, even though good levels of anti-LCMV (autoreactive) CTLs were generated and infiltrated the pancreas. However, the MHC class I upregulation on β cells usually seen in RIP-LCMV transgenic mice after LCMV infection was not detectable. The conclusion from these studies was that (i) INF-γ has a strong inflammatory effect by upregulating MHC molecules and that (ii) this cytokine is a prerequisite for β-cell destruction, since CTLs against a self-antigen on β cells cannot successfully destroy them in the absence of INF-γ-induced MHC class I upregulation. These results also illustrate the built-in safety mechanisms that an organism must have to prevent autoimmune disease: basal MHC class I levels on β cells appear to be insufficient to support β-cell destruction by autoreactive CTLs. Moreover, INF-γ can indirectly protect mice from IDDM due to its ability to induce islet cell regeneration and proliferation, a process described by Gianani and Sarvetnick (21).

IL-2 was evaluated by using a double transgenic RIP-LCMV × RIP–IL-2 model (85). An interesting contrast to the effect of INF-γ is that the IL-2 model did not support spontaneous IDDM, despite the presence of local insulitis in the absence of LCMV infection. However, virus-induced IDDM was enhanced in these double transgenic mice. Thus, IL-2 induced lymphocyte proliferation, but it was not sufficient to break unresponsiveness to self. Possible reasons are that the IL-2 produced locally only partially activated lymphocytes in the pancreas, or, alternatively or additionally, not enough self-reactive lymphocytes were activated (see also the previous paragraph noting the implication of autoreactive lymphocyte levels). The increased apoptosis of autoreactive lymphocytes, described in other models, was not observed in our RIP-LCMV × RIP-IL-2 transgenic mice; consequently, no amelioration of IDDM was seen.

Several Th2 cytokines were also evaluated. For example, IL-10 promoted IDDM, showing clearly that the Th1/Th2 paradigm is not always applicable (44). In contrast, IL-4 prevented IDDM in preliminary studies with the RIP-LCMV model. These findings are similar to those described by use of NOD mice. IL-4 was shown to prevent islet cell destruction under certain but not all conditions (60), whereas IL-10 did not prevent IDDM. Without question, the regulation of local autoimmunity is most complex.

In addition to the amount of the cytokine produced, the timing in relation to the stage of the autoimmune process appears to be important to the pathogenic outcome. Along this line, studies with the NOD mouse model have shown that elimination or overproduction of certain cytokines such as TNF-α can have differential effects in the early or late phases of IDDM development (16). Studies with switchable (tetracycline-dependent) promoters are under way to dissect the differential effects of various cytokines during the lag phase prior to clinical IDDM.

A third class of cytokines, termed Th3, of which the main candidate is transforming growth factor β (TGF-β), is produced primarily by lymphocytes in the gut and may have regulatory function, particularly in response to orally administered antigens. In several studies by Howard Weiner's group (97) and my laboratory (89), mice treated orally with insulin had a predominance of TGF-β-producing lymphocytes in the pancreas that prevented IDDM (see also the paragraph on the treatment of autoimmune diseases in this chapter). In contrast, overexpression of TGF-β in β cells of double transgenic RIP-LCMV × RIP–TGF-β mice did not prevent virus-induced IDDM. However, the high levels of TGF-β produced in these mice led to considerable fibrosis of the islets (46). Possibly, then, only lower levels of TGF-β can prevent islet cell destruction, and the local concentration of TGF-β is very important to determining its regulatory function. From studies performed in vitro, TGF-β is now known to suppress both Th1 and Th2 responses. Thus, TGF-β is a molecule that has general immunosuppressive properties and is therefore a likely downstream factor in regulating immune responses. However, upstream factors such as IL-4 (40) probably precede the involvement of TGF-β in regulating autoimmunity.

When direct immunohistochemical or ELISPOT analyses of cytokines were performed with the islets, an interesting scheme emerged: mice that were prediabetic or that had received successful immune intervention therapy (89) had more IL-4 than INF-γ in their islets. In contrast, mice with IDDM showed a predominance of INF-γ (90). This outcome indicates a potential regulatory or islet-protective role for certain cytokines such as IL-4 and provides a mechanism by which immune intervention can influence the local inflammatory process in the islets, a possibility that is discussed further in the following chapter (see also Fig. 3 and 4).

MODULATION OF AUTOIMMUNE DIABETES BY REGULATORY LYMPHOCYTES: ORAL ANTIGEN ADMINISTRATION AND DNA VACCINATION

The RIP-LCMV transgenic mouse model is ideally suited for the testing of novel approaches to antigen-specific immune therapy (42, 83). The induction of IDDM by viral infection can be controlled and is directed to a known and well-characterized self-antigen, the LCMV transgene expressed by the β cells. As shown in Fig. 3, 4, and 5, antigen-specific interventions can be successful at several stages during the development of IDDM. Described in detail here are two such strategies: oral antigen administration (89, 98, 107) and immunization with plasmid DNA expressing self-antigens.

Oral Self-Antigens Prevent Autoimmune Diabetes

When RIP-LCMV transgenic mice, which develop slow-onset IDDM, received insulin orally twice per week, starting 10 days after LCMV infection and lasting until 6

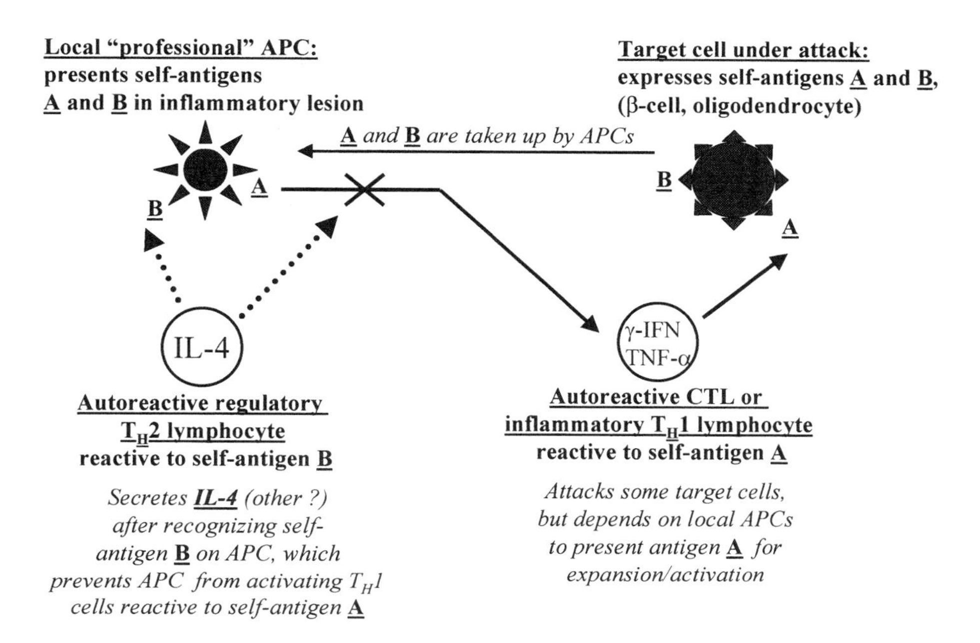

Figure 4. Concept of bystander suppression in treating autoimmune diseases. Destructive lymphocytes reactive with self-antigen A can be negatively influenced in the target organ by regulatory (bystander suppressor) lymphocytes reactive with a different self-antigen B.

weeks postinfection, 40 to 60% of the mice remained free of IDDM (time of observation of protection from diabetes was 6 months) (89, 98, 107). Systemically, anti-self (viral) CTLs were generated at equal levels in these insulin-treated mice with and without IDDM. However, within the pancreas, markedly less insulitis evolved in insulin-treated transgenic mice without IDDM. Specifically, lymphocytes infiltrated around but not into the islets, and β cells were not destroyed. Pancreata from insulin-treated RIP-NP transgenic mice without IDDM contained sixfold more IL-4-, IL-10-, or TGF-β-producing lymphocytes and twofold less IL-2- or IFN-γ-producing lymphocytes than pancreata from insulin-treated mice with IDDM. Thus, oral administration of insulin as a self-antigen, which is not involved in the primary induction of IDDM in RIP-LCMV transgenic mice, prevented autoimmune disease. Our findings suggest that a bystander effect of regulatory T cells that changed the ratio of Th2/Th1 type cytokines in the target tissue can be of value in preventing and treating IDDM. In similar studies with the NOD model of IDDM and other animal models of autoimmune disease, Howard Weiner and colleagues (97, 107) and other investigators (89) also identified regulatory lymphocytes that secreted TGF-β.

This protective effect of oral porcine insulin was completely abrogated by either a one- or two-amino-acid substitution in the beta chain of the insulin molecule (93). Oral administration of an insulin analog (PheB$_{25}$ [Asn]; AlaB$_{30}$ [Thr]) with a 100-fold reduced intrinsic activity and of human insulin (AlaB$_{30}$ [Thr]), with intrinsic activity similar to that of porcine insulin did not abort the IDDM. These results suggest, first, that the efficacy of insulin in preventing IDDM was not from a direct effect of the hormone. Second, a one-amino-acid substitution completely abrogated the ability of insulin to induce regulatory

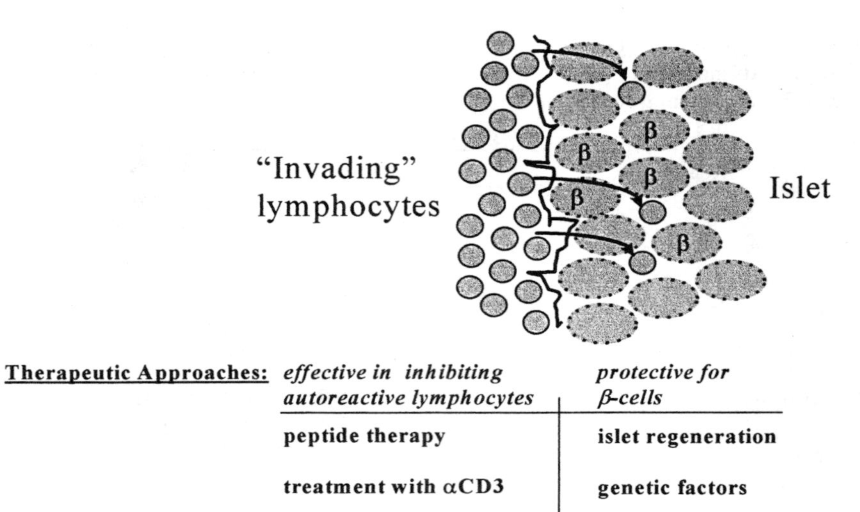

Figure 5. "The battle for the islet": therapeutic approaches to prevention of autoimmune diabetes. Infiltration and destruction of islet cells by autoreactive lymphocytes and APCs constitute a complex process that can be reversed by various mechanisms even after its initiation. Not clear, at this point, is whether "attack" of an islet cell results in increased regeneration of β cells as a way for the islet cells to fight back and prevent their own destruction. Apparently, from present evidence, islet cell destruction should be viewed as a dynamic process that largely depends on the amount of local inflammatory mediators and destructive versus protective lymphocytes.

autoreactive lymphocytes that prevented IDDM. Third, selection of the "wrong" antigen could potentiate IDDM; therefore, oral antigen administration must be evaluated carefully or it may harm the patient.

Immunization with Plasmid DNA Expressing Self-Antigens

The second approach to antigen-specific intervention was vaccination with DNA plasmids that expressed self-antigens. Mice that expressed LCMV-NP as a transgene in their β cells developed IDDM after LCMV infection. Subsequent inoculation of plasmid DNA that encoded the insulin B chain reduced the incidence of this virally induced autoimmune diabetes by 50%. The protection provided by insulin B-chain DNA proceeded through induction of anti-self, regulatory CD4 lymphocytes that reacted with the insulin B chain, secreted IL-4, and locally reduced the activities of LCMV-NP autoreactive CTLs in the pancreatic draining lymph nodes. In contrast, similar vaccination with plasmids that expressed the LCMV ("self") protein did not prevent IDDM, because no such regulatory cells were induced. Thus, immunization with DNA plasmids that express self-antigens might constitute a novel and attractive therapeutic approach to the prevention of autoimmune diseases, but only if the antigens are carefully preselected for their ability to induce regulatory lymphocytes in vivo. Our findings suggest that self-antigens (such as LCMV self-antigen) involved primarily in the destructive inflammatory autoimmune process are not suited for induction of regulatory lymphocytes. In contrast, antigens not involved in this primary autoreactive process, such as insulin, are good candidates for induction of bystander suppressor or regulatory cells (see also Fig. 4).

Our findings that oral administration of insulin or plasmid DNA vaccination both induce insulin-specific regulatory lymphocytes are noteworthy, because autoreactive lymphocytes are usually thought of as destructive mediators of autoimmune disease. However, some autoreactive lymphocytes can ameliorate or prevent autoimmune disease, and they have been designated regulatory cells. Potentially, such cells could be activated by viral infection either directly or through mimicry-related APL effects (73). In experiments with type I diabetes, splenic cells (β-cell-reactive $CD4^+$ T cells) as well as thymic regulatory cells prevented the disease after adoptive transfer into prediabetic NOD mice (2, 29, 76). A group of promising strategies for therapeutic induction and expansion of regulatory cells in vivo has been collectively called "oral tolerance" and refers to the prevention and treatment of autoimmune diseases by oral administration of antigens related to autoimmune target tissue-specific antigens. High-dose regimens have resulted in deletion (8) or anergy or autoreactive lymphocytes specific for the administered antigen; alternatively, low- to medium-dose treatments represent a form of mucosal immunization, since their success relies on the induction of antigen-specific regulatory cells (96, 97). In several animal models of autoimmune disease, including type I diabetes, such regulatory cells have been generated. Initially, these cells were deemed CD8 T cells (49, 56, 62); however, subsequent reports documented the suppression of experimental autoimmune encephalitis (EAE) (14, 15) by CD8 and CD4 T cells as well as blocking of collagen-induced arthritis (82) and diabetes (9) and induction of tolerance to ovalbumin (18, 19) or bovine alpha s1-casein (105) by regulatory CD4 T cells. As reviewed recently (35), TCR T cells mediated oral tolerance to ovalbumin (39, 51–53) in experimental autoimmune uveitis (100). Elsewhere, CD8-γδTCR T cells prevented diabetes in NOD mice given insulin by aerosol spray (26).

On the basis of our findings with DNA vaccination, the special immune environment for the priming of lymphocytes at large mucosal surfaces may not be critical for the generation of regulatory cells. In general, such lymphocytes can be characterized by their ability to secret cytokines such as IL-4, IL-10, and TGF-β and by their activities as suppressive regulatory cells. If such cells home to an autoimmune target site, they can prevent disease. This mechanism has been termed "bystander suppression" (Fig. 4) when the specificity of regulatory cells differs from that of the autoreactive destructive lymphocytes. Bystander suppressor cells, which by definition must recognize an autoantigen different from the initiating self-antigen, have been suggested to be participants in EAE (3, 15, 55, 56) and arthritis (106, 108); additionally, their existence was shown in our model of virus-induced diabetes (89, 93). Although in some cases phenotypic analyses have been performed to seek bystander suppressor cells (56), other studies have documented the production of TGF-β, IL-4, and IL-10 (89, 96). These and other observations in a transgenic TCR system (13) argue that antigen-specific autoreactive T cells can differentiate in vivo into autoaggressive or regulatory cells, depending on the site and context of the initial priming event. However, only in our RIP-LCMV transgenic model of experimental diabetes has bystander suppression been clearly demonstrated by using two distinct and well-defined self-antigens.

SUMMARY AND CONCLUSIONS

The concept of molecular mimicry as an explanation of how microbes can induce autoimmune disease is most intriguing. Support for the role of this and many contributory factors in the RIP-LCMV transgenic mouse model of IDDM have been reviewed in this chapter. It is possible that studies like those conducted by Strominger's laboratory (101, 102), in which the conformational cross-reactivities of human T cells with microbial ligands were defined, will yield direct evidence for cross-reactive T cells that recognize viral and self-ligands. Certainly, this approach will become more straightforward when data banks that can be used to seek conformational matches among such molecules are available. These data, along with animal models, are required to study many pathogenic scenarios that engage researchers and clinicians. From these findings, intervention strategies can be devised and tested. Thus, proof for the involvement of viruses in autoimmune disease will likely come forth indirectly. For example, therapeutic interventions tested with animal models could prove successful in humans. Furthermore, antiviral treatments that reduce the incidence and/or severity of autoimmune diseases would indirectly provide evidence for links between certain viral infectious diseases. Important issues that should continue to be addressed by the use of animal models are the multiple related or unrelated viral infections that cause autoimmunity and the ability of antiviral T cells to cross-react with self-ligands in vivo.

Acknowledgments. This work was supported by National Institutes of Health grants R29 DK51091 and RO1 AI44451 to M.G.v.H. and AG04342 to Michael B. A. Oldstone. M.G.v.H. is the recipient of a Juvenile Diabetes Foundation International Career Development Award (JDFI 296120). The author thanks Diana Frye for assistance with the manuscript and Phyllis Minick for helpful editorial comments.

REFERENCES

1. **Adams, T. E., S. Alpert, and D. Hanahan.** 1987. Non tolerance and autoantibodies to a transgenic self-antigen expressed in pancreatic beta-cells. *Nature* **325:**223–228.
2. **Akhtar, I., J. P. Gold, L. Y. Pan, J. L. Ferrara, X. D. Yang, J. I. Kim, and K. N. Tan.** 1995. $CD4^+$ beta islet cell-reactive T cell clones that suppress autoimmune diabetes in nonobese diabetic mice. *J. Exp. Med.* **182:**87–97.
3. **al-Sabbagh, A., A. Miller, L. M. Santos, and H. L. Weiner.** 1994. Antigen-driven tissue-specific suppression following oral tolerance: orally administered myelin basic protein suppresses proteolipid protein-induced experimental autoimmune encephalomyelitis in the SJL mouse. *Eur. J. Immunol.* **24:**2104–2109.
4. **Bach, J.-F., and D. Mathis.** 1997. The NOD mouse. *Res. Immunol.* **148:**281–370.
5. **Bachmaier, K., N. Neu, L. M. de la Maza, S. Pal, A. Hessel, and J. M. Penninger.** 1999. *Chlamydia* infections and heart disease linked through antigenic mimicry. *Science* **283:**1335–1339.
6. **Baekkeskov, S., H. J. Anastoot, S. Christgau, A. Reetz, M. Solimena, M. Cascalho, F. Folli, H. Richter-Dlesen, P. DeCamillin, and P. D. Camilli.** 1990. Identification of the 64K autoantigen in insulin-dependent diabetes as the GABA-synthesizing enzyme glutamic acid decarboxylase. *Nature* **347:**151–156.
7. **Baekkeskov, S., and B. Hansen.** 1990. Human diabetes. *Curr. Top. Microbiol. Immunol.* **164:**1–193.
8. **Benson, J. M., and C. C. Whitacre.** 1997. The role of clonal deletion and anergy in oral tolerance. *Res. Immunol.* **148:**533–541.
9. **Bergerot, I., N. Fabien, V. Maguer, and C. Thivolet.** 1994. Oral administration of human insulin to NOD mice generates $CD4^+$ T cells that suppress adoptive transfer of diabetes. *J. Autoimmunity* **7:**655–663.
10. **Borrow, P., and M. B. A. Oldstone.** 1997. Lymphocytic choriomeningitis virus, p. 593–627. *In* N. Nathanson (ed.), *Viral Pathogenesis.* Lippincott-Raven, Philadelphia, Pa.
11. **Buchmeier, M., R. Welsh, F. Dutko, and M. B. A. Oldstone.** 1980. The virology and immunobiology of LCMV infection. *Adv. Immunol.* **30:**275–331.

12. **Chan, S., M. Correia-Neves, A. Dierich, C. Benoist, and D. Mathis.** 1998. Visualization of CD4/CD8 T cell commitment. *J. Exp. Med.* **188:**2321–2333.
13. **Chen, Y., J. Inobe, V. K. Kuchroo, J. L. Baron, C. A. Janeway, Jr., and H. L. Weiner.** 1996. Oral tolerance in myelin basic protein T-cell receptor transgenic mice: suppression of autoimmune encephalomyelitis and dose-dependent induction of regulatory cells. *Proc. Natl. Acad. Sci. USA* **93:**388–391.
14. **Chen, Y., J. Inobe, and H. L. Weiner.** 1995. Induction of oral tolerance to myelin basic protein in CD8-depleted mice: both $CD4^+$ and $CD8^+$ cells mediate active suppression. *J. Immunol.* **155:**910–916.
15. **Chen, Y., V. K. Kuchroo, J. I. Inobe, D. A. Hafler, and H. L. Weiner.** 1994. Regulatory T cell clones induced by oral tolerance: suppression of autoimmune encephalomyelitis. *Science* **265:**1237–1240.
16. **Cope, A., R. Ettinger, and H. McDivitt.** 1997. The role of TNF alpha and related cytokines in the development and function of the autoreactive T-cell repertoire. *Res. Immunol.* **148:**307–312.
17. **Forster, I., R. Hirose, J. M. Arbeit, B. E. Clausen, and D. Hanahan.** 1995. Limited capacity for tolerization of $CD4^+$ T cells specific for a pancreatic beta cell neo-antigen. *Immunity* **2:**573–585.
18. **Garside, P., M. Steel, F. Y. Liew, and A. M. Mowat.** 1995. $CD4^+$ but not $CD8^+$ T cells are required for the induction of oral tolerance. *Int. Immunol.* **7:**501–504.
19. **Garside, P., M. Steel, E. A. Worthey, A. Satoskar, J. Alexander, H. Bluethmann, F. Y. Liew, and A. M. Mowat.** 1995. T helper 2 cells are subject to high dose oral tolerance and are not essential for its induction. *J. Immunol.* **154:**5649–5655.
20. **Genain, C., D. Lee-Parritz, M. Nguyen, L. Massacesi, N. Joshi, R. Ferrante, K. Hoffman, K. Moseley, N. Letvin, and S. Hauser.** 1994. In healthy primates, circulating autoreactive T-cells mediate autoimmune disease. *J. Clin. Investig.* **94:**1339–1345.
21. **Gianani, R., and N. Sarvetnick.** 1996. Viruses, cytokines, antigens, and autoimmunity. *Proc. Natl. Acad. Sci. USA* **93:**2257–2259.
22. **Guerder, S., J. Meyerhoff, and R. Flavell.** 1994. The role of the T cell costimulator B7-1 in autoimmunity and the induction and maintenance of tolerance to peripheral antigen. *Immunity* **1:**155–166.
23. **Guerder, S., D. E. Picarella, P. S. Linsley, and R. A. Flavell.** 1994. Costimulator B7-1 confers antigen-presenting-cell function to parenchymal tissue and in conjunction with tumor necrosis factor α leads to autoimmunity in transgenic mice. *Proc. Natl. Acad. Sci. USA* **91:**5138–5142.
24. **Hanahan, D.** 1998. Peripheral-antigen-expressing cells in thymic medulla: factors in self-tolerance and autoimmunity. *Curr. Opin. Immunol.* **10:**656–662.
25. **Harding, F. A., and J. P. Allison.** 1993. CD28-B7 interactions allow the induction of $CD8^+$ cytotoxic T-lymphocytes in the absence of exogenous help. *J. Exp. Med.* **177:**1791–1796.
26. **Harrison, L. C., M. Dempsey-Collier, D. R. Kramer, and K. Takahashi.** 1996. Aerosol insulin induces regulatory CD8 gamma delta T cells that prevent murine insulin-dependent diabetes. *J. Exp. Med.* **184:**2167–2174.
27. **Haskins, K., and D. Wegmann.** 1996. Diabetogenic T-cell clones. *Perspect. Diabetes* **45:**1299–1305.
28. **Healey, D., P. Ozegbe, S. Arden, P. Chandler, J. Hutton, and A. Cooke.** 1995. In vivo activity and in vitro specificity of $CD4^+$ Th1 and Th2 cells derived from the spleens of diabetic NOD mice. *J. Clin. Investig.* **95:**2979–2985.
29. **Herbelin, A., J.-M. Gombert, F. Lepault, J.-F. Bach, and L. Chatenoud.** 1998. Mature mainstream $TCR\alpha\beta^+CD4^+$ thymocytes expressing L-selectin mediate "active tolerance" in the nonobese diabetic mouse. *J. Immunol.* **161:**2620–2628.
30. **Homo-Delarche, F., and C. Boitard.** 1996. Autoimmune diabetes: the role of the islets of Langerhans. *Immunol. Today* **17:**456–460.
31. **Horwitz, M. S., L. M. Bradley, J. Harbertson, T. Krahl, J. Lee, and N. Sarvetnick.** 1998. Diabetes induced by coxsackie virus: initiation by bystander damage and not molecular mimicry. *Nat. Med.* **4:**781–785.
32. **Horwitz, M. S., K. T. Krahl, C. Fine, J. Lee, and N. Sarvetnick.** 1999. Protection from lethal coxsackievirus-induced pancreatitis by expression of gamma interferon. *J. Virol.* **73:**1756–1766.
33. **Jenson, A. B., H. S. Rosenberg, and A. L. Notkins.** 1980. Pancreatic islet cell damage in children with fatal viral infections. *Lancet* **2:**354–358.
34. **Kagi, D., B. Ledermann, K. Burki, P. Seiler, B. Odermatt, K. J. Olsen, E. R. Podack, R. M. Zinkernagel, and H. Hengartner.** 1994. Cytotoxicity mediated by T cells and natural killer cells greatly impaired in perforin-deficient mice. *Nature* **369:**1–7.

35. **Kapp, J. A., and Y. Ke.** 1997. The role of gamma delta TCR-bearing T cells in oral tolerance. *Res. Immunol.* **148:**561–567.
36. **Katz, J., C. Benoist, and D. X. Mathis.** 1995. T helper cell subsets in IDDM. *Science* **268:**1185–1188.
37. **Kaufman, D. L., M. G. Clare-Salzler, J. Tian, T. Forsthuber, G. Ting, P. Robinson, M. A. Atkinson, E. E. Sercarz, A. J. Tobin, and P. V. Lehmann.** 1993. Spontaneous loss of T-cell tolerance to glutamic acid decarboxylase in murine insulin-dependent diabetes. *Nature* **366:**69–72.
38. **Kaufman, D. L., M. G. Erlander, M. J. Clare-Salzler, M. A. Atkinson, N. K. Maclaren, and A. J. Tobin.** 1992. Autoimmunity to two forms of glutamate decarboxylase in insulin-dependent diabetes mellitus. *J. Clin. Investig.* **89:**283–292.
39. **Ke, Y., K. Pearce, J. P. Lake, H. K. Ziegler, and J. A. Kapp.** 1997. Gamma delta T lymphocytes regulate the induction and maintenance of oral tolerance. *J. Immunol.* **158:**3610–3618.
40. **King, C., J. Davies, R. Mueller, M. S. Lee, T. Krahl, B. Yeung, E. O'Connor, and N. Sarvetnick.** 1998. TGF-beta1 alters APC preference, polarizing islet antigen responses toward a Th2 phenotype. *Immunity* **5:**601–603.
41. **King, M. L., A. Shaikh, D. Bidwell, A. Voller, and J. E. Banatvala.** 1983. Coxsackie B virus specific IgM responses in children with insulin-dependent (juvenile-onset, type 1) diabetes mellitus. *Lancet* **i:**1397–1399.
41a**Kurts, C., W.R. Heath, H. Kosada, J. F. Miller, and F. R. Carbone.** 1998. The peripheral deletion of autoreactive CD8+ T cells induced by cross-presentation of self-antigens involves signaling through CD95 (FAS, Apo-1). *J. Exp. Med.* **188:**415–420.
42. **Lamont, A.** 1994. Are we closer to selective immunotherapy for autoimmune diseases? *Immunol. Today* **15:**45–47.
43. **Lee, M. S., R. Mueller, L. S. Wicker, L. B. Peterson, and N. Sarvetnick.** 1996. IL-10 is necessary and sufficient for autoimmune diabetes in conjunction with NOD MHC homozygosity. *J. Exp. Med.* **183:**2663–2668.
44. **Lee, M.-S., R. Mueller, L. S. Wicker, L. B. Peterson, and N. Sarvetnick.** 1996. IL-10 is necessary and sufficient for autoimmune diabetes in conjunction with NOD MHC homozygosity. *J. Exp. Med.* **183:**2663–2668.
45. **Lee, M.-S., M. G. von Herrath, H. Reiser, M. B. A. Oldstone, and N. Sarvetnick.** 1995. Sensitization to self antigens by in situ expression of interferon-y. *J. Clin. Investig.* **95:**486–492.
46. **Lee, M.-S., M. G. von Herrath, S. Sawyer, M. Arnush, T. Krahl, M. B. A. Oldstone, and N. Sarvetnick.** 1996. TGF-beta fails to inhibit allograft rejection in transgenic mice. *Transplantation* **71:**1–10.
47. **Li, Q., A. E. Borovitskaya, M. G. DeSilva, C. Wasserfall, N. K. Maclaren, A. L. Notkins, and M. S. Lan.** 1997. Autoantigens in insulin-dependent diabetes mellitus: molecular cloning and characterization of human IA-2β. *Proc. Assoc. Am. Physicians* **109:**429–439.
48. **Liblau, R. S., S. M. Singer, and H. McDevitt.** 1995. Th1 and Th2 $CD4^+$ T cells in the pathogenesis of organ-specific autoimmune diseases. *Immunol. Today* **16:**34–38.
49. **Lider, O., L. M. Santos, C. S. Lee, P. J. Higgins, and H. L. Weiner.** 1989. Suppression of experimental autoimmune encephalomyelitis by oral administration of myelin basic protein. II. Suppression of disease and in vitro immune responses is mediated by antigen-specific $CD8^+$ T lymphocytes. *J. Immunol.* **142:**748–752.
50. **Matzinger, P.** 1994. Tolerance, danger, and the extended family. *Annu. Rev. Immunol.* **12:**991–1045.
51. **McMenamin, C., M. McKersey, P. Kuhnlein, T. Hunig, and P. G. Holt.** 1995. Gamma delta T cells down-regulate primary IgE responses in rats to inhaled soluble protein antigens. *J. Immunol.* **154:**4390–4394.
52. **McMenamin, C., C. Pimm, M. McKersey, and P. G. Holt.** 1994. Regulation of IgE responses to inhaled antigen in mice by antigen-specific gamma delta T cells. *Science* **265:**1869–1871.
53. **Mengel, J., F. Cardillo, L. S. Aroeira, O. Williams, M. Russo, and N. M. Vaz.** 1995. Anti-gamma delta T cell antibody blocks the induction and maintenance of oral tolerance to ovalbumin in mice. *Immunol. Lett.* **48:**97–102.
54. **Menser, M. A., J. Forrest, and R. Bransby.** 1978. Rubella infection and diabetes mellitus. *Lancet* **i:**57–60.
55. **Miller, A., A. al-Sabbagh, L. M. Santos, M. P. Das, and H. L. Weiner.** 1993. Epitopes of myelin basic protein that trigger TGF-beta release after oral tolerization are distinct from encephalitogenic epitopes and mediate epitope-driven bystander suppression. *J. Immunol.* **151:**7307–7315.

56. **Miller, A., O. Lider, and H. L. Weiner.** 1991. Antigen-driven bystander suppression after oral administration of antigens. *J. Exp. Med.* **174:**791–798.
57. **Miller, J. F., W. R. Heath, J. Allison, G. Morahan, M. Hoffmann, C. Kurts, and H. Kosaka.** 1997. T cell tolerance and autoimmunity. *Ciba Found. Symp.* **204:**159–168.
58. **Modlin, R., and T. B. Nutman.** 1993. Type 2 cytokines and negative immune regulation in human infections. *Curr. Opin. Immunol.* **5:**511–517.
59. **Morgan, D., R. Liblau, B. Scott, S. Fleck, H. O. McDevitt, N. Sarvetnick, D. Lo, and L. Sherman.** 1996. $CD8^+$ T cell-mediated spontaneous diabetes in neonatal mice. *J. Immunol.* **157:**978–984.
60. **Mueller, R. T., and N. Sarvetnick.** 1996. Pancreatic expression of IL-4 abrogates insulitis and diabetes in NOD mice. *J. Exp. Med.* **184:**1093–1099.
61. **Notkins, A. L., and J. W. Yoon.** 1984. Virus-induced diabetes mellitus, p. 241–247. *In* A. L. Notkins and M. B. A. Oldstone (ed.), *Concepts in Viral Pathogenesis.* Springer-Verlag, New York, N.Y.
62. **Nussenblatt, R. B., R. R. Caspi, R. Mahdi, C. C. Chan, F. Roberge, O. Lider, and H. L. Weiner.** 1990. Inhibition of S-antigen induced experimental autoimmune uveoretinitis by oral induction of tolerance with S-antigen. *J. Immunol.* **144:**1689–1695.
63. **Ohashi, P., S. Oehen, K. Buerki, H. Pircher, C. Ohashi, B. Odermatt, B. Malissen, R. Zinkernagel, and H. Hengartner.** 1991. Ablation of tolerance and induction of diabetes by virus infection in viral antigen transgenic mice. *Cell* **65:**305–317.
64. **Oldstone, M. B. A.** 1987. Molecular mimicry and autoimmune disease. *Cell* **50:**819–820.
65. **Oldstone, M. B. A.** 1989. Molecular mimicry as a mechanism for the cause and as a probe uncovering etiologic agent(s) of autoimmune disease. *Curr. Top. Microbiol. Immunol.* **145:**127–135.
66. **Oldstone, M. B. A., M. Nerenberg, P. Southern, J. Price, and H. Lewicki.** 1991. Virus infection triggers insulin-dependent diabetes mellitus in a transgenic model: role of anti-self (virus) immune response. *Cell* **65:**319–331.
67. **Oldstone, M. B. A., M. G. von Herrath, C. F. Evans, and M. S. Horwitz.** 1996. Virus-induced autoimmune disease: transgenic approach to mimic insulin-dependent diabetes mellitus and multiple sclerosis. *Curr. Top. Microbiol. Immunol.* **206:**67–83.
68. **Patterson, K., R. Chandra, and A. Jenson.** 1981. Congenital rubella, insulitis and diabetes mellitus in an infant. *Lancet* **i:**1048–1049.
69. **Plebanski, M., E. A. M. Lee, C. M. Hannan, K. L. Flanagan, S. C. Gilbert, M. B. Gravenor, and A. V. S. Hill.** 1999. Altered peptide ligands narrow the repertoire of cellular immune responses by interfering with T-cell priming. *Nat. Med.* **5:**565–571.
70. **Prince, G., A. B. Jenson, L. Billups, and A. L. Notkins.** 1978. Infection of human pancreatic beta cell cultures with mumps virus. *Nature* **27:**158–161.
71. **Refaeli, Y., L. Van Parijs, C. A. London, J. Tschopp, and A. K. Abbas.** 1998. Biochemical mechanisms of IL-2-regulated fas-mediated T cell apoptosis. *Immunity* **8:**615–623.
72. **Roman, L., L. F. Simons, R. E. Hammer, J. F. Sambrook, and M. J. Gething.** 1990. The expression of influenza virus hemagglutinin in the pancreatic beta-cells of tg mice results in autoimmune diabetes. *Cell* **61:**383–396.
73. **Ruiz, P. J., H. Garren, D. L. Hirschberg, A. M. Langer-Gould, M. Levite, M. V. Karpuj, S. Southwood, A. Sette, P. Conlon, and L. Steinman.** 1999. Microbial epitopes act as altered peptide ligands to prevent experimental autoimmune encephalomyelitis. *J. Exp. Med.* **189:**1275–1283.
74. **Sarvetnick, N., J. Shizuru, D. Liggitt, L. Martin, B. McIntyre, A. Gregory, T. Parslow, and T. Stewart.** 1990. Loss of pancreatic islet tolerance induced by B-cell expression of interferon-gamma. *Nature* **346:**844–847.
75. **Scott, B., R. Liblau, S. Degermann, L. A. Marconi, L. Ogata, A. J. Caton, H. O. McDevitt, and D. Lo.** 1994. A role for non-MHC genetic polymorphism in susceptibility to spontaneous autoimmunity. *Immunity* **1:**73–82.
76. **Sempe, P., M. F. Richard, J. F. Bach, and C. Boitard.** 1994. Evidence of $CD4^+$ regulatory T cells in the non-obese diabetic male mouse. *Diabetologia* **37:**337–343.
77. **Shrinivasappa, J., J. Saegusa, B. Prabhakar, M. Gentry, M. Buchmeier, T. Wiktor, H. Koprowski, M. Oldstone, and A. Notkins.** 1986. Frequency of reactivity of monoclonal antiviral antibodies with normal tissues. *J. Virol.* **57:**397–401.
78. **Smith, K. M., D. C. Olson, R. Hirose, and D. Hanahan.** 1997. Pancreatic gene expression in rare cells of thymic medulla: evidence for functional contribution to T cell tolerance. *Int. Immunol.* **9:**1355–1365.

79. **Sonderstrup, G., and H. McDevitt.** 1998. Identification of autoantigen epitopes in MHC class II transgenic mice. *Immunol. Rev.* **164:**129–138.
80. **Steinman, L.** 1996. A few autoreactive cells in an autoimmune infiltrate control a vast population of nonspecific cells: a tale of smart bombs and the infantry. *Proc. Natl. Acad. Sci. USA* **93:**2253–2256.
81. **Swain, S. L.** 1994. Generation and in vivo persistence of polarized Th1 and Th2 memory cells. *Immunity* **1:**543–552.
82. **Tada, Y., A. Ho, D. R. Koh, and T. W. Mak.** 1996. Collagen-induced arthritis in CD4- or CD8-deficient mice: $CD8^+$ T cells play a role in initiation and regulate recovery phase of collagen-induced arthritis. *J. Immunol.* **156:**4520–4526.
83. **Tisch, R., and H. McDevitt.** 1994. Antigen specific immunotherapy: is it a real possibility to combat T-cell medicated autoimmunity? *Proc. Natl. Acad. Sci. USA* **91:**437–438.
84. **Vella, A., T. K. Teague, J. Ihle, J. Kappler, and P. Marrack.** 1997. Interleukin 4 (IL-4) or IL-7 prevents the death of resting T cells: stat6 is probably not required for the effect of IL-4. *J. Exp. Med.* **186:**325–330.
85. **von Herrath, M. G., J. Allison, J. F. Miller, and M. B. Oldstone.** 1995. Focal expression of interleukin-2 does not break unresponsiveness to "self" (viral) antigen expressed in beta cells but enhances development of autoimmune disease (diabetes) after initiation of an anti-self immune response. *J. Clin. Investig.* **95:**477–485.
86. **von Herrath, M. G., B. Coon, H. Lewicki, H. Mazarguil, J. E. Gairin, and M. B. A. Oldstone.** 1998. In vivo treatment with a MHC class I-restricted blocking peptide can prevent virus-induced autoimmune diabetes. *J. Immunol.* **161:**5087–5096.
87. **von Herrath, M. G., J. Dockter, M. Nerenberg, J. E. Gairin, and M. B. A. Oldstone.** 1994. Thymic selection and adaptability of cytotoxic T lymphocyte responses in transgenic mice expressing a viral protein in the thymus. *J. Exp. Med.* **180:**1901–1910.
88. **von Herrath, M. G., J. Dockter, and M. B. A. Oldstone.** 1994. How virus induces a rapid or slow onset insulin-dependent diabetes mellitus in a transgenic model. *Immunity* **1:**231–242.
89. **von Herrath, M. G., T. Dyrberg, and M. B. A. Oldstone.** 1996. Oral insulin treatment suppresses virus-induced antigen-specific destruction of beta cells and prevents autoimmune diabetes in transgenic mice. *J. Clin. Investig.* **98:**1324–1331.
90. **von Herrath, M. G., S. Guerder, H. Lewicki, R. Flavell, and M. B. A. Oldstone.** 1995. Coexpression of B7.1 and viral (self) transgenes in pancreatic β-cells can break peripheral ignorance and lead to spontaneous autoimmune diabetes. *Immunity* **3:**727–738.
91. **von Herrath, M. G., and A. Holz.** 1997. Pathological changes in the islet milieu precede infiltration of islets and destruction of β-cells by autoreactive lymphocytes in a transgenic model of virus-induced IDDM. *J. Autoimmun.* **10:**231–238.
92. **von Herrath, M. G., A. Holz, D. Homann, and M. B. A. Oldstone.** 1998. Role of viruses in type I diabetes. *Semin. Immunol.* **10:**87–100.
93. **von Herrath, M. G., and D. Homann.** 1997. Treatment of virus-induced autoimmune diabetes by oral administration of insulin: study on the mechanism by which oral antigens can abrogate autoimmunity. *Endocrinol. Diabetes* **105:**24–25.
94. **von Herrath, M. G., and M. B. Oldstone.** 1996. Virus-induced autoimmune disease. *Curr. Opin. Immunol.* **8:**878–885.
95. **von Herrath, M. G., and M. B. A. Oldstone.** 1997. IFN-gamma is essential for β-cell destruction by CTL. *J. Exp. Med.* **185:**531–539.
96. **Weiner, H. L.** 1997. Oral tolerance: immune mechanisms and treatment of autoimmune diseases. *Immunol. Today* **18:**335–343.
97. **Weiner, H. L., A. Friedman, A. Miller, S. J. Khoury, A. Al-Sabbagh, L. Santos, M. Sayegh, R. B. Nussenblatt, D. E. Trentham, and D. A. Hafler.** 1994. Oral tolerance: immunologic mechanisms and treatment of animal and human organ-specific autoimmune diseases by oral administration of autoantigens. *Annu. Rev. Immunol.* **12:**809–837.
98. **Weiner, H. L., A. Friedman, A. Miller, S. J. Khoury, A. Al-Sabbagh, L. Santos, M. Sayegh, R. B. Nussenblatt, D. E. Trentham, and D. A. Hafler.** 1994. Oral tolerance: immunologic mechanisms and treatment of animal and human organ-specific autoimmune diseases by oral administration of autoantigens. *Annu. Rev. Immunol.* **12:**809–837.
99. **Wicker, L. S., J. A. Todd, and L. B. Peterson.** 1995. Genetic control of autoimmune diabetes in the NOD mouse. *Annu. Rev. Immunol.* **13:**179–200.

100. **Wildner, G., and S. R. Thurau.** 1995. Orally induced bystander suppression in experimental autoimmune uveoretinitis occurs only in the periphery and not in the eye. *Eur. J. Immunol.* **25:**1292–1297.
100a. **Wong, F. S., J. Karttunen, C. Dumont, L. Wen, I. Visintin, I. M. Pilip, N. Shastri, E. G. Pamer, and C. A. Janeway, Jr.** 1999. Identification of an MHC class I-restricted autoantigen in type 1 diabetes by screening an organ-specific cDNA library. *Nat Med.* **5:**1026–1031.
101. **Wucherpfennig, K. W., and J. L. Strominger.** 1995. Molecular mimicry in T-cell mediated autoimmunity: viral peptides activate human T-cell clones specific for myelin basic protein. *Cell* **80:**695–705.
102. **Wucherpfennig, K. W., B. Yu, K. Bhol, D. S. Monos, E. Argyris, R. W. Karr, A. R. Ahmed, and J. L. Strominger.** 1995. Structural basis for major histocompatibility complex (MHC)-linked susceptibility to autoimmunity: charged residues of a single MHC binding pocket confer selective presentation of self-peptides in pemphigus vulgaris. *Proc. Natl. Acad. Sci. USA* **92:**11935–11939.
103. **Yoon, J. W., M. Austin, T. Onodera, and A. L. Notkins.** 1979. Virus-induced diabetes mellitus: isolation of a virus from the pancreas of a child with diabetic ketoacidosis. *N. Engl. J. Med.* **300:**1173–1179.
104. **Yoon, J.-W., C.-S. Yoon, H.-W. Lim, Q. Q. Huang, Y. Kang, K. H. Pyun, K. Hirasawa, R. S. Sherwin, and H.-S. Jun.** 1999. Control of autoimmune diabetes in NOD mice by GAD expression or suppression in beta cells. *Science* **284:**1183–1187.
105. **Yoshida, H., S. Hachimura, K. Hirahara, T. Hisatsune, K. Nishijima, A. Shiraishi, and S. Kaminogawa.** 1998. Induction of oral tolerance in splenocyte-reconstituted SCID mice. *Clin. Immunol. Immunopathol.* **87:**282–291.
106. **Yoshino, S., E. Quattrocchi, and H. L. Weiner.** 1995. Suppression of antigen-induced arthritis in Lewis rats by oral administration of type II collagen. *Arthritis Rheum.* **38:**1092–1096.
107. **Zang, J. A., L. Davidson, G. Eisenbarth, and H. Weiner.** 1991. Suppression of diabetes in NOD mice by oral administration of porcine insulin. *Proc. Natl. Acad. Sci. USA* **88:**10252–10256.
108. **Zhang, Z. Y., C. S. Lee, O. Lider, and H. L. Weiner.** 1990. Suppression of adjuvant arthritis in Lewis rats by oral administration of type II collagen. *J. Immunol.* **145:**2489–2493.

Molecular Mimicry, Microbes, and Autoimmunity
Edited by M. W. Cunningham and R. S. Fujinami

Chapter 18

Molecular Mimicry and Chagas' Disease

Edecio Cunha-Neto and Jorge Kalil

Chronic Chagas' disease cardiomyopathy (CCC) is one of the few well-defined examples of human postinfectious autoimmunity in which an infectious episode with an established pathogen—the protozoan parasite *Trypanosoma cruzi*—clearly triggers autoimmune phenomena, most of which are related to documented molecular mimicry and organ-specific damage. The time-scale dissociation between primary infection with high levels of parasites in tissue and blood and tissue pathology, allied with the scarcity of *T. cruzi* in the heart lesions of patients with CCC, prompted investigators as early as 60 years ago (130) to suggest that the mononuclear cell infiltrate should directly damage the heart in an autoimmune fashion. This life-threatening heart disorder affects ca. 6 million of the 20 million patients infected with *T. cruzi* in Latin America; in the absence of immunoprophylaxis and effective antiparasite drug treatment, it is a major public health problem in that region. Unraveling of the relationship by which molecular mimicry between an infectious agent and self-components can trigger organ-specific autoimmunity may lead to reverse strategies for the identification of infectious agents that putatively trigger autoimmune diseases of suspected infectious etiology. In addition, the testing of current concepts on the molecular pathogenesis of proven postinfectious autoimmune diseases like CCC and rheumatic fever (58) may allow the early identification of susceptible individuals and the treatment of affected patients.

A PRIMER ON CHAGAS' DISEASE

Ninety years ago, the Brazilian physician scientist Carlos Chagas described American trypanosomiasis (later dubbed Chagas' disease) and identified its etiologic agent, the flagelate protozoan *T. cruzi,* its full life cycle, and the clinical spectrum of the disease (30). Human *T. cruzi* infection occurs after a blood meal by infected hematophagous hemipteran bugs (family *Triatominae*), in which metacyclic trypomastigotes, the infective form of *T. cruzi,* from the insect's fecal material invade the cytoplasm of many different types of host cells. Once inside the cytoplasm, trypomastigote forms change into the replicative amastigote forms, which multiply by binary fission, changing back to trypomastigote forms just prior to host cell lysis and releasing hundreds of infective forms of *T. cruzi* ready

Edecio Cunha-Neto • Transplantation Immunology Laboratory, Heart Institute (InCor), University of São Paulo School of Medicine, São Paulo, Brazil. ***Jorge Kalil*** • Howard Hughes Medical Institute and Department of Medicine and Heart Institute (InCor), University of São Paulo School of Medicine, São Paulo, Brazil.

to invade new target cells. After several cycles of invasion-replication-release, parasitism is widespread in several tissues and blood; acute asymptomatic myocarditis is thought to occur nearly always (24).

The acute phase of infection, which lasts about 2 months, is usually asymptomatic, but flu-like symptoms and, more rarely, fulminant myocarditis may occur in 10 to 30% of infected individuals. The high parasite load ensures a strong cellular and humoral immune response against *T. cruzi,* leading to the control but not the complete elimination of tissue and blood parasitism. Thus, a low-grade persistent infection is established, regardless of the clinical progression of the disease, as indicated by the invariable reactivation of parasitemia in chronically infected individuals after immunodeficiency or pharmacological immunosuppression (20, 52, 75, 80, 137). For reasons that are yet unknown, there is no "sterile immunity," probably due to multiple immune escape mechanisms employed by the parasite (89).

Chagas' disease is endemic in many countries of Latin America, where 20 million people may be infected. Since there are no anti-*T. cruzi* vaccines or highly effective chemotherapeutic agents, disease control is based on vector control in areas of endemicity and serological screening of donor blood. Up to 6 million patients are afflicted with CCC, a dilated cardiomyopathy with a T-cell-rich inflammatory infiltrate that can be accompanied by severe heart conduction defects, arrhythmia, and thromboembolism which often lead to a fatal course and which develop in ca. 30% of infected individuals 5 to 30 years after primary infection (71, 77). CCC is a particularly lethal form of dilated cardiomyopathy, as the length of survival after presentation is two- to fourfold shorter than that after presentation with idiopathic dilated cardiomyopathy (19, 87). The remaining 60 to 70% of chronically *T. cruzi*-infected individuals either remain asymptomatic (ASY; "indeterminate" patients) or develop denervation of parietal smooth muscle in the digestive system, generally the esophagus or colon (5 to 10%). Functional damage to the autonomic nervous system is also observed and affects a subgroup of symptomatic and indeterminate ASY patients (9, 55). In patients with CCC, dilated cardiomyopathy, arrhythmia, and systemic (arterial) thromboembolism may occur as single forms or in combination, and patients usually die of refractory heart failure or sudden arrhythmic death. Cardiac or digestive syndromes of chronic Chagas' disease may also present in isolated or overlapping forms (11).

Heart-Specific Inflammatory Lesions in CCC: Parasite Antigen-Driven Immunopathology or Autoimmunity?

The major histopathological feature that accompanies dilated cardiomyopathy in patients with CCC is the presence of a diffuse myocarditis, with intense heart fiber damage and significant fibrosis in the presence of very scarce *T. cruzi* forms (63, 64). On the other hand, a focal, nondestructive myocarditis can be observed in ASY individuals (85); focal myocarditis has been associated with the presence of *T. cruzi* antigen (62). Focal inflammatory infiltrates associated with parasite foci can also be observed in several different organs (17). Histiocytes and endothelial cells in the heart tissue of patients with CCC display increased levels of expression of human leukocyte antigen (HLA) class I and class II molecules; CCC cardiomyocytes display increased levels only of HLA class I (110). The inflammatory infiltrate is composed of macrophages (50%), T cells (40%), and B cells (10%) (96); among the T cells, a 2:1 predominance of the $CD8^+$ subset over the $CD4^+$ subset was observed (63, 131). The demonstration of restricted heterogeneity of T-cell receptor Vα transcripts in heart biopsy specimens from CCC patients (40) is in line with similar findings in established autoimmune diseases (60). Heart-infiltrating T-cell lines obtained from CCC patients produce gamma interferon (IFN-γ) and tumor necrosis factor alpha (TNF-α) in the absence of interleukin 4 (IL-4) upon stimulus with phytohemagglutinin (PHA) (41), in line with the predominant detection of IFN-γ and TNF-α in immunohistochemical studies of heart tissue from patients with CCC (111, 112). This also appears to be in line with the apparent ability of *T. cruzi* to induce a systemic shift in the cytokine profile in infected patients toward T1-type cytokines, with suppression of T2-type cytokines (35, 41), which may be related to the ability of mucin-like glycoconjugates from *T. cruzi* to induce the production of IL-12 (28).

Since it is known that *T. cruzi* establishes a lifelong, low-grade infection, the possibility that heart tissue damage in patients with CCC is a mere result of recognition of parasite antigen on target tissue, with the accompanying inflammation, must be entertained (62, 71). Even though heart-infiltrating T cells have been implicated as the ultimate effectors of heart tissue damage, a direct role for heart parasitism has been proposed after the identification of *T. cruzi* antigen and DNA in the hearts of patients with CCC by immunohistochemical and PCR techniques (62, 68). Several findings, however, fail to lend support to local recognition of *T. cruzi* as the trigger of heart tissue damage, as follows (i) *T. cruzi* DNA has been detected in the hearts of both individuals with CCC (with diffuse myocarditis) and ASY individuals (with focal, nondestructive myocarditis) (15, 104); (ii) $CD4^+$ T-cell clones obtained from the heart tissue of a CCC patient failed to recognize recombinant and crude *T. cruzi* antigens (38); (iii) low-grade parasitism is widespread in several organs apart from the heart (136, 137), where parasitism is associated with inflammatory foci (17) but where functional damage to the organ is absent; and (iv) systemic parasitism among CCC and ASY patients is indistinguishable, as assessed by direct parasitological assays or PCR for detection of *T. cruzi* DNA (25). The bulk of the evidence indicates that patients with chronic Chagas' disease undergo low-grade tissue parasitism associated with scarce inflammatory foci in all organs, including the heart; this apparently does not cause tissue damage. Diffuse myocarditis with the accompanying tissue-damaging inflammation is found in the hearts of CCC patients but not in the hearts of ASY patients, and the overwhelming majority of microscope fields of slides with samples of hearts from patients with CCC is devoid of *T. cruzi,* even when immunohistochemical detection of *T. cruzi* antigen is used. This may indicate that while direct *T. cruzi* tissue parasitism might induce focal inflammatory foci, it is apparently unable to evoke sufficient heart damage to lead to dilated cardiomyopathy. Thus, some other factor must be operating along with parasite persistence to lead to heart damage in a subgroup of *T. cruzi*-infected individuals. The recent identification of persistent virus infection in patients with bona fide human autoimmune diseases like multiple sclerosis (3) and insulin-dependent diabetes mellitus (46) may indicate that this might be a common theme among such diseases.

IMMUNOPATHOGENESIS OF CHAGAS' DISEASE

The time-scale dissociation between high levels of parasitemia and tissue pathology, allied with the scarcity of *T. cruzi* in the heart lesions of patients with CCC, prompted early investigators (130) to suggest that the mononuclear cell infiltrate in the heart should recognize and mount delayed-type hypersensitivity responses toward a tissue-specific heart component as a result of chronic *T. cruzi* infection, the so-called autoimmune hypothesis of pathogenesis. As discussed in the previous section, although chronic *T. cruzi* infection generates focal, non-tissue-damaging inflammatory infiltrates in most organs in CCC and ASY "indeterminate" patients, only in the hearts of CCC patients and not in the hearts of ASY patients does the inflammatory infiltrate become tissue damaging, leading to dilated cardiomyopathy. The existence of nonpathogenic and pathogenic inflammatory infiltrates has been described in animal models of autoimmune diabetes (14) and thyroiditis (6). The mechanisms that underlie this transition to the pathogenic potential of the inflammatory infiltrate are incompletely understood even in highly controlled experimental transgenic models of autoimmune disease (14). The susceptibility factors that lead 30% of *T. cruzi*-infected patients to undergo such a transition and develop CCC are largely unknown, but immunological, genetic, environmental, and parasite-related factors may play a role. This may be the single most important step in answering the riddle of autoimmune disease.

An array of reports of autoreactivity among Chagas' disease patients and experimentally infected animals has been published over the last three decades. One of the critical points in validating autoimmunity and molecular mimicry in Chagas' disease is the lack of animal models that can reproduce all the components of the disease in humans. Concerning CCC in particular, experimentally *T. cruzi*-infected mice may develop myocarditis but never develop dilated cardiomyopathy, the life-threatening component of Chagas' disease in humans. Experimental models of dilated cardiomyopathy have been developed with larger outbred animals, like rabbits and dogs (10, 12, 13, 124). Here we critically review reports on such models, with an emphasis on those that report on molecular mimicry between *T. cruzi* and host target organs.

Early Evidence for Autoimmunity and Molecular Mimicry

Studies performed in the 1970s were characterized by the lack of molecular definition of the antigen systems used; most used tissue or *T. cruzi* homogenates (Table 1). Cardiac tissue homogenates induced lymphokine production (43, 129) but not proliferative responses among peripheral blood T cells from patients with CCC (97, 128). Noninfected cardiomyocytes were targets of cytotoxicity by peripheral blood mononuclear cells (PBMCs) from patients with CCC (125). Interestingly, tests related to effector function (lymphokine production, cytotoxicity) yielded positive results for the patients, whereas T-cell proliferation assays apparently had uniformly negative results. Studies with experimentally infected animals disclosed similar findings. Peripheral T cells from rabbits experimentally infected with *T. cruzi* also displayed cytotoxicity to uninfected cardiomyocytes (117). Several investigators reported on the induction of myocardial inflammatory lesions after repeated injection of *T. cruzi* subcellular fractions into rabbits (124) and mice (5, 115). Peripheral T cells from experimentally infected mice displayed proliferative responses to heart homogenates (45, 49). Finally, the passive transfer of inflammatory lesions with T-cell populations from chronically infected mice to naive recipients (45, 66, 76) supported a pathogenic role for autoreactive T cells (Table 2). The autoreactive nature of those pathogenic T cells was disputed recently by using an infection model similar to that used by dos Santos et al. (45) but with different *T. cruzi* strains (123). Using PCR amplification to detect *T. cruzi* DNA, those investigators have shown that, in their system, transfer of lesions occurs only in the presence

Table 1. Autoreactivity after *T. cruzi* infection[a]

Host component	Host	Disease related	Molecular definition	Reference
Cardiac myosin	M	—	$CD4^+$ T cells	114
Cardiac myosin, p150	M	Yes	Serum IgG	127
Heart homogenate	M	—	T cells	45, 49
43-kDa muscle glycoprotein	M	—	Serum IgG	93
Nervous tissue, heart and skeletal muscle	M	—	Serum IgG	126
Second extracellular loop, M2 cholinergic receptor	M	—	Serum IgG	95
Second extracellular loop, β1 adrenergic receptor	M	—	Serum IgG	95
M2 cholinergic receptor	H	Arrhythmia	Serum IgG	44
M2 cholinergic receptor	H	Dysautonomy	Serum IgG	53, 56, 81
Second extracellular loop, M2 cholinergic receptor	H	Dysautonomy	Serum IgG	55
Neurons	H	—	Serum IgG	113
Sciatic nerve homogenate	H	—	Serum IgG	51
Small nuclear ribonucleoprotein	H	No	Serum IgG	16
Heart homogenate	H	Yes	T cells	43, 129
Cardiomyocytes	H	—	T cells	125
Cardiomyocytes	Rb	—	T cells	117

[a] Abbreviations: M, mouse; Rb, rabbit; H, human; IgG, immunoglobulin G. —, not tested or not applicable.

of *T. cruzi*. Dos Santos et al. (45) recently used the same PCR detection in their system and failed to detect *T. cruzi* DNA in inflammatory lesions (94).

Regarding antibody recognition, Cossio et al. (33) described antibodies that were present in the serum of CCC patients and that were capable of binding to vascular endothelium and interstitium in murine heart sections, the so-called EVI antibodies; such antibodies could be absorbed with *T. cruzi* epimastigotes (Table 3). Later, these antibodies were demonstrated to recognize α-galactosyl glycoproteins absent from human tissue, and the initial findings of correlation with cardiac symptoms were not reproducible (74). Anti-sarcoplasmic reticulum (anti-SRA) antibody identified in sera from patients with CCC recognized a conserved ion-translocating enzyme present in *T. cruzi* as well as heart and skeletal striated muscle membranes that were also recognized by sera from patients with degenerative muscle diseases (116). Several investigators identified antibodies cross-reactive

Table 2. Passive transfer of lesions or functional damage with antibodies or T cells from chronically *T. cruzi*-infected individuals[a]

Immunological effectors	Origin	Effect	Disease related	Reference
Splenocytes	M	Focal myocarditis	—	76
$CD4^+$ T cell lines	M	Demyelination	—	66
$CD4^+$ T splenocytes	M	Focal myocarditis	—	45, 118
Anti-*T. cruzi* MAb	M	Receptor agonist	—	34, 140
Mouse antireceptor Ab	M	Ca^{2+} channels	—	95
Anti-M2 muscarinic Ab from arrhythmic patients	H	Conduction defect in rabbit hearts	Yes	44, 91
Anti-M2 receptor O2 loop Ab from Chagasic patients	H	Decreased contractility of rat atria	Dysautonomy	53, 55, 81
Ab against *T. cruzi* P protein, β1-adrenoreceptor	H	Accelerate beating on rat cardiomyocytes	No	47, 72

[a] Abbreviations: M, mouse; H, human; Ab, antibody; MAb, monoclonal antibody.

Table 3. Molecular mimicry after *T. cruzi* infection: human disease[a]

T. cruzi molecule	Host component	Disease related	Molecular definition	Reference
α-Galactosyl residues	α-Galactosyl EVI Ab	No	Sugar moieties	33, 74
SRA	Skeletal muscle SRA	Yes	AS	5
FL-160	47-kDa neuron protein	No	rDNA, AS	132–134
MAP	MAP (brain)	No	rDNA, AS	73
23-kDa ribosomal protein	23-kDa ribosomal protein	No	Ab	21
Ribosomal P protein	Ribosomal P protein	Yes	rDNA, Ab, SP	83
Ribosomal P0 and P2β proteins	β1-adrenoreceptor M2 muscarinic receptor	—	rDNA, Ab, SP	47, 72, 91
B13 protein	Human cardiac myosin heavy chain	Yes	rDNA, Ab, T-cell clones	2, 38, 39

[a] Abbreviations: AS, antiserum; Ab, patient antibody; MAb, monoclonal antibody; rDNA, recombinant DNA; SP, synthetic peptides; —, not tested.

between *T. cruzi* and heart muscle in sera or monoclonal antibodies obtained from experimentally infected mice (78, 79, 92, 126, 140). Conversely, it has been reported that mice with experimental autoimmune myocarditis induced by immunization with heart homogenates developed anti-*T. cruzi* antibodies (31).

Molecular Mimicry: in Search of Defined Antigenic Targets

In the 1980s and 1990s monoclonal antibody and recombinant DNA technologies aided in the definition of several molecular mimicry systems. There have been several reports of immunological cross-reactivity or antigenic mimicry between more or less defined *T. cruzi* and host self-antigens. The cross-reactive antigen pairs are classified here as to the nature of the host self-antigen: evolutionarily conserved structures, neuroantigens, adrenergic or cholinergic cardiovascular receptors, and cardiac myosin.

Evolutionarily Conserved Structures

Given the evolutionary conservation of primary sequences of many key structural proteins or enzymes from protists to humans, it is not surprising that these kinds of cross-reactive antigens can be detected. Data are reported in Tables 3 and 4. Serum antibodies that recognize Ca^{2+}-dependent ATPase from SRA cross-reactively recognized microsomal membranes from *T. cruzi* (4). This antibody was also present in sera from patients with degenerative muscle diseases (116). Sera from *T. cruzi*-infected mice contained cross-reactive antibodies that recognized microtubule-associated proteins (MAPs) from *T. cruzi*, rat fibroblasts, and bovine brain; such antibodies were also contained in patient sera, but without an association with clinical symptoms (73). It was reported that sera from CCC patients, more so than sera from ASY patients, possessed antibodies against a C-terminal epitope of *T. cruzi* ribosomal P2β protein (R13; EEEDDDMGFGLFD) which is very similar in primary sequence to the corresponding C terminus of mammalian ribosomal P protein (H13; EESDDDMGFGLFD) (82, 83). It was claimed that such antibodies were indeed autoreactive and able to bind to mammalian P protein, recognizing its acidic portion (72), but some investigators could not reproduce this finding (22). The demonstration that sera from *T. cruzi*-infected C3H/He mice possess significant amounts of antibodies which bind to a broad class of antigens that contain runs of acidic amino acids (70) might explain the origin and nature of reactivity toward the acidic epitopes of ribosomal P proteins.

Table 4. Molecular mimicry after *T. cruzi* infection: murine models[a]

T. cruzi antigen	Host component	Molecular definition	Reference
Microsomal fraction	Heart and skeletal muscle	MAb	78, 79
?	Neurons, liver, kidney, testis	MAb	121
?	Neurons	MAb	139
?	Heart tissue	Serum IgG	92
SRA	Skeletal muscle Ca^{2+}-dependent SRA	AS	4
Sulfated glycolipids	Neurons	MAb	106–109
Glycosphingolipids	Glycosphingolipids	Serum IgG	138
MAP	MAP (brain)	rDNA, AS	73
T. cruzi soluble extract	Myelin basic protein	Serum IgG, T cells	8
150-kDa protein	Smooth and striated muscle	Serum IgG	140
150-kDa protein	β 1-adrenoreceptor M2 cholinergic receptor	MAb	34
55-kDa membrane protein	28-kDa lymphocyte membrane protein	MAb	61

[a] Abbreviations: AS, antiserum; MAb, monoclonal antibody; rDNA, recombinant DNA; IgG, immunoglobulin G.

Cross-reactive antibodies between a 23-kDa ribosomal protein and a corresponding mammalian protein were observed in ca. 30% of patients with chronic Chagas' disease, independent of their clinical presentation (21). However, as ribosomal proteins are present in all cell types, they should be unlikely targets for a heart-specific autoimmune attack. Sulfated glycolipids and neutral glycosphingolipids found in *T. cruzi* are essentially the same as those found in mammalian hosts and are cross-reactively recognized by antibodies formed during infection (106, 138).

Neuroantigens

In human Chagas' disease, there is a net loss of neurons from the autonomic system along the hollow viscera and the heart (113) (Table 1), bearing an obvious causal relationship with the denervation syndromes of the esophagus and sigmoid colon. Furthermore, the autonomic nervous system dysfunction observed in symptomatic and asymptomatic patients may be related to such a denervation (69, 88). Early studies indicated that sera from more than 80% of Chagas' disease patients contained antineuron autoantibodies (113). In order to further investigate the hypothesis that the denervation seen in Chagas' disease has immunological causes, several groups produced monoclonal antibodies that display cross-reactivity between *T. cruzi* and mammalian nervous tissue (Table 4). Monoclonal antibody CE5 raised against the dorsal root ganglia recognized *T. cruzi* as well as neurons and heart muscle (139). Anti-*T. cruzi* monoclonal antibodies 5H7 and 3H3 recognized antigens with molecular weights of 58,000 and 37,000 in mouse brain and spinal cord, including neurons and glial cells, and antigens with molecular weights of 58,000 and 35,000 in *T. cruzi* (121). Monoclonal antibodies that were raised against bat *Trypanosoma* species and that cross-reacted with *T. cruzi* also recognized mouse cerebellar neurons and astrocytes (108). Later, it was shown that sulfated glycolipids were the target antigens; administration of such monoclonal antibodies induced immediate paralysis and death by respiratory insufficiency (106, 107). Van Voorhis, Eisen, and colleagues (132–134) (Table 3) described a cross-reactive epitope (TPQRKTTEDRPQ) between FL-160, a 160-kDa *T. cruzi* flagellar protein, and a neuronal 47-kDa protein; the peptide was capable of inhibiting the binding of anti-FL-160 antibodies to the neuronal antigen. However, the presence of cross-reactive antibodies in patients' sera did not correlate with any clinical form of Chagas' disease (29). A cross-reactivity between *T. cruzi* and myelin basic protein was observed at the levels of both antibodies and T cells in experimentally infected mice (8) (Table 4).

Cardiovascular Receptors

The autonomic nervous system dysfunction observed in patients with Chagas' disease (69, 88) prompted investigators to study the possible involvement of antibodies against adrenergic or muscarinic cholinergic receptors. Antibodies against the β-adrenoreceptor are found in patients with severe idiopathic dilated cardiomyopathy and may be functionally active (32, 67). Functional antibodies against adrenergic G-protein-coupled receptors were first demonstrated in serum from Chagas' disease patients as early as 1984 (23, 105, 122), and agonistic antibodies against muscarinic (M2) cholinergic receptors have received more attention in the last decade (53, 54, 56); recently, it was shown that functional antireceptor antibodies bind to the second extracellular loop of the M2 muscarinic receptor (55) (Tables 2, 3, and 4). Investigators have demonstrated that the presence of such functionally active antireceptor antibodies correlates not with symptomatic patients but rather with those who have a dysfunction of the autonomous nervous system (55). A recent report showed molecular mimicry at the epitope level between ribosomal protein P0 of *T. cruzi* (ASEEE) and an epitope (AESDE) at the external domains of the cardiovascular β1-adrenergic receptor (47, 72). This cross-reactive antibody had agonistic functional activity, in that it could accelerate the beating of rat cardiomyocytes which could be inhibited by a specific peptide. However, up to now it has not been demonstrated that antibodies that recognize the epitope shared between the P0 protein and β_1-adrenoreceptor are more prevalent among patients with CCC than among ASY *T. cruzi*-seropositive patients. A functionally active monoclonal antibody against M2 muscarinic receptors obtained from *T. cruzi*-infected mice was shown to be cross-reactive with a 150-kDa polypeptide from *T. cruzi* (34). A recent study demonstrated that peptides from ribosomal proteins P0 and P2β, which bear acidic epitopes, could block antibodies with muscarinic activity (91), probably by an interaction with the second extracellular loop of the receptor. Again, as mentioned above, the recognition of acidic epitopes in muscarinic and adrenergic receptors by cross-reactive anti-P-protein antibodies appeared to be dependent on a negative charge interaction, which may suggest a common origin (70). Two interesting experimental studies involved the immunization of mice with *T. cruzi* ribosomal protein P2β (86) and the R13 peptide from ribosomal P protein (98) (Table 5). In both studies, electrocardiogram alterations were seen, but no signs of myocardial inflammation were obtained.

Cardiac Myosin

Myosin is the most abundant heart protein, making up to 50% of muscle protein by weight (59). It is a major antigen in several instances of heart-specific autoimmunity (27, 42, 99, 135); moreover, immunization with cardiac myosin in Freund's adjuvant induces severe T-cell-dependent myocarditis in genetically susceptible mice (84, 100, 101, 119). Cardiac myosin is recognized by $CD4^+$ T cells in chronically *T. cruzi*-infected mice (114). Anti-cardiac myosin antibodies (127) and delayed-type hypersensitivity responses to cardiac myosin (J. S. Leon and D. M. Engman, personal communication) have been found in *T. cruzi*-infected mice with intense myocarditis (Table 1). Taken together, these reports on cardiac myosin autoimmunity in murine models of Chagas' disease suggested the possible

Table 5. Induction of heart disorders after immunization with *T. cruzi* antigens[a]

T. cruzi antigen	Host	Disorder	Reference
T. cruzi microsomal fraction	Rb	Myocarditis	124
T. cruzi SRA	M	Myocarditis	5
T. cruzi microsomal and cytoplasmic fractions	M	Myocarditis	115
Recombinant ribosomal protein P2β	M	ECG alteration	86
R13 peptide from ribosomal protein P0	M	ECG alteration	98

[a] Abbreviations: M, mouse; Rb, rabbit; ECG, electrocardiogram.

relevance of myosin recognition in the pathogenesis of CCC in humans. Our group detected anti-human ventricular cardiac myosin heavy-chain immunoglobulin G antibodies at similar levels among sera from individuals with CCC, ASY individuals, and healthy individuals (39). Affinity-selected anti-human ventricular cardiac myosin heavy-chain antibodies from Chagas' disease patients' sera specifically recognized a defined *T. cruzi* antigen (39), the recombinant tandemly repetitive protein B13 (57) (Table 3). Cardiac myosin-B13 cross-reactive antibodies were present in sera from 100% of CCC patients but only 14% of ASY patients (Fig. 1) (39). In the absence of exogenous antigen, (PHA-expanded) $CD4^+$ T-cell clones from heart tissue of a CCC patient cross-reactively recognized the ventricular cardiac (but not skeletal) myosin heavy chain and *T. cruzi* protein B13 (Table 3; Fig. 2) (38). However, none of the 17 clones tested responded to the immunodominant recombinant *T. cruzi* antigens CRA, FRA, JL5, and B12 or to *T. cruzi* trypomastigote lysate (38). Such results may suggest that autoimmunity or molecular mimicry targets, rather than the direct antigenic stimulus of *T. cruzi*, are the primary stimuli of the heart tissue-damaging T-cell infiltrate.

PBMCs from CCC, ASY, or healthy individuals recognize the B13 protein but are nonresponsive to cardiac myosin (1, 35). However, in vitro sensitization of lymphocytes from a *T. cruzi*-seronegative individual with B13 protein elicits cardiac myosin-cross-

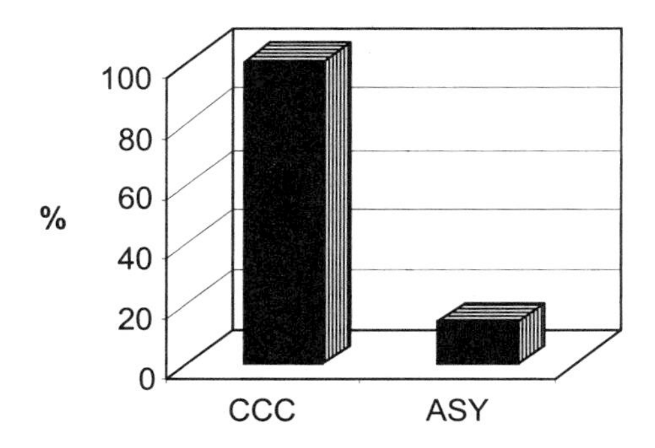

Figure 1. Prevalence of immunoglobulin G antibodies cross-reactive with cardiac myosin and *T. cruzi* B13 protein in sera from CCC patients and ASY individuals (percent positive) (modified from reference 39).

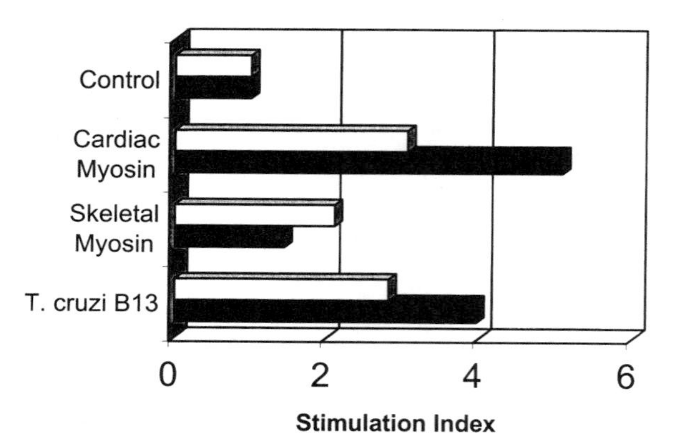

Figure 2. Cross-reactive recognition of cardiac myosin and *T. cruzi* protein B13 by T-cell clones E2O5 (black bars) and E2.17 (white bars) derived from heart tissue from a CCC patient (modified from reference 38).

reactive T-cell clones (Fig. 3) (2). The investigators hypothesized that in vivo challenge with B13 antigen along with *T. cruzi* infection could break immunological tolerance to cardiac myosin and elicit cardiac myosin-responsive T cells in vivo. Given the fact that tandemly repetitive units of B13 protein show sequence variation (26, 57, 65), we recently tested the recognition of all the variant epitopes by patients with CCC and ASY individuals. The variant epitopes recognized by CCC patients were different from those recognized by ASY individuals (1). Although this was not tested directly, the differential anti-B13 T-cell recognition repertoire of CCC patients could hypothetically be more prone than that of

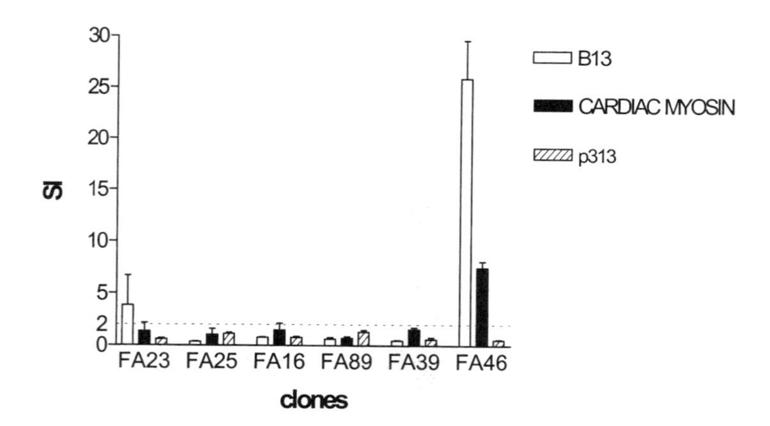

Figure 3. Cross-reactive recognition of cardiac myosin (black bars) and *T. cruzi* protein B13 (white bars) by T-cell clones obtained from a PBMC T-cell line sensitized with B13 protein. The synthetic peptide p313 (hatched bars) is a negative control (modified from reference 2).

ASY patients to recognition of cardiac myosin in a cross-reactive fashion. It is known that the fine recognition of epitopes can determine whether an immune response will be protective or pathogenic, as has been described before for other diseases (102).

Thus, in parallel to our findings, it can be hypothesized that in natural *T. cruzi* infection of a susceptible individual, macrophages that have endocytosed *T. cruzi* will produce IL-12 (50) and express the costimulatory molecule B7 (48), presenting a heart-cross-reactive *T. cruzi* antigen (e.g., B13 protein). Such antigen-loaded, activated macrophages will sensitize cross-reactive $CD4^+$ T-cell clones that will become experienced T1-type T cells (1, 37, 41). Upon recirculation to the heart tissue and an encounter with heart interstitial macrophages that constitutively present heart antigen (e.g., cardiac myosin) epitopes complexed to class II MHC molecules (120), such cells would become activated (2, 38). From then on, $CD4^+$ T cells might initiate and maintain a typical delayed-type hypersensitivity reaction in heart tissue by the release of inflammatory cytokines, as reported previously (41), and cellular recruitment.

SUMMARY AND CONCLUSION

Chagas' disease is a conundrum of several clinical syndromes triggered by *T. cruzi* infection in a group of susceptible individuals. Expression of clinical syndromes can be nonoverlapping. It is therefore not surprising that several different systems of molecular mimicry have been identified. Inasmuch as several of those may simply be secondary to degeneracy in immune recognition (90), and thus the result is inconsequential to pathogenesis, it is likely that some of them may play an important role in disease causation. Functional autoantibodies may have a pathogenic role on autonomic system disorders and may also be relevant in heart conduction disorders. Heart inflammatory lesions that are related to T1-type cytokine production and that ultimately lead to dilated cardiomyopathy and life-threatening heart failure are most likely related to cross-reactive recognition of heart-specific epitopes (e.g., cardiac myosin) by T cells in situ. The persistence of a parasite which induces IL-12 production (7) and upregulation of B7 molecules (48) is likely to boost the production of potentially pathogenic, experienced, T1-type T cells. Even though no significant differences were observed in the HLA class II profiles of CCC and ASY patients in an extensive survey by Abel (1, 46a), the differential T-cell recognition repertoire for recognition of a cross-reactive antigen, as described in CCC patients in the previous section, may be one of the underlying factors of the differential susceptibility to the development of CCC (35). In fact, the T-cell recognition repertoire has been acknowledged as one of the pathogenesis "checkpoints" in autoimmune diseases (6, 14). The study of fine recognition and immune repertoires in diseases in which molecular mimicry is thought to play a role is probably going to become an important focus of research in the near future. Furthermore, the identification of relevant heart-specific antigenic targets may allow the future use of antigen-specific immunomodulation (e.g., oral tolerance induction) to avoid progression of CCC, as has been done for autoimmune uveitis (103) and rheumatoid arthritis (18).

REFERENCES

1. **Abel, L.** 1999. Reposta imune celular periférica a proteína B13 de *T. cruzi:* estudo do reconhecimento antigênico e perfil de citocinas. Doctoral thesis. University of São Paulo, São Paulo, Brazil.

2. **Abel, L. C., J. Kalil, and E. Cunha-Neto.** 1997. Molecular mimicry between cardiac myosin and *Trypanosoma cruzi* antigen B13: identification of a B13-driven human T cell clone that recognizes cardiac myosin. *Braz. J. Med. Biol. Res.* **30:**1305–1308.
3. **Ablashi, D. V., W. Lapps, M. Kaplan, J. E. Whitman, J. R. Richert, and G. R. Pearson.** 1998. Human herpesvirus-6 (HHV-6) infection in multiple sclerosis: a preliminary report. *Mult. Scler.* **4:**490–496.
4. **Acosta, A. M., M. Sadigursky, and C. A. Santos-Buch.** 1983. Anti-striated muscle antibody activity produced by *Trypanosoma cruzi. Proc. Soc. Exp. Biol. Med.* **172:**364–369.
5. **Acosta, A. M., and C. A. Santos-Buch.** 1985. Autoimmune myocarditis induced by *Trypanosoma cruzi. Circulation* **71:**1255–1261.
6. **Akkaraju, S., W. Y. Ho, D. Leong, K. Canaan, M. M. Davis, and C. C. Goodnow.** 1997. A range of CD4 T cell tolerance: partial inactivation to organ-specific antigen allows nondestructive thyroiditis or insulitis. *Immunity* **7:**255–271.
7. **Aliberti, J. C., M. A. Cardoso, G. A. Martins, R. T. Gazzinelli, L. Q. Vieira, and J. S. Silva.** 1996. Interleukin-12 mediates resistance to *Trypanosoma cruzi* in mice and is produced by murine macrophages in response to live trypomastigotes. *Infect. Immun.* **64:**1961–1967.
8. **Al-Sabbagh, A., C. A. Garcia, B. M. Diaz-Bardales, C. Zaccarias, J. K. Sakurada, and L. M. Santos.** 1998. Evidence for cross-reactivity between antigen derived from *Trypanosoma cruzi* and myelin basic protein in experimental Chagas disease. *Exp. Parasitol.* **89:**304–311.
9. **Amorim, D. D., and N. J. Marin.** 1995. Functional alterations of the autonomic nervous system in Chagas' heart disease. *Rev. Paulista Med.* **113:**772–784.
10. **Amorim, D. S.** 1985. Chagas' heart disease: experimental models. *Heart Vessels Suppl.* **1:**236–239.
11. **Andrade, Z. A.** 1983. Mechanisms of myocardial damage in *Trypanosoma cruzi* infection. *Ciba. Found. Symp.* **99:**214–233.
12. **Andrade, Z. A., S. G. Andrade, and M. Sadigursky.** 1987. Enhancement of chronic *Trypanosoma cruzi* myocarditis in dogs treated with low doses of cyclophosphamide. *Am. J. Pathol.* **127:**467–473.
13. **Andrade, Z. A., S. G. Andrade, M. Sadigursky, R. J. J. Wenthold, S. L. Hilbert, and V. J. Ferrans.** 1997. The indeterminate phase of Chagas' disease: ultrastructural characterization of cardiac changes in the canine model. *Am. J. Trop. Med. Hyg.* **57:**328–336.
14. **Andre, I., A. Gonzalez, B. Wang, J. Katz, C. Benoist, and D. Mathis.** 1996. Checkpoints in the progression of autoimmune disease: lessons from diabetes models. *Proc. Natl. Acad. Sci. USA* **93:**2260–2263.
15. **Anez, N., H. Carrasco, H. Parada, G. Crisante, A. Rojas, C. Fuenmayor, N. Gonzalez, G. Percoco, R. Borges, P. Guevara, and J. L. Ramirez.** 1999. Myocardial parasite persistence in chronic chagasic patients. *Am. J. Trop. Med. Hyg.* **60:**726–732.
16. **Bach-Elias, M., D. Bahia, D. C. Teixeira, and R. M. Cicarelli.** 1998. Presence of autoantibodies against small nuclear ribonucleoprotein epitopes in Chagas' patients' sera. *Parasitol. Res.* **84:**796–799.
17. **Barbosa, A. A., Jr., and Z. A. Andrade.** 1984. Identificacao do *Trypanosoma cruzi* nos tecidos extracardiacos de portadores de miocardite cronica chagásica. *Rev. Soc. Bras. Med. Trop.* **17:**123–126.
18. **Barnett, M. L., J. M. Kremer, E. W. St. Clair, D. O. Clegg, D. Furst, M. Weisman, M. J. Fletcher, S. Chasan-Taber, E. Finger, A. Morales, C. H. Le, and D. E. Trentham.** 1998. Treatment of rheumatoid arthritis with oral type II collagen. Results of a multicenter, double-blind, placebo-controlled trial. *Arthritis Rheum.* **41:**290–297. (Erratum, **41:**938.)
19. **Bestetti, R. B., and G. Muccillo.** 1997. Clinical course of Chagas' heart disease: a comparison with dilated cardiomyopathy. *Int. J. Cardiol.* **60:**187–193.
20. **Bocchi, E. A., M. L. Higuchi, M. L. Vieira, N. Stolf, G. Bellotti, A. Fiorelli, D. Uip, A. Jatene, and F. Pileggi.** 1998. Higher incidence of malignant neoplasms after heart transplantation for treatment of chronic Chagas' heart disease. *J. Heart Lung Transplant.* **17:**399–405.
21. **Bonfa, E., V. S. Viana, A. C. Barreto, N. H. Yoshinari, and W. Cossermelli.** 1993. Autoantibodies in Chagas' disease. An antibody cross-reactive with human and *Trypanosoma cruzi* ribosomal proteins. *J. Immunol.* **150:**3917–3923.
22. **Bonfa, E., A. V. Tavares, M. V. Regenmortel, W. Cossermelli, and M. T. Levin.** 1992. The importance of one aminoacid substitution in the anti-P response in Chagas' disease. *Mem. Inst. Oswaldo Cruz* **87**(Suppl)**:**172.
23. **Borda, E., J. Pascual, P. Cossio, M. De La Vega, R. Arana, and L. Sterin-Borda.** 1984. A circulating IgG in Chagas' disease which binds to beta-adrenoceptors of myocardium and modulates their activity. *Clin. Exp. Immunol.* **57:**679–686.

24. **Brener, Z.** 1973. Biology of *Trypanosoma cruzi*. *Annu. Rev. Microbiol.* **27:**347–382.
25. **Britto, C., M. A. Cardoso, C. M. Vanni, A. Hasslocher-Moreno, S. S. Xavier, W. Oelemann, A. Santoro, C. Pirmez, C. M. Morel, and P. Wincker.** 1995. Polymerase chain reaction detection of *Trypanosoma cruzi* in human blood samples as a tool for diagnosis and treatment evaluation. *Parasitology* **110:**241–247.
26. **Buschiazzo, A., O. E. Campetella, R. A. Macina, S. Salceda, A. C. Frasch, and D. O. Sanchez.** 1992. Sequence of the gene for a *Trypanosoma cruzi* protein antigenic during the chronic phase of human Chagas disease. *Mol. Biochem. Parasitol.* **54:**125–128.
27. **Caforio, A. L., M. Grazzini, J. M. Mann, P. J. Keeling, G. F. Bottazzo, W. J. McKenna, and S. Schiaffino.** 1992. Identification of alpha- and beta-cardiac myosin heavy chain isoforms as major autoantigens in dilated cardiomyopathy. *Circulation* **85:**1734–1742.
28. **Camargo, M. M., I. C. Almeida, M. E. Pereira, M. A. Ferguson, L. R. Travassos, and R. T. Gazzinelli.** 1997. Glycosylphosphatidylinositol-anchored mucin-like glycoproteins isolated from *Trypanosoma cruzi* trypomastigotes initiate the synthesis of proinflammatory cytokines by macrophages. *J. Immunol.* **158:**5890–5901.
29. **Cetron, M. S., R. Hoff, S. Kahn, H. Eisen, and W. C. Van Voorhis.** 1992. Evaluation of recombinant trypomastigote surface antigens of *Trypanosoma cruzi* in screening sera from a population in rural northeastern Brazil endemic for Chagas' disease. *Acta Trop.* **50:**259–266.
30. **Chagas, C.** 1909. Nova tripanozomiaze humana. Estudos sobre a morfolojia e o ciclo evolutivo do Schizotripanum cruzi n.gen n. sp, ajente etiolojico de nova entidade morbida do homen. *Mem. Inst. Oswaldo Cruz* **1:**159–218.
31. **Chambo, J. G., M. P. Cabeza, and R. P. Laguens.** 1990. Presence of anti-*Trypanosoma cruzi* antibodies in the sera of mice with experimental autoimmune myocarditis. *Experientia* **46:**977–979.
32. **Chiale, P. A., M. B. Rosenbaum, M. V. Elizari, A. Hjalmarson, Y. Magnusson, G. Wallukat, and J. Hoebeke.** 1995. High prevalence of antibodies against beta 1- and beta 2-adrenoceptors in patients with primary electrical cardiac abnormalities. *J. Am. Coll. Cardiol.* **26:**864–869.
33. **Cossio, P. M., C. Diez, A. Szarfman, E. Kreutzer, B. Candiolo, and R. M. Arana.** 1974. Chagasic cardiopathy. Demonstration of a serum gamma globulin factor which reacts with endocardium and vascular structures. *Circulation* **49:**13–21.
34. **Cremaschi, G., N. W. Zwirner, G. Gorelik, E. L. Malchiodi, M. G. Chiaramonte, C. A. Fossati, and L. Sterin-Borda.** 1995. Modulation of cardiac physiology by an anti-*Trypanosoma cruzi* monoclonal antibody after interaction with myocardium. *FASEB J.* **9:**1482–1488.
35. **Cunha-Neto, E., L. Abel, L. V. Rizzo, A. Goldberg, C. Mady, B. Ianni, J. Hammer, F. Sinigaglia, and J. Kalil.** 1998. Checkpoints for autoimmunity-induced heart tissue damage in human chagas' disease: a multistep process? *Mem. Inst. Oswaldo Cruz* **93**(Suppl. I)**:**40–41.
36. **Cunha-Neto, E.** 1999. MHC-restricted antigen presentation and recognition: constraints on gene, recombinant and peptide vaccines in humans *Braz. J. Med. Biol. Res.* **32:**199–205.
37. **Cunha-Neto, E., L. C. Abel, L. V. Rizzo, A. Goldberg, C. Mady, B. Ianni, J. Hammer, F. Sinigaglia, and J. Kalil.** 1997. Autoimmunity in human Chagas' disease cardiomyopathy: cytokine production and antigen recognition by T cells. *Mem. Inst. Oswaldo Cruz* **92**(Suppl. I)**:**40–41.
38. **Cunha-Neto, E., V. Coelho, L. Guilherme, A. Fiorelli, N. Stolf, and J. Kalil.** 1996. Autoimmunity in Chagas' disease. Identification of cardiac myosin-B13 *Trypanosoma cruzi* protein crossreactive T cell clones in heart lesions of a chronic Chagas' cardiomyopathy patient. *J. Clin. Investig.* **98:**1709–1712.
39. **Cunha-Neto, E., M. Duranti, A. Gruber, B. Zingales, I. DeMessia, N. Stolf, G. Bellotti, M. E. Patarroyo, F. Pilleggi, and J. Kalil.** 1995. Autoimmunity in Chagas disease cardiopathy: biological relevance of a cardiac myosin-specific epitope crossreactive to an immunodominant *Trypanosoma cruzi* antigen. *Proc. Natl. Acad. Sci. USA* **92:**3541–3545.
40. **Cunha-Neto, E., R. Moliterno, V. Coelho, L. Guilherme, E. Bocchi, M. L. Higuchi, N. Stolf, F. Pileggi, L. Steinman, and J. Kalil.** 1994. Restricted heterogeneity of T cell receptor variable alpha chain transcripts in hearts of Chagas' disease cardiomyopathy patients. *Parasite Immunol.* **16:**171–179.
41. **Cunha-Neto, E., L. V. Rizzo, F. Albuquerque, L. Abel, L. Guilherme, E. Bocchi, F. Bacal, D. Carrara, B. Ianni, C. Mady, and J. Kalil.** 1998. Cytokine production profile of heart-infiltrating T cells in Chagas' disease cardiomyopathy. *Braz. J. Med. Biol. Res.* **31:**133–137.
42. **Cunningham, M. W., S. M. Antone, M. Smart, R. Liu, and S. Kosanke.** 1997. Molecular analysis of human cardiac myosin-cross-reactive B- and T-cell epitopes of the group A streptococcal M5 protein. *Infect. Immun.* **65:**3913–3923.

43. **de la Vega, M. T., G. Damilano, and C. Diez.** 1976. Leukocyte migration inhibition test with heart antigens in American trypanosomiasis. *J. Parasitol.* **62:**129–130.
44. **De Oliveira, S., R. C. Pedrosa, J. H. Nascimento, A. C. Campos de Carvalho, and M. O. Masuda.** 1997. Sera from chronic chagasic patients with complex cardiac arrhythmias depress electrogenesis and conduction in isolated rabbit hearts. *Circulation* **96:**2031–2037.
45. **dos Santos, R. R., M. A. Rossi, J. L. Laus, J. S. Silva, W. Savino, and J. Mengel.** 1992. Anti-CD4 abrogates rejection and reestablishes long-term tolerance to syngeneic newborn hearts grafted in mice chronically infected with *Trypanosoma cruzi. J. Exp. Med.* **175:**29–39.
46. **el-Zayadi, A. R., O. E. Selim, H. Hamdy, H. Dabbous, A. Ahdy, and S. A. Moniem.** 1998. Association of chronic hepatitis C infection and diabetes mellitus. *Trop. Gastroenterol.* **19:**141–144.
46a. **Faé, K. C., S. A. Drigo, E. Cunha-Neto, B. Ianni, C. Mady, J. Kalil, and A. C. Goldberg.** Role of genetic susceptibility to cardiomyopathy in Chagas' disease. *Microbes Infect.,* in press.
47. **Ferrari, I., M. J. Levin, G. Wallukat, R. Elies, D. Lebesgue, P. Chiale, M. Elizari, M. Rosenbaum, and J. Hoebeke.** 1995. Molecular mimicry between the immunodominant ribosomal protein P0 of *Trypanosoma cruzi* and a functional epitope on the human beta 1-adrenergic receptor. *J. Exp. Med.* **182:**59–65.
48. **Frosch, S., D. Kuntzlin, and B. Fleischer.** 1997. Infection with *Trypanosoma cruzi* selectively upregulates B7-2 molecules on macrophages and enhances their costimulatory activity. *Infect. Immun.* **65:**971–977.
49. **Gattass, C. R., M. T. Lima, A. F. Nobrega, M. A. Barcinski, and A. G. Dos Reis.** 1988. Do self-heart-reactive T cells expand in *Trypanosoma cruzi*-immune hosts? *Infect. Immun.* **56:**1402–1405.
50. **Gazzinelli, R. T., A. Talvani, M. M. Camargo, H. C. Santiago, M. A. Oliveira, L. Q. Vieira, G. A. Martins, J. C. Aliberti, and J. S. Silva.** 1998. Induction of cell-mediated immunity during early stages of infection with intracellular protozoa. *Braz. J. Med. Biol. Res.* **31:**89–104.
51. **Gea, S., P. Ordonez, F. Cerban, D. Iosa, C. Chizzolini, and E. Vottero-Cima.** 1993. Chagas' disease cardioneuropathy: association of anti-*Trypanosoma cruzi* and anti-sciatic nerve antibodies. *Am. J. Trop. Med. Hyg.* **49:**581–588.
52. **Gluckstein, D., F. Ciferri, and J. Ruskin.** 1992. Chagas' disease: another cause of cerebral mass in the acquired immunodeficiency syndrome. *Am. J. Med.* **92:**429–432.
53. **Goin, J. C., E. Borda, C. P. Leiros, R. Storino, and L. Sterin-Borda.** 1994. Identification of antibodies with muscarinic cholinergic activity in human Chagas' disease: pathological implications. *J. Auton. Nerv. Syst.* **47:**45–52.
54. **Goin, J. C., E. Borda, A. Segovia, and L. Sterin-Borda.** 1991. Distribution of antibodies against beta-adrenoceptors in the course of human *Trypanosoma cruzi* infection. *Proc. Soc. Exp. Biol. Med.* **197:**186–192.
55. **Goin, J. C., C. P. Leiros, E. Borda, and L. Sterin-Borda.** 1997. Interaction of human chagasic IgG with the second extracellular loop of the human heart muscarinic acetylcholine receptor: functional and pathological implications. *FASEB J.* **11:**77–83.
56. **Goin, J. C., L. C. Perez, E. Borda, and L. Sterin-Borda.** 1994. Modification of cholinergic-mediated cellular transmembrane signals by the interaction of human chagasic IgG with cardiac muscarinic receptors. *Neuroimmunomodulation* **1:**284–291.
57. **Gruber, A., and B. Zingales.** 1993. *Trypanosoma cruzi:* characterization of two recombinant antigens with potential application in the diagnosis of Chagas' disease. *Exp. Parasitol.* **76:**1–12.
58. **Guilherme, L., E. Cunha-Neto, V. Coelho, R. Snitcowsky, P. M. Pomerantzeff, R. V. Assis, F. Pedra, J. Neumann, A. Goldberg, and M. E. Patarroyo.** 1995. Human heart-infiltrating T-cell clones from rheumatic heart disease patients recognize both streptococcal and cardiac proteins. *Circulation* **92:**415–420.
59. **Harrington, W. F., and M. E. Rodgers.** 1984. Myosin. *Annu. Rev. Biochem.* **53:**35–73.
60. **Heber-Katz, E., and H. Acha-Orbea.** 1989. The V region hypothesis: evidence from autoimmune encephalomyelitis. *Immunol. Today* **10:**164–169.
61. **Hernandez-Munain, C., J. L. De Diego, A. Alcina, and M. Fresno.** 1992. A *Trypanosoma cruzi* membrane protein shares an epitope with a lymphocyte activation antigen and induces crossreactive antibodies. *J. Exp. Med.* **175:**1473–1482.
62. **Higuchi, M. L., M. M. Reis, V. D. Aiello, L. A. Benvenuti, P. S. Gutierrez, G. Bellotti, and F. Pileggi.** 1997. Association of an increase in $CD8^+$ T cells with the presence of *Trypanosoma cruzi* antigens in chronic, human, chagasic myocarditis. *Am. J. Trop. Med. Hyg.* **56:**485–489.
63. **Higuchi, M. L., P. S. Gutierrez, V. D. Aiello, S. Palomino, E. Bocchi, J. Kalil, G. Bellotti, and F. Pileggi.** 1993. Immunohistochemical characterization of infiltrating cells in human chronic chagasic myocardi-

tis: comparison with myocardial rejection process. *Virchows Arch. A. Pathol. Anat. Histopathol.* **423:**157–160.

64. **Higuchi, M. L., M. C. De, B. A. Pereira, E. A. Lopes, N. Stolf, G. Bellotti, and F. Pileggi.** 1987. The role of active myocarditis in the development of heart failure in chronic Chagas' disease: a study based on endomyocardial biopsies. *Clin. Cardiol.* **10:**665–670.
65. **Hoft, D. F., K. S. Kim, K. Otsu, D. R. Moser, W. J. Yost, J. H. Blumin, J. E. Donelson, and L. V. Kirchhoff.** 1989. *Trypanosoma cruzi* expresses diverse repetitive protein antigens. *Infect. Immun.* **57:**1959–1967.
66. **Hontebeyrie-Joskowicz, M., G. Said, G. Milon, G. Marchal, and H. Eisen.** 1987. L3T4^{+} T cells able to mediate parasite-specific delayed-type hypersensitivity play a role in the pathology of experimental Chagas' disease. *Eur. J. Immunol.* **17:**1027–1033.
67. **Jahns, R., V. Boivin, C. Siegmund, G. Inselmann, M. J. Lohse, and F. Boege.** 1999. Autoantibodies activating human beta 1-adrenergic receptors are associated with reduced cardiac function in chronic heart failure. *Circulation* **99:**649–654.
68. **Jones, E. M., D. G. Colley, S. Tostes, E. R. Lopes, C. L. Vnencak-Jones, and T. L. McCurley.** 1993. Amplification of a *Trypanosoma cruzi* DNA sequence from inflammatory lesions in human chagasic cardiomyopathy. *Am. J. Trop. Med. Hyg.* **48:**348–357.
69. **Junqueira, J. L., P. S. Beraldo, E. Chapadeiro, and P. C. Jesus.** 1992. Cardiac autonomic dysfunction and neuroganglionitis in a rat model of chronic Chagas' disease. *Cardiovasc. Res.* **26:**324–329.
70. **Kahn, S., M. Kahn, and H. Eisen.** 1992. Polyreactive autoantibodies to negatively charged epitopes following *Trypanosoma cruzi* infection. *Eur. J. Immunol.* **22:**3051–3056.
71. **Kalil, J., and E. Cunha-Neto.** 1996. Autoimmunity in Chagas' disease cardiomyopathy: fulfilling the criteria at last? *Parasitol. Today* **12:**396–399.
72. **Kaplan, D., I. Ferrari, P. L. Bergami, E. Mahler, G. Levitus, P. Chiale, J. Hoebeke, M. H. Van Regenmortel, and M. J. Levin.** 1997. Antibodies to ribosomal P proteins of *Trypanosoma cruzi* in Chagas disease possess functional autoreactivity with heart tissue and differ from anti-P autoantibodies in lupus. *Proc. Natl. Acad. Sci. USA* **94:**10301–10306.
73. **Kerner, N., P. Liegeard, M. J. Levin, and M. Hontebeyrie-Joskowicz.** 1991. *Trypanosoma cruzi:* antibodies to a MAP-like protein in chronic Chagas' disease cross-react with mammalian cytoskeleton. *Exp. Parasitol.* **73:**451–459.
74. **Khoury, E. L., C. Diez, P. M. Cossio, and R. M. Arana.** 1983. Heterophil nature of EVI antibody in *Trypanosoma cruzi* infection. *Clin. Immunol. Immunopathol.:* **27:**283–288.
75. **Kohl, S., L. K. Pickering, L. S. Frankel, and R. G. Yaeger.** 1982. Reactivation of Chagas' disease during therapy of acute lymphocytic leukemia. *Cancer* **50:**827–828.
76. **Laguens, R. P., P. C. Meckert, G. Chambo, and R. J. Gelpi.** 1981. Chronic Chagas disease in the mouse. II. Transfer of the heart disease by means of immunocompetent cells. *Medicina (Buenos Aires)* **41:**40–43.
77. **Laranja, F., E. Dias, G. Nobrega, and A. Miranda.** 1956. Chagas' disease. A clinical, epidemiologic and pathologic study. *Circulation* **14:**1035.
78. **Laucella, S. A., H. E. de Titto, and E. L. Segura.** 1996. Epitopes common to *Trypanosoma cruzi* and mammalian tissues are recognized by sera from Chagas' disease patients: prognosis value in Chagas disease. *Acta Trop.* **62:**151–162.
79. **Laucella, S. A., E. Velazquez, M. Dasso, and E. de Titto.** 1996. *Trypanosoma cruzi* and mammalian heart cross-reactive antigens. *Acta Trop.* **61:**223–238.
80. **Leiguarda, R., A. Roncoroni, A. L. Taratuto, L. Jost, M. Berthier, M. Nogues, and H. Freilij.** 1990. Acute CNS infection by *Trypanosoma cruzi* (Chagas' disease) in immunosuppressed patients. *Neurology* **40:**850–851.
81. **Leiros, C. P., L. Sterin-Borda, E. S. Borda, J. C. Goin, and M. M. Hosey.** 1997. Desensitization and sequestration of human m2 muscarinic acetylcholine receptors by autoantibodies from patients with Chagas' disease. *J. Biol. Chem.* **272:**12989–12993.
82. **Levin, M. J., E. Mesri, R. Benarous, G. Levitus, A. Schijman, P. Levy-Yeyati, P. A. Chiale, A. M. Ruiz, A. Kahn, and M. B. Rosenbaum.** 1989. Identification of major *Trypanosoma cruzi* antigenic determinants in chronic Chagas' heart disease. *Am. J. Trop. Med. Hyg.* **41:**530–538.
83. **Levitus, G., M. Hontebeyrie-Joskowicz, M. H. Van Regenmortel, and M. J. Levin.** 1991. Humoral autoimmune response to ribosomal P proteins in chronic Chagas heart disease. *Clin. Exp. Immunol.* **85:**413–417.
84. **Liao, L., R. Sindhwani, L. Leinwand, B. Diamond, and S. Factor.** 1993. Cardiac alpha-myosin heavy

chains differ in their induction of myocarditis. Identification of pathogenic epitopes. *J. Clin. Investig.* **92:**2877–2882.

85. **Lopes, E. R., E. Chapadeiro, Z. A. Andrade, H. O. Almeida, and A. Rocha.** 1981. Pathological anatomy of hearts from asymptomatic Chagas disease patients dying in a violent manner. *Mem. Inst. Oswaldo Cruz.* **76:**189–197. (In Portuguese.)
86. **Lopez, B. P., M. P. Cabeza, D. Kaplan, G. Levitus, F. Elias, F. Quintana, M. Van Regenmortel, R. Laguens, and M. J. Levin.** 1997. Immunization with recombinant *Trypanosoma cruzi* ribosomal P2beta protein induces changes in the electrocardiogram of immunized mice. *FEMS Immunol. Med. Microbiol.* **18:**75–85.
87. **Mady, C., R. H. Cardoso, A. C. Barretto, P. L. da Luz, G. Bellotti, and F. Pileggi.** 1994. Survival and predictors of survival in patients with congestive heart failure due to Chagas' cardiomyopathy. *Circulation* **90:**3098–3102.
88. **Marin-Neto, J. A., G. Bromberg-Marin, A. Pazin-Filho, M. V. Simoes, and B. C. Maciel.** 1998. Cardiac autonomic impairment and early myocardial damage involving the right ventricle are independent phenomena in Chagas' disease. *Int. J. Cardiol.* **65:**261–269.
89. **Martin, U. O., D. Afchain, A. de Marteleur, O. Ledesma, and A. Capron.** 1987. Circulating immune complexes in different developmental stages of Chagas' disease. *Medicina (Buenos Aires)* **47:**159–162. (In Portuguese).
90. **Mason, D.** 1998. A very high level of crossreactivity is an essential feature of the T-cell receptor. *Immunol. Today* **19:**395–404.
91. **Masuda, M. O., M. Levin, S. F. De Oliveira, P. C. Dos Santos Costa, P. L. Bergami, N. A. Dos Santos Almeida, R. C. Pedrosa, I. Ferrari, J. Hoebeke, and A. C. Campos de Carvalho.** 1998. Functionally active cardiac antibodies in chronic Chagas' disease are specifically blocked by *Trypanosoma cruzi* antigens. *FASEB J.* **12:**1551–1558.
92. **McCormick, T. S., and E. C. Rowland.** 1989. *Trypanosoma cruzi:* cross-reactive anti-heart autoantibodies produced during infection in mice. *Exp. Parasitol.* **69:**393–401.
93. **McCormick, T. S., and E. C. Rowland.** 1993. *Trypanosoma cruzi:* recognition of a 43-kDa muscle glycoprotein by autoantibodies present during murine infection. *Exp. Parasitol.* **77:**273–281.
94. **Mengel, J., and R. Ribeiro-dos-Santos.** 1998. Autoreactive $CD4^+$ T cells in the pathogenesis of chronic myocarditis found in experimental *Trypanosoma cruzi* infection. *Mem. Inst. Oswaldo Cruz* **93**(Suppl. I)**:**28–29.
95. **Mijares, A., L. Verdot, N. Peineau, B. Vray, J. Hoebeke, and J. Argibay.** 1996. Antibodies from *Trypanosoma cruzi* infected mice recognize the second extracellular loop of the beta 1-adrenergic and M2-muscarinic receptors and regulate calcium channels in isolated cardiomyocytes. *Mol. Cell Biochem.* **163–164:**107–112.
96. **Milei, J., R. Storino, A. G. Fernandez, R. Beigelman, S. Vanzulli, and V. J. Ferrans.** 1992. Endomyocardial biopsies in chronic chagasic cardiomyopathy. Immunohistochemical and ultrastructural findings. *Cardiology* **80:**424–437.
97. **Mosca, W., J. Plaja, R. Hubsch, and R. Cedillos.** 1985. Longitudinal study of immune response in human Chagas' disease. *J. Clin. Microbiol.* **22:**438–441.
98. **Motran, C. C., F. M. Cerban, W. Rivarola, D. Iosa, and E. Vottero de Cima.** 1998. *Trypanosoma cruzi:* immune response and functional heart damage induced in mice by the main linear B-cell epitope of parasite ribosomal P proteins. *Exp. Parasitol.* **88:**223–230.
99. **Neu, N., S. W. Craig, N. R. Rose, F. Alvarez, and K. W. Beisel.** 1987. Coxsackievirus induced myocarditis in mice: cardiac myosin autoantibodies do not cross-react with the virus. *Clin. Exp. Immunol.* **69:**566–574.
100. **Neu, N., B. Ploier, and C. Ofner.** 1990. Cardiac myosin-induced myocarditis. Heart autoantibodies are not involved in the induction of the disease. *J. Immunol.* **145:**4094–4100.
101. **Neu, N., N. R. Rose, K. W. Beisel, A. Herskowitz, G. Gurri-Glass, and S. W. Craig.** 1987. Cardiac myosin induces myocarditis in genetically predisposed mice. *J. Immunol.* **139:**3630–3636.
102. **Nicholson, L. B., H. Waldner, A. M. Carrizosa, A. Sette, M. Collins, and V. K. Kuchroo.** 1998. Heteroclitic proliferative responses and changes in cytokine profile induced by altered peptides: implications for autoimmunity. *Proc. Natl. Acad. Sci. USA* **95:**264–269.
103. **Nussenblatt, R. B., I. Gery, H. L. Weiner, F. L. Ferris, J. Shiloach, N. Remaley, C. Perry, R. R. Caspi, D. A. Hafler, C. S. Foster, and S. M. Whitcup.** 1997. Treatment of uveitus by oral administration of retinal antigens: results of a phase I/II randomized masked trial. *Am. J. Ophthalmol.* **123:**583–592.

104. **Olivares-Villagomez, D., T. L. McCurley, C. L. Vnencak-Jones, R. Correa-Oliveira, D. G. Colley, and C. E. Carter.** 1998. Polymerase chain reaction amplification of three different *Trypanosoma cruzi* DNA sequences from human chagasic cardiac tissue. *Am. J. Trop. Med. Hyg.* **59:**563–570.
105. **Pascual, J., E. Borda, and L. Sterin-Borda.** 1987. Chagasic IgG modifies the activity of sarcolemmal ATPases through a beta adrenergic mechanism. *Life Sci.* **40:**313–319.
106. **Petry, K., and H. Eisen.** 1989. Chagas disease: a model for the study of autoimmune diseases. *Parasitol. Today* **5:**111–115.
107. **Petry, K., E. Nudelman, H. Eisen, and S. Hakomori.** 1988. Sulfated lipids represent common antigens on the surface of *Trypanosoma cruzi* and mammalian tissues. *Mol. Biochem. Parasitol.* **30:**113–121.
108. **Petry, K., P. Voisin, and T. Baltz.** 1987. Complex lipids as common antigens to *Trypanosoma cruzi, T. dionisii, T. vespertilionis* and nervous tissue (astrocytes, neurons). *Acta Trop.* **44:**381–386.
109. **Petry, K., P. Voisin, T. Baltz, and J. Labouesse.** 1987. Epitopes common to trypanosomes [*T. cruzi, T. dionisii* and *T. vespertilionis* (Schizotrypanum)]: astrocytes and neurons. *J. Neuroimmunol.* **16:**237–252.
110. **Reis, D. D., E. M. Jones, S. Tostes, E. R. Lopes, E. Chapadeiro, G. Gazzinelli, D. G. Colley, and T. L. McCurley.** 1993. Expression of major histocompatibility complex antigens and adhesion molecules in hearts of patients with chronic Chagas' disease. *Am. J. Trop. Med. Hyg.* **49:**192–200.
111. **Reis, D. D., E. M. Jones, S. J. Tostes, E. R. Lopes, G. Gazzinelli, D. G. Colley, and T. L. McCurley.** 1993. Characterization of inflammatory infiltrates in chronic chagasic myocardial lesions: presence of tumor necrosis factor-alpha+ cells and dominance of granzyme A+, $CD8^+$ lymphocytes. *Am. J. Trop. Med. Hyg.* **48:**637–644.
112. **Reis, M. M., M. D. L. Higuchi, L. A. Benvenuti, V. D. Aiello, P. S. Gutierrez, G. Bellotti, and F. Pileggi.** 1997. An in situ quantitative immunohistochemical study of cytokines and IL-2R+ in chronic human chagasic myocarditis: correlation with the presence of myocardial *Trypanosoma cruzi* antigens. *Clin. Immunol. Immunopathol.* **83:**165–172.
113. **Ribeiro dos Santos, R. R., J. O. Marquez, C. C. Von Gal Furtado, J. C. Ramos de Oliveira, A. R. Martins, and F. Koberle.** 1979. Antibodies against neurons in chronic Chagas' disease. *Tropenmed. Parasitol.* **30:**19–23.
114. **Rizzo, L. V., E. Cunha-Neto, and A. R. Teixeira.** 1989. Autoimmunity in Chagas' disease: specific inhibition of reactivity of $CD4^+$ T cells against myosin in mice chronically infected with *Trypanosoma cruzi. Infect. Immun.* **57:**2640–2644.
115. **Ruiz, A. M., M. Esteva, M. P. Cabeza, R. P. Laguens, and E. L. Segura.** 1985. Protective immunity and pathology induced by inoculation of mice with different subcellular fractions of *Trypanosoma cruzi. Acta Trop.* **42:**299–309.
116. **Santos-Buch, C. A., A. M. Acosta, H. J. Zweerink, M. Sadigursky, O. F. Andersen, B. F. von Kreuter, C. I. Brodskyn, C. Sadigursky, and R. J. Cody.** 1985. Primary muscle disease: definition of a 25-kDa polypeptide myopathic specific chagas antigen. *Clin. Immunol. Immunopathol.* **37:**334–350.
117. **Santos-Buch, C. A., and A. R. Teixeira.** 1974. The immunology of experimental Chagas' disease. 3. Rejection of allogeneic heart cells in vitro. *J. Exp. Med.* **140:**38–53.
118. **Silva-Barbosa, S. D., V. Cotta-de-Almeida, I. Riederer, J. De Meis, M. Dardenne, A. Bonomo, and W. Savino.** 1997. Involvement of laminin and its receptor in abrogation of heart graft rejection by autoreactive T cells from *Trypanosoma cruzi*-infected mice. *J. Immunol.* **159:**997–1003.
119. **Smith, S. C., and P. M. Allen.** 1991. Myosin-induced acute myocarditis is a T cell-mediated disease. *J. Immunol.* **147:**2141–2147.
120. **Smith, S. C., and P. M. Allen.** 1992. Expression of myosin-class II major histocompatibility complexes in the normal myocardium occurs before induction of autoimmune myocarditis. *Proc. Natl. Acad. Sci. USA* **89:**9131–9135.
121. **Snary, D., J. E. Flint, J. N. Wood, M. T. Scott, M. D. Chapman, J. Dodd, T. M. Jessell, and M. A. Miles.** 1983. A monoclonal antibody with specificity for *Trypanosoma cruzi,* central and peripheral neurones and glia. *Clin. Exp. Immunol.* **54:**617–624.
122. **Sterin-Borda, L., L. C. Perez, M. Wald, G. Cremaschi, and E. Borda.** 1988. Antibodies to beta 1 and beta 2 adrenoreceptors in Chagas' disease. *Clin. Exp. Immunol.* **74:**349–354.
123. **Tarleton, R. L., L. Zhang, and M. O. Downs.** 1997. "Autoimmune rejection" of neonatal heart transplants in experimental Chagas disease is a parasite-specific response to infected host tissue. *Proc. Natl. Acad. Sci. USA* **94:**3932–3937.

124. **Teixeira, A. R., and C. A. Santos-Buch.** 1975. The immunology of experimental Chagas' disease. II. Delayed hypersensitivity to *Trypanosoma cruzi* antigens. *Immunology* **28:**401–410.
125. **Teixeira, A. R., G. Teixeira, V. Macedo, and A. Prata.** 1978. *Trypanosoma cruzi*-sensitized T-lymphocyte mediated 51CR release from human heart cells in Chagas' disease. *Am. J. Trop. Med. Hyg.* **27:**1097–1107.
126. **Tekiel, V. S., G. A. Mirkin, and C. S. Gonzalez.** 1997. Chagas' disease: reactivity against homologous tissues induced by different strains of *Trypanosoma cruzi. Parasitology* **115:**495–502.
127. **Tibbetts, R. S., T. S. McCormick, E. C. Rowland, S. D. Miller, and D. M. Engman.** 1994. Cardiac antigen-specific autoantibody production is associated with cardiomyopathy in *Trypanosoma cruzi*-infected mice. *J. Immunol.* **152:**1493–1499.
128. **Todd, C. W., N. R. Todd, and A. C. Guimaraes.** 1983. Do lymphocytes from chagasic patients respond to heart antigens? *Infect. Immun.* **40:**832–835.
129. **Toledo, B. M., N. V. Amato, E. Mendes, and I. Mota.** 1979. In vitro cellular immunity in Chagas' disease. *Clin. Exp. Immunol.* **38:**376–380.
130. **Torres, C. M.** 1929. Patologia de la miocarditis crónica en la enfermedad de Chagas. An.5a *Reun. Soc. Argent. Pat. Reg. Norte* **2:**902–916.
131. **Tostes, J. S., E. R. Lopes, F. E. Pereira, and E. Chapadeiro.** 1994. Human chronic chagasic myocarditis: quantitative study of $CD4^+$ and $CD8^+$ lymphocytes in inflammatory exudates. *Rev. Soc. Bras. Med. Trop.* **27:**127–134. (In Portuguese).
132. **Van Voorhis, W., L. Barrett, R. Koelling, and A. G. Farr.** 1993. FL-160 proteins of *Trypanosoma cruzi* are expressed from a multigene family and contain two distinct epitopes that mimic nervous tissues. *J. Exp. Med.* **178:**681–694.
133. **Van Voorhis, W., and H. Eisen.** 1989. Fl-160. A surface antigen of *Trypanosoma cruzi* that mimics mammalian nervous tissue. *J. Exp. Med.* **169:**641–652.
134. **Van Voorhis, W., L. Schlekewy, and H. L. Trong.** 1991. Molecular mimicry by *Trypanosoma cruzi:* the Fl-160 epitope that mimics mammalian nerve can be mapped to a 12-amino acid peptide. *Proc. Natl. Acad. Sci. USA* **88:**5993–5997.
135. **Vashishtha, V., and V. A. Fischetti.** 1993. Surface-exposed conserved region of the streptococcal M protein induces autoantibodies cross-reactive with denatured forms of myosin. *J. Immunol.* **150:**4693–4701.
136. **Vazquez, M. C., A. Riarte, M. Pattin, and M. Lauricella.** 1993. Chagas' disease can be transmitted through kidney transplantation. *Transplant. Proc.* **25:**3259–3260.
137. **Vazquez, M. C., R. Sabbatiello, R. Schiavelli, E. Maiolo, N. Jacob, M. Pattin, and A. Rearte.** 1996. Chagas disease and transplantation. *Transplant. Proc.* **28:**3301–3303.
138. **Vermelho, A. B., M. D. N. de Meirelles, M. C. Pereira, G. Pohlentz, and E. Barreto-Bergter.** 1997. Heart muscle cells share common neutral glycosphingolipids with *Trypanosoma cruzi. Acta Trop.* **64:**131–143.
139. **Wood, J. N., L. Hudson, T. M. Jessell, and M. Yamamoto.** 1982. A monoclonal antibody defining antigenic determinants on subpopulations of mammalian neurones and *Trypanosoma cruzi* parasites. *Nature* **296:**34–38.
140. **Zwirner, N. W., E. L. Malchiodi, M. G. Chiaramonte, and C. A. Fossati.** 1994. A lytic monoclonal antibody to *Trypanosoma cruzi* bloodstream trypomastigotes which recognizes an epitope expressed in tissues affected in Chagas' disease. *Infect. Immun.* **62:**2483–2489.

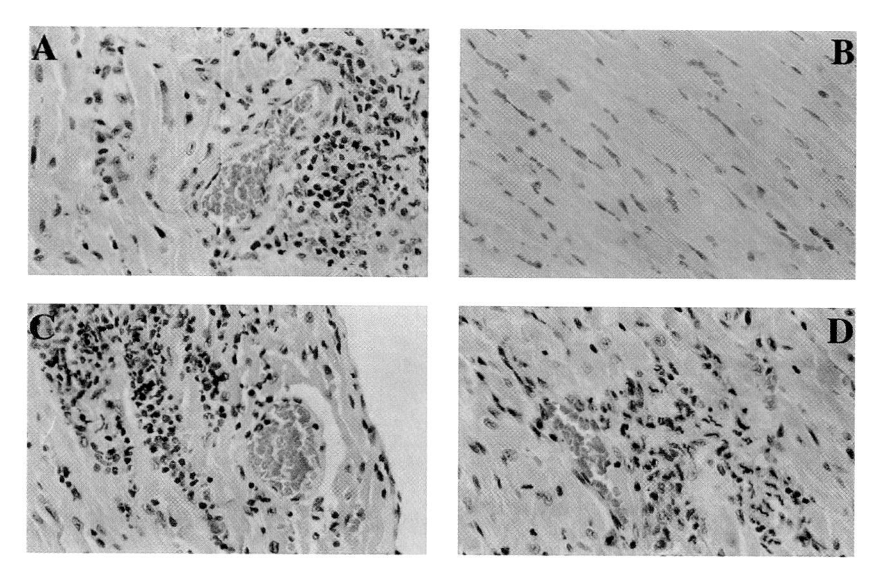

Figure 1. (See p. 75.) Inflammatory heart disease in BALB/c mice that were immunized with the endogenous mouse M7Aα peptide from the α myosin heavy chain (A), the control endogenous M7Aβ peptide from the homologous region of the β-myosin heavy chain (B), the 60-kDa CRP-derived peptide from *C. trachomatis* (ChTR1) (C), and the 60-kDa CRP-derived peptide from *C. pneumoniae* (ChPN) (D) (1). Hearts were analyzed 21 days after the initial immunization. Hematoxylin-eosin staining was used. Magnifications, $\times$320.

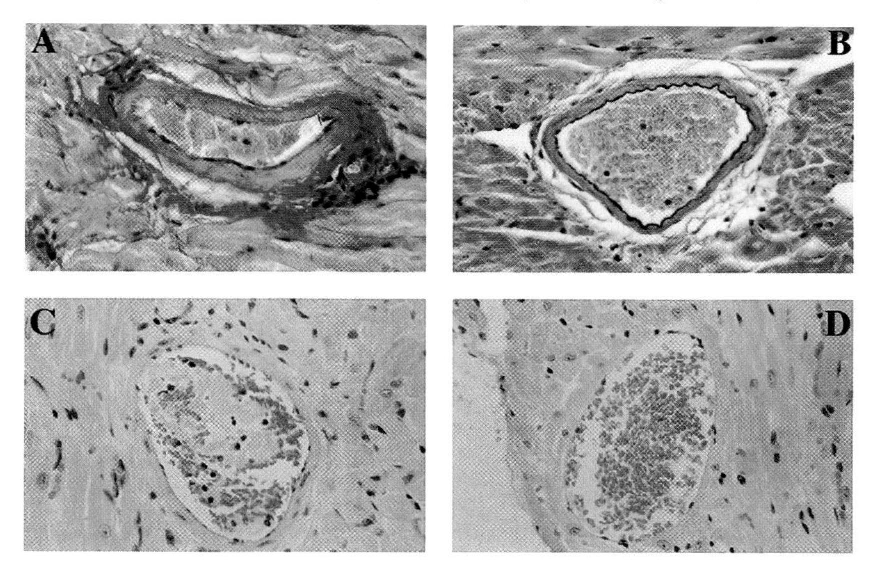

Figure 2. (See p. 79.) Blood vessels in mice immunized with *C. trachomatis* 60-kDa CRP-derived peptide. (A) Thickening of the arterial wall and perivascular fibrotic changes in mice immunized with ChTR1. Note the perivascular mononuclear inflammatory cells. (B) Normal morphology of the cardiac artery in mice immunized with Freud's complete adjuvant (FCA) alone. (C) Occlusion of cardiac blood vessels in mice immunized with ChTR1. (D) No occlusions in cardiac blood vessel were seen in control mice immunized with FCA alone. (A and B) Elastica staining for collagen (red) for detection of fibrotic changes. (C and D) Hematoxylin-eosin staining was used. Magnifications, $\times$320.

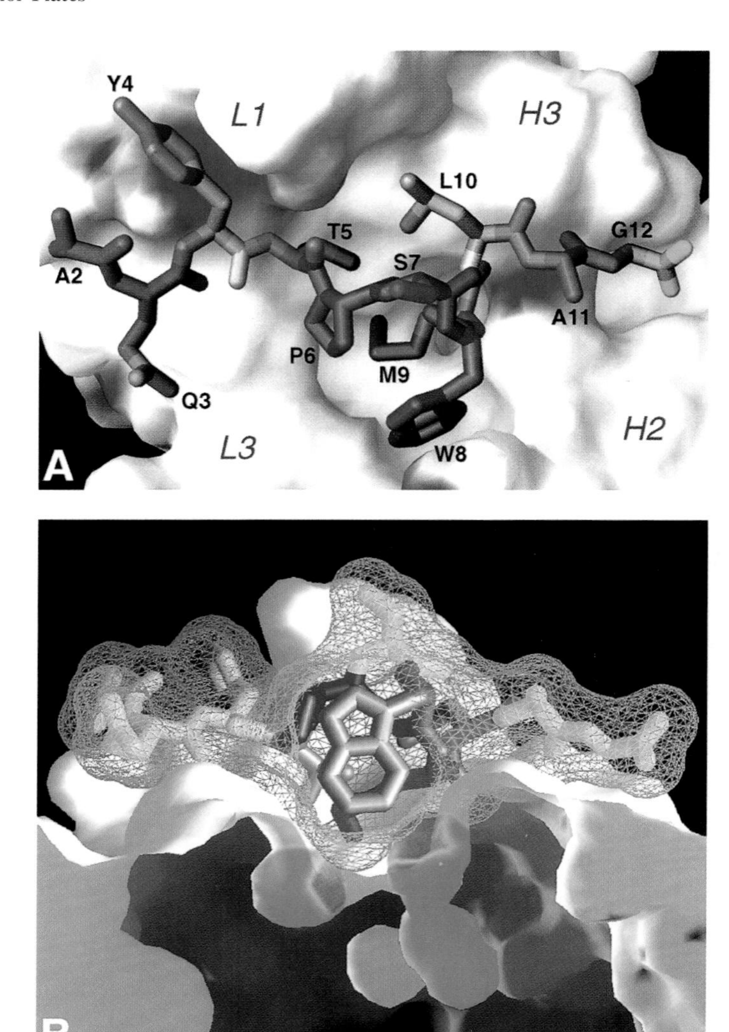

Figure 6. (See p. 154.) Crystal structure of PA1 bound to MAb 2H1. (A) Looking down on PA1 resting in the binding site of 2H1, which is shown in white, with positively charged regions in blue and negatively charged areas in red. Antibody heavy-chain CDR2 and CDR3 and light-chain CDR1 and CDR3 are denoted by H2, H3, L1, and L3, respectively. (B) Side view showing the cutaway surface of MAb 2H1 in white and the molecular surface of PA1 as a lilac mesh. The orientation of the peptide is similar to that in panel A, and the residues corresponding to the PA1 motif are colored as follows: T5, blue-green; P6, purple; W8, pink; M9, orange; L10, green. The remainder of the peptide is shown in yellow. The surfaces of two cavities between the antibody and the bound peptide are colored yellow, green, and orange according to their proximity to the antibody light chain, antibody heavy chain, and peptide, respectively.

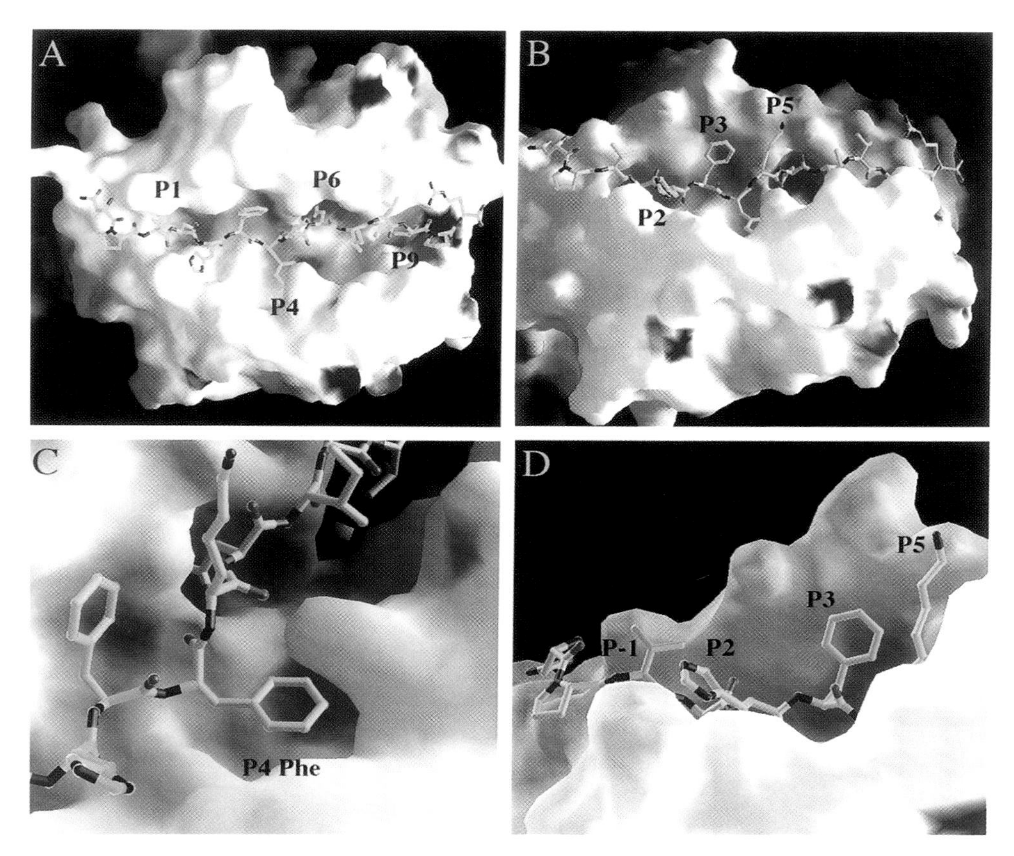

Figure 1. (See p. 203.) Crystal structure of the complex of HLA-DR2 and the MBP peptide (residues 85 to 99). (A) Overview of the structure. MBP peptide residues V89, F92, I95, and T97 occupy the P1, P4, P6, and P9 pockets of the HLA-DR2 binding site, respectively. (B) Solvent-exposed residues that are important for TCR recognition of the MBP peptide (residues 85 to 99). MBP residues H90, F91, and K93 were identified as important TCR contact residues. These are located at the P2, P3, and P5 positions, respectively, and are available for interaction with the TCR. (C) P4 pocket of the HLA-DR2 binding site. This pocket is occupied by F92 of the MBP peptide. The necessary room for this aromatic side chain is created by the DRβ71 polymorphism. (D) Close-up view of MBP peptide residues that are important for TCR recognition. Preferences at positions P-1, P2, P3, and P5 were considered in the search criteria for cross-reactive microbial peptides. Reprinted from *The Journal of Experimental Medicine* (57) with permission of the publisher.

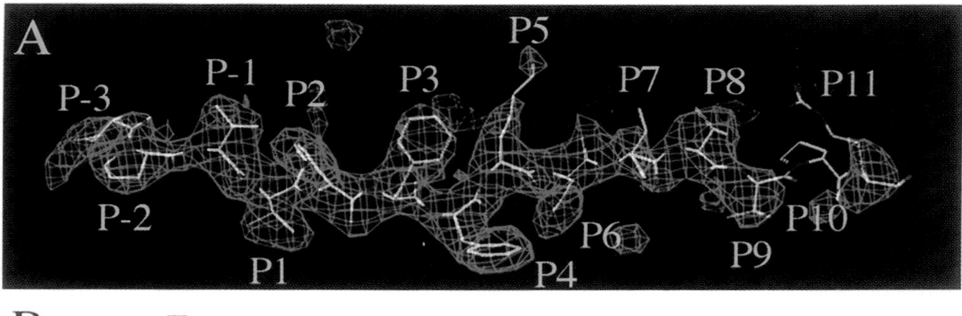

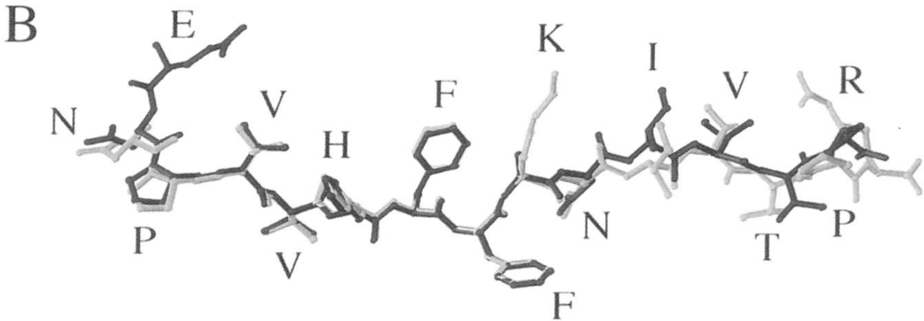

Figure 2. (See p. 205.) Electron density and model of the MBP peptide in the binding site of HLA-DR2. (A) Electron density of the MBP peptide bound to HLA-DR2. The C terminus of the peptide (P10, P11) is partially disordered. (B) Superposition of the two MBP peptides in the asymmetric unit. The DR2-MBP peptide complex crystallized as a dimer of dimers, like other HLA-DR molecules (4, 61). The model for the MBP peptide includes residues P-3 to P11 and P-4 to P10 for the two copies in the asymmetric unit, yellow and blue, respectively. The peptide backbones superimpose in the P-1 to P4 segment and are more divergent in the C-terminal segment due to different crystal contacts. A crystal contact between peptide residue P-3 in one molecule and P5 from a symmetrically related molecule stabilizes the N terminus of one peptide, enabling P-4 to be included in the model for this peptide and P5 Lys to be included in the model for the other peptide. Reprinted from *The Journal of Experimental Medicine* (57) with permission of the publisher.

INDEX